Designing the Just Transition

Just Sustainabilities

Just Sustainabilities contributes to understanding, theorizing and ultimately developing strategies towards the development of more just and sustainable communities in both the global North and South. Through a collection of solutions orientated books, the series looks at policy and planning themes that improve people's quality of life and well-being, both now and into the future; that are carried out with an intentional focus on just and equitable processes, outputs and outcomes in terms of people's access to environmental, social, political and economic space(s); and that aim to achieve a high quality of life and well-being within environmental limits.

Series Editor
Julian Agyeman

Titles already published

Julian Agyeman, *Introducing Just Sustainabilities: Policy, Planning, and Practice*

Karen Bickerstaff, Gordon Walker, Harriet Bulkeley, *Energy Justice in a Changing Climate: Social Equity and Low-carbon Energy*

Peter Utting, *Social and Solidarity Economy: Beyond the Fringe*

Jenny Pickerill, *Eco-Homes: People, Place and Politics*

Dean Saitta, *Intercultural Urbanism: Planning from the Ancient World to the Modern Day*

Phoebe Godfrey and Mary Buchanan, *Global Im-Possibilities: Exploring the Paradoxes of Just Sustainabilities*

Peter Adey, Tim Cresswell, Jane Yeonjae Lee, Anna Nikolaeva, André Nóvoa and Cristina Temenos, *Moving Towards Transition*

Designing the Just Transition

Design-Politics, Labor, and the Battle for Post-Carbon Futures

Damian White and Nicholas Pevzner

BLOOMSBURY ACADEMIC
LONDON • NEW YORK • OXFORD • NEW DELHI • SYDNEY

BLOOMSBURY ACADEMIC
Bloomsbury Publishing Plc, 50 Bedford Square, London, WC1B 3DP, UK
Bloomsbury Publishing Inc, 1359 Broadway, New York, NY 10018, USA
Bloomsbury Publishing Ireland, 29 Earlsfort Terrace, Dublin 2, D02 AY28, Ireland

BLOOMSBURY, BLOOMSBURY ACADEMIC and the Diana logo are trademarks of Bloomsbury Publishing Plc

First published in Great Britain 2026

Cover design: Burgess & Beech, Grace Ridge
Cover image © Andriy Onufriyenko via Getty Images

A catalogue record for this book is available from the British Library.

Library of Congress Cataloging-in-Publication Data available

ISBN: HB: 978-1-3503-3255-3
ePDF: 978-1-3503-3257-7
eBook: 978-1-3503-3256-0

Typeset by Newgen KnowledgeWorks Pvt. Ltd., Chennai, India

For product safety related questions contact productsafety@bloomsbury.com.

To find out more about our authors and books visit www.bloomsbury.com and sign up for our newsletters.

Designing the Just Transition

Design-Politics, Labor, and the Battle for Post-Carbon Futures

Damian White and Nicholas Pevzner

BLOOMSBURY ACADEMIC
LONDON • NEW YORK • OXFORD • NEW DELHI • SYDNEY

BLOOMSBURY ACADEMIC
Bloomsbury Publishing Plc, 50 Bedford Square, London, WC1B 3DP, UK
Bloomsbury Publishing Inc, 1359 Broadway, New York, NY 10018, USA
Bloomsbury Publishing Ireland, 29 Earlsfort Terrace, Dublin 2, D02 AY28, Ireland

BLOOMSBURY, BLOOMSBURY ACADEMIC and the Diana logo are trademarks of Bloomsbury Publishing Plc

First published in Great Britain 2026

Cover design: Burgess & Beech, Grace Ridge
Cover image © Andriy Onufriyenko via Getty Images

A catalogue record for this book is available from the British Library.

Library of Congress Cataloging-in-Publication Data available

ISBN: HB: 978-1-3503-3255-3
ePDF: 978-1-3503-3257-7
eBook: 978-1-3503-3256-0

Typeset by Newgen KnowledgeWorks Pvt. Ltd., Chennai, India

For product safety related questions contact productsafety@bloomsbury.com.

To find out more about our authors and books visit www.bloomsbury.com and sign up for our newsletters.

Praise

"This fascinating book treats the transition ahead with the nuance and complexity it deserves, but also the brio and confidence it demands. Read it and then go make it real."

Bill McKibben, author of *Here Comes the Sun*

"I have waited for this book. Pulling together as it does the particulars of labor/ecology relations while taking to task the design professions' historical complicity with neoliberalism, it uniquely offers a set of tactics for building a worker-forward transition to a healthy life on this planet."

Peggy Deamer, Professor Emerita of Architecture, Yale University, author of *Architecture and Labor*

"Our societies need massive and rapid transitions. White and Pevzner provide three badly missing things to that challenge. They bring labor politics to the practice of transition design, design to talk of just transitions, and critical histories to visioning more sustainable future societies. *Designing the Just Transition* will remind systems change-oriented designers that their agency lies in acknowledging that they also are laborers who must act in solidarity across scales."

Cameron Tonkinwise, Professor of Design Studies, University of Technology Sydney

"As it has entered the mainstream, the strategy of just transition has been separated from its roots in labor environmentalism and political agency and often seen as the equivalent of energy justice. *Designing the Just Transition* applies the theoretical and practical promise of the strategy to the political economy of design while keeping it grounded, intentionally and creatively, in the world of work and workers."

Dimitris Stevis, Professor of Political Science, Colorado State University, and author of *Just Transitions: Promise and Contestation*

"White and Pevzner's *Designing the Just Transition* is a rare gift of a book: at once richly historical and yet compellingly pertinent to today's predicaments. It offers a comprehensive reassessment of design and architecture's diverse and problematic relationship to environmental politics. More than this, the book also presents a visionary position on where to go from here: how design might effectively reengage with progressive political agendas that genuinely enable environmental responsibility and decarbonization, while simultaneously ensuring solidarity with all forms of labor. It is path-breaking and deeply relevant across many disciplines, continents, and political subject positions, and should inspire designers, scholars, environmental advocates, and policymakers to delve confidently into designing just forms of systems change."

Jesse Adams Stein, Senior Lecturer and Director of Design Studies, University of Technology Sydney, editor of *Designing through Planetary Breakdown: Locating Material Knowledge and Practical Skill*

Designing the Just Transition is a comprehensive argument for designers becoming more reflexive to the many moments in design history where labor, political economy, ecology, and systemic change have been taken up not as issues marginal to design activity but essential to it in the realization of more just and equitable societies. A counterargument against disillusionment in the face of the many challenges facing a field in crisis and a world in transition, this book makes a powerful case for why designers should become more involved in progressive politics and build power through collective organizing and action.

Ahmed Ansari, Associate Professor, Integrated Design & Media, Tandon School of Engineering, New York University

"These are tough times for positive, forward-looking social and environmental visioning. But Damian White and Nicholas Pevzner unflinchingly embody hope for a better future, while giving us theoretical and historical richness and piercing critique where it's long been overdue. Colleagues and comrades: read this one."

Timmons Roberts, Professor of Environment and Society, Brown University, and Director, Climate Social Science Network

Contents

Figures

Prologue and Acknowledgments

This book explores the relationship between design, labor, and political ecology in brittle, turbulent times. We clearly find ourselves in a moment marked by multiple intersecting economic, political, democratic and ecological crises, a fragmenting elite consensus on climate change, a geopolitical order framed increasingly around "might is right", and, an escalating populist backlash that excels in stoking all manner of resentments but offers little in terms of possible solutions to the problems it diagnoses. *Designing the Just Transition* is nevertheless, an unfashionable attempt to think about the "spaces of hope" (Harvey, 2000) that might help us head in rather different directions to business as usual. It does so by focusing on a set of fields, knowledges, and makerly practices that have occupied a fairly marginal space in the radical imagination of recent decades. Notably, this book attempts to consider what design, planning, and architecture might contribute to a politics of democratic worldmaking, that, at the level of ideas, might help us contest the politics of grievance and division and, at a more material and cultural level, help us start to identify the many creative agencies that will be needed to mitigate and adapt to life lived on a warmer, more restless planet. We were drawn together by a range of further shared concerns and assumptions.

First, it strikes us that the defining question that *haunts* all manner of environment-political discussions at the present time is the status of "the transition." Specifically, the question arises of whether we can—with any credibility—argue we are on some kind of pathway to decarbonization, whether we are, more accurately, (and more dismally), facing the consolidation and lock-in of fossil capitalism and carbon modernity, or whether we actually find ourselves in a far more uncertain, contingent and chaotic moment (Wallace, 2026). We might describe this third space as the proposition that we are in the midst of some kind of transition, but this is a transition proceeding in a *combined and uneven* fashion in an unsettled and unpredictable world, a transition that is deeply fragmented across different geographies, polities and sectors, marked by progression and regression. It is a transition that is intensifying global and national rivalries between "petro-states" and "electro-states" (Tooze, 2025c) as well as triggering regional, local and sectoral conflicts, between those who can gain from transitions and those who have most to lose (Oatley and Blyth, 2021; Colgan, Green, and Hale, 2021). This book's starting point is located in this messy third space. Rather than provide one more catalog of reasons to embrace doom, which increasingly seem to foreclose political thought, we attempt to "stay with the trouble" (Haraway, 2016) of the current moment, to consider what opening might exist for a design and labor friendly politics of transition that can revitalize the public imagination and the public realm.

Second, the intuition of this book is that while policy and activism are essential if we are to have any hope of outflanking and ultimately transcending fossil capital and carbon modernity, a politics of *making* and *unmaking, designing* and *planning* that takes both the dignity of labor *and* our ecological crisis seriously, is also going to play some significant role in shaping the socio-ecological futures we inhabit.

Third, we argue that *the just transition*—a thoroughly compromised and contested political framework, used by the great and good and radical activists for all its limitations remains an important space for thinking about the possibilities and dilemmas that attempts to enact just social and ecological futures will encounter in the future. We see the just transition not as a pre-decided plan, route map, template, or recipe that needs to be enacted to "get everything right." As we will see in this book, the issues, demands, and desires that have been raised by currents of labor-environmentalism and advocates of just transitions have ebbed and flowed across the decades, and been co-opted by all manner of actors. However, despite this, we suggest the demand for just transitions, in their most thoughtful expression, is perhaps the only politics situated within the environmental and climate space that emerges from the labor movement and has, from its beginnings, foregrounded the relationship between labor, working people and ecological problems. It is the only politics that has consistently argued that progress on climate and broader environmental issues need to be enfolded in broader struggles for meaningful work and good jobs, affordable housing, and livable communities to have any chance of success. As such, and in contrast to the "environmentalisms of the rich" that have endlessly foregrounded an ecology of scarcity and limits, austerity and sacrifice, advocates of labor-environmentalisms and just transitions have persistently argued that environmental and climate politics have to be pursued in ways that can also bring real material benefits to people's lives. This book takes its lead from such movements, thinkers and activists. It is informed by the view that designs for a just transition have to speak to the aspirations and concerns of working people around the planet to live better, richer, more creative lives, to contribute to climate solutions and build regenerative social ecologies, without impoverishing themselves or their children, or climate and environmental politics is a lost cause.

Bringing the socio-ecological, classed, labor, and design aspects of this crisis together, then, we suggest could generate productive openings, alignments, and realignments. For design fields that are often drawn to ways of thinking, mapping, and diagraming that often tidy up or contain reality, the *political ecology* and broader social science of the just transition brings the mess of the history, sociology and geography of transition, of conflict, trade offs and vested interests back in. For sustainable design discussions that have long been entranced by homeostatic ideas of ecology that assume we can return to a world of balance and harmony (an idea that the science of ecology itself let go of a good half-century ago), the just transition forces us to more fully confront the fact that transitions, where they are occurring, are already proving to be deeply fraught and uneven affairs. Indeed, if we start to think through the scale of further disruptions that is likely to be unleashed by decarbonization and energy transition, we can begin to see ripples of conflict occurring in ever-widening concentric circles. For the environmental social sciences, where discussions over the past decade have rather calcified around first-world debates that assume "we" need to choose singular imaginaries and singular pathways, such as either ecomodernism or degrowth, the social fix or the technological fix, the just transition implies that if we think about transitions across the vastly uneven and unequal landscapes that shape life chances and politics across the Global North and South, more agile thinking and practice will be necessary.

Finally, in evoking and exploring the contributions that forms of design, planning, and architecture and the resources that these fields and social practices might offer to the just transition, we argue that such an inquiry highlights that designing the just transition, at its root, is a *battle* for the future. This battle for the future is giving rise to clashes between contested political-ecological, social, *and design imaginaries*. And at a time now where "futures" seem to be everywhere foreclosing, fascism rising, and a politics of disenchantment, fatalism, and nihilism is coursing through not only the politics of the populist Right but increasingly found in many parts of the ecological Left, the question of how we might link political demands with forms of creative agency that seek to design, imagine, and materialize desirable, sustainable, and just post-carbon futures seems more urgent to explore than ever.

Designing the Just Transition proceeds through exploring how traditions of design, planning, and architecture have engaged with—and sometimes tried to avoid—questions of ecology, "nature," and climate, and questions of justice, labor, and work over the past 100 years. To disrupt the presentism of those forms of design discourse and practice that take the neoliberalization of design over the past four decades as some inevitable given and settled matter we take the long view here. Rather, we read design, planning, and architecture *historically* as fields that have been institutionalized in many ways over the past hundred years and produced rich debates about how they could and should act in society. We read the design fields *sociologically* as important but fallible, power-ridden, political, and material practices, that have always been in the business of enrolling many kinds of visible and invisible, credentialed and non-credentialed, exploited and better rewarded labor to engage in *worldmaking*. We also read design *anthropologically* as a set of practices that long predate the rise of the professional designer/architect, and practices that continue today to be moved forward by the knowledge, insights, and racialized, gendered, and classed labor of all manner of design workers, design adjacent workers, and non-credentialed lay designers.

To be clear, our argument in this book is not that design will save us. Such a proposition would be ridiculous and *Designing the Just Transition* is an exploration of only one thread of the many material agencies that will need to be redirected to build more just social and ecological futures. As scholars who work in the design academy, we are well aware that neoliberal design "as-is" is very much inscribed in the crisis that we currently face. There are many deep-seated sources of resistance within professional design, planning, and architecture for moving beyond the status quo, and this also needs to be recognized, confronted, and addressed. *Designing the Just Transition* nevertheless seeks to make the argument that an exploration of the ways in which many credentialed and non-credentialed design currents have wrestled with questions of labor, ecology, and justice potentially offer a critical repository of schemes and dreams, of insights and possibilities for how we could live differently that could have much to say to the present.

So, that's the "big picture elevator pitch," but all writing occurs in particular times and places. All books are the product of the dynamic forces of history, geography, positionality, and circumstance. We should be transparent about these forces to better situate the strengths and limits of our inquiry at the outset.

Making the claim to write about design, labor, political ecology, and a planetary climate crisis, of course, is one thing. Grappling with and making sense of these vast categories (which all contain multitudes) is another. As Peggy Deamer and Guy Julier have emphasized, design, architecture, and planning are shaped by different institutional, political, economic, and national design culture contexts, which have changed over time (Deamer 2020; Julier 2017). An explosion of global conversations around labor-environmental studies, political ecologies of labor and gender, political sociologies, and political geographies of the just transition have also demonstrated how climate/labor/design relations are playing out in many ways in different spaces and places with different geographies and histories (Morena et al., 2020; Räthzel et al., 2021; Stevis 2023).

This book is certainly informed by and in solidarity with these differing struggles for just transitions. Still, it is also an intervention that is situated in particular geographies (mostly concerned with the design traditions and debates that have dominated in the urban cores of the North Atlantic rim societies of the past hundred years), the product of certain vantage points (notably design academia in the United States, where we are located) and preoccupied with climate questions in particular.

Designing the Just Transition is also the product of a dialogue between a political sociologist and political theorist who teaches the sociology of architecture and design, urban political ecology, and climate politics in a design school (Damian), and a landscape architect and ecological designer (Nick) who teaches courses on energy, ecology, climate change, and design of public spaces and public infrastructures in a design school with a strong tradition of landscape architecture and regional planning.

Our explorations of labor, climate, political ecology, and design reach out beyond these horizons at different points but make no claim to grasping the totality. Our modest hope, though, is that *Designing the Just Transition* opens up some opportunities for conversation and comparison between colleagues and comrades working through these issues in different contexts and in different geographies to stir further conversations, scholarship, and political struggles.

Discussions of climate futures at the moment—particularly in North America—rarely evoke anything other than dismay and gloom. While this is quite understandable, *Designing the Just Transition* attempts to focus on the long game. The futures we are going to inhabit are not determined by singular events, the actions of single administrations, nor are they adequately captured by modes of analysis that offer snapshot readings of the contemporary moment that center the United States as the be-all and end-all. Our socio-ecological futures are going to be produced through a complex interplay of a range of short- and long-term political-economic, technological, demographic, geopolitical, cultural, and social forces and movements, ecological and climate forcing agents, as well as sudden "events" and surprises, which are going to play out over decades and across continents. Similarly, our capacities to act, mitigate, and adapt to climate change are dynamic and changing and are going to play out in unexpected ways. We should be suspicious of prophets and seers—on the ecological Left and the populist Right—who tell us otherwise. The struggle to build just climate futures are always struggles that occur over the long game, because as climate scientists emphasize, every fraction of a degree matters, and furthermore as the climate scientist Kate Marvell has constantly underlined, "… Climate change is not pass-fail …. There is a real continuum of futures, a continuum of possibilities" (Marvel quoted in Schwartz, 2018; but also see Marvel, 2025). This book could be seen as an attempt in dark times to keep open discussions about the "continuum of possibilities" about our climate-designed futures.

Many books are the product of both isolation and irritation and a deep desire for dialogue and engagement. Writing a book takes a long time, so you have to reach a fairly critical point of frustration, a sense that things have been missed, that people are talking past each other, that there could be other paths taken, and that there are conversations to move toward to make the expenditure of effort worthwhile. This intervention is no exception, but as the writing proceeded and we had a better sense of what we were up to, we started talking to many more friends and colleagues to test ideas, fill out blank spots in our knowledge, and get different vantage points. For many, many productive conversations about climate, design, planning, and architecture, we are enormously in debt to the incomparable expertise of Stephanie Carlisle. Anne Tate and Charlie Cannon patiently and carefully read through very scrappy early drafts of chapters two and three and provided vital feedback. Cameron Tonkinwise generously suggested many fruitful lines of inquiry that we took up in this book. Fletcher Linder, Jennifer Coffman, Chris Wilbert, Kurt Teichert, Hannah Teicher, Pedro Gadanho, Laura Briggs, Markus Berger, Ijlal Muzaffar, Ahmed Ansari, Gideon Kossoff, Terry Irwin, Eric Anderson, Liliane Wong, Idil Gaziulusoy, Billy Fleming, Daniel Aldana Cohen, Daniel Barber, Tony Fry, Jennifer Lawrence, Jesse Goldstein, Dustin Mulvaney, Samuel Zipp, Jae Shin, Timmons Roberts, and Neil Brenner all got lured into discussions about the politics of architecture, design, and urbanism at different points in the writing of this book. Pelle Ehn provided tea, cake, and hospitality in Copenhagen as he helped us understand the history of worker-centered design in Scandinavia. Les Leopold at the Workers Institute graciously spent time helping us understand the political context in the United States, which informed the rise of the just transition narrative in the labor movement and the context under which Tony Mazzocchi developed his thinking. We have learnt a great deal about the global complexities of the just transition and labor-environmentalism from Dimitris Stevis, Dunja Krause, Edouard Morena, and colleagues in the Just Transitions Research Group, Sean Sweeney at Trade Unions for Energy Democracy, and colleagues in the Climate and Community Project. We would like to thank Peggy Deamer and Michael Toste for helping us think about the history of labor and unionization in architecture and design. Deep appreciation to the Architecture Lobby's Green New Deal Working Group (Ryan Ludwig, Elisa Iturbe, Caitlin Watson, Jack Rusk, Adare Brown) for spending time with us and helping us understand some of the complexities that are involved in building coalitions of design workers across architecture, engineering, and the building trades. We would like to thank everyone who participated and attended the two conferences held at the Rhode Island School of Design in 2018 and 2019 on "Climate Futures, Design and the Just Transition."

Julian Agyeman saw the potential in a scrappy book idea we pitched to him many years ago. We would like to thank our editor Nick Wolterman, Anna Eggers, Nadine Staes-Polet, Oliver Dellow, our copy editor Kirthika Priya and all the colleagues at Bloomsbury for bringing this project over the line. We would like to thank Jennifer Prewitt-Freilino and Dean Patricia Barbeito for providing support to reproduce the visual images for this book. Damian would like to acknowledge the daily joy and love that Sarah Friel, Xavier White, Cormac White, and Finbar White bring into his life and his friends, family, and colleagues who sustain him. Nick would like to again thank Stephanie Carlisle for her constant support, as well as her immense knowledge of climate, architecture, embodied carbon, and how stuff is made—her countless insights over many conversations on these topics that formed the foundations of many of the thoughts captured in this text.

The fight to align labor demands, environmental justice, and design justice, the fight for designs for the just transition, and the fight for a livable planet are all part of an ongoing project that will take on many different contours in different spaces and places. There are many other stories to be told and many struggles we have missed. If you come across this book, are working on common projects, and feel inclined to reach out, please do so.

Damian White, dwhite01@risd.edu, The Rhode Island School of Design, Providence, RI.

Nicholas Pevzner pevzner@design.upenn.edu, The University of Pennsylvania, Weitzman School of Design, Philadelphia, PA.

February 2026.

Introduction

Resisting the It's Too Late-o-cene ...by Design?

Inertia, delay, and despair may well define our current responses to the climate and broader socio-ecological crisis. Still, it is increasingly clear that socio-technical struggles around the capacities *to embark on* sustainable transitions *or the capacities to mobilize resistance to such demands* are playing a critical, disruptive role in unsettling the domestic politics of many nations as well as shaping the increasingly fraught and unpredictable geopolitical relations between the dominant global powers (Gilman, 2025). If a few years ago, elite discussions of sustainable and low carbon transitions often proceeded with a steady confidence, which, while recognizing setbacks and difficulties could proclaim that moves to design and build low-carbon futures, at least within high-income countries, and perhaps further afield, were becoming a central force of politics, culture, design, architecture, planning, and innovation, recent developments suggest a far more uneven terrain.

To be sure, China is not only surging as a green tech manufacturing superpower but now exporting clean tech around the world at extraordinary rates that approximate and exceed the investments that the United States made in the Marshall Plan at the end of the Second World War (Xue and Larsen, 2025). We can point to a solar revolution that is moving forward faster than any energy source in history. India has emerged as the fastest growing solar market in the world (Bond and Sinha, 2026) and markets in rooftop solar are booming from Australia to Pakistan (McKibben, 2025) Over the past decade, the cost of grid-scale battery storage systems has fallen by nearly 90 percent (Ember, 2025), while their total installed capacity worldwide has expanded roughly tenfold (IEA, 2026). Circular economy action plans have been embraced by a range of political actors from Brazil to the EU. Over one quarter of the cars sold around the world in 2024 were electric vehicles or hybrid cars (Martinez-Diaz et al., 2025).

At the same time, much of the elite consensus that has informed climate politics over the last two decades has clearly fragmented (Fertik, 2025) and a backlash against climate and environmental measures is now in full force (Patterson, 2023). Where transitions have advanced the furthest, the technical, political, financial, and engineering complexities and trade-offs involved in replacing carbon intensive energy and mobility systems has come into full view (Grubert and Hastings-Simon, 2022). Climate skeptic Right populists are on the march in virtually every elected democracy, and climate, environmental, and energy questions have become increasingly ensnared in seemingly interminable culture wars with all manner of better and less well-informed claims circulating around the potential economic costs and social and economic impacts of decarbonized technologies, policies, and politics (Buck, 2023; Ritchie, 2025). Perhaps most strikingly, in the United States, we have seen talk of a future shaped by clean tech, green jobs, and green manufacturing systematically replaced by an administration committed to actively sabotaging climate science and the domestic solar and wind industries, undercutting global climate treaties, and propping up fossil fueled "energy dominance." The climate, energy and labor impacts of AI and power-hungry data centers cast a further shadow on our capacities to make any confident predictions of the future.

To be sure, the winds of politics can change and change again in fast order. However, our inquiry needs to start with clear-eyed recognition that the power of the forces committed to maintaining business as usual are ascendant in many places. Fossil capital (Malm, 2016) and carbon modernity (Iturbe, 2019) are resilient, well-financed, and still immensely profitable (Christophers, 2024). The fate of many communities, regions, and nations remains deeply tethered to the extraction, production, and consumption of fossil fuels (Oatley and Blyth, 2021; Cha, 2024), and such dependencies are increasingly shaping political support or political resistance to decarbonization (Colgan, Green, and Hale 2021). The technical complexity, cost, and thorny politics of decarbonization generates further potential for all kinds of surprises, resistance, reversals, and backsliding (Buck, 2021). Even in future scenarios where the sheer physics of the climate system forces even the most skeptical currents to act, we do not know whether such moves to mitigate and adapt to climate change will come into being through some ordered process or chaos and pure panic. It seems reasonable to surmise, though, that given the existing balance of forces in the contemporary world, all things remaining equal, dealing with climate change in future decades could well amplify existing class, gendered, and racial inequalities and exclusions, patterns of colonial violence, and injustice (Estevez and Riofrancos, 2025; Riofrancos, 2025b).

So, this is what we know. Perhaps the more interesting question to think about given that our social, political, and ecological relations are not pre-determined, is what can be done to make things otherwise? If we are in the midst of some kind of transition, but this is a transition currently proceeding in a *combined, uneven, contingent* and often *chaotic* fashion in an unsettled, unequal, and increasingly unpredictable world, a transition that is deeply fragmented across different geographies, polities and sectors, marked by progression, regression, and reversals (Fertik, 2025; Tooze, 2025c; Wallace, 2026), what kinds of practices, alliances, and struggles might open up alternative possibilities?

This book follows the lead of the more enlightened labor unions, environmental justice movements, Indigenous activists, and climate justice campaigners that have long argued that the *just transition* has to stand as the horizon for a labor-friendly politics that can seek to address climate and environmental concerns while also foregrounding the pressing broader material concerns and aspirations of working people. As we shall see in the next chapter, the just transition is a foundational concept of labor-environmentalism (Morena et al., 2020; Räthzel et al., 2021; Stevis, 2023). It has been reworked through environmental justice (see Mulvaney, 2019; Cha, 2024; Lennon, 2025), feminist (Barca, 2024), and indigenous struggles (Whyte, 2017; Estes, 2019b; Indigenous Peoples Principles and Protocols for Just Transition, 2024), and while ebbing and flowing, it has continually reemerged as a central point of reference in discussions of green new deals (Aronoff et al., 2019), just sustainabilities (Agyeman et al., 2003), Latin American/Andean traditions of *buen vivir*/sumak kawsay/suma qamaña (Escobar, 2018), struggles for Indigenous land and food sovereignty (Estes et al., 2021), and even heated debates around the viability of Left eco-modernisms (Huber, 2022), ecofeminisms (Barca, 2024), anti-imperial political ecologies (Ajl, 2021), and degrowth (Schmelzer et al., 2022). Given that the language of "just transitions" has been used both by the great and good inside climate summits to carefully manage incremental steps to decarbonization and by radical movements outside the summit that are demanding fundamental system change, the value of the term is certainly open to question. Nevertheless, we suggest in this book that a *labor-centered* just transition focused on the material agencies of design offers an important horizon for political engagement in dark times.

We might say here that at its best and in contrast to the surging politics of doom and nihilism that increasingly define climate conversations in many high-income countries in the global North, the just transition is a demand *for a future* that ties *the social question* and *the ecology question* together. In contrast to those forms of climate advocacy that can slip into *climate reductionism* or *climate determinism*,[1] it is a discussion that has long recognized, as IPCC co-chair Diana Ürge-Vorsatz has bluntly observed, that "sustainability is not the top priority of everyone on the planet and probably never will be."[2] Sustainable or low-carbon transition proposals will always have to be advanced while simultaneously and directly addressing desires for meaningful work, labor, well-being, economic security, and shared prosperity, or they will not occur. In contrast to what Nancy Fraser refers to as "the environmentalisms of the rich" (Fraser, 2022), more radical advocates of just transitions have long argued that transition proposals have to be transparent about *who* exactly is going to do the classed, raced, and gendered labor of transition and who benefits? Who will be the winners? Who will be the loser? And how can we make things different?

This book, though, is primarily interested in the useful work that the just transition might do for grounding and reorienting design, architecture, and planning *and* the useful work that a design imaginary focused on people's agency might offer for climate, environmental, and labor struggles. Why focus, though, on such modest fields and practices in the face of the scale of the climate crisis and the gigantic task of decarbonization?

Designing and Planning Transitions

The climate crisis is presently narrated to us as an epochal defining moment. As we are now regularly told, many large levers need to be pulled and big moves need to occur to facilitate rapid decarbonization: domestic policy, geopolitics, energy policy, finance, agriculture, transport, engineering, culture, and so on. To be sure, the current state of climate science clearly indicates we do need to make big moves and recognize the daunting scale of the task ahead. With most recent median estimates of future warming suggesting that if we follow current policies, we could be heading well beyond 2°C global warming by 2100 (with bulk of estimates spanning from 2.3°C to 3°C, with a median estimate of 2.7°C) (Hausfather, 2025), it is clear that we are departing from the Holocene envelope that settled human societies have resided within for the past 12,000 years. In the best possible scenarios, now we will have to adjust to life on a hotter, more restless, more unstable, and less predictable planet[3] (Jamieson, 2014; Marvel, 2025). While the hard-won knowledge of climate science and the social science of decarbonization is vital for acting on this challenge, it's not the whole story. Transitions toward low-carbon futures require smart energy, industrial and agricultural policies, new financial codes, new terms of trade, new investment agreements, international treaties, and so on. This is all critically important to attend to. However, this book is also interested in how transitions will also have to be *materialized*, built and rebuilt, designed and planned, fabricated and realized, coded and created, conceptualized, imagined and re-imagined, and enacted by particular forms of *design labor*. Political and policy struggles around decarbonization are vital. However, such struggles are also intertwined in important ways with a broader series of *ongoing* socio-technical and labor struggles over the design of the architectural, infrastructural and urban forms, systems, and services we might seek to build, the design choices we might make in order to invent, redirect, or repair our existing material and digital cultures, our infrastructures, urbanscapes, and ruralscapes. Designing and planning the *phasing in* of zero-carbon systems and the *phasing out* of fossil-fuel systems (Grubert and Hastings-Simon, 2022) is going to be essential for managing the complexities of decarbonization across the next decades. Discussions of the material, cultural, and imaginative aspects of the decarbonization project have, to date, been largely missing from conversations about the social science of just transitions.

But what do we mean by *design* here? How might we define these terms if we are concerned with drawing design, design labor, and design politics into engagement with the just transition?

What Is Design?

> Design is how we can be dominated by instrumental rationality, and love it, too.
>
> Ken Liu (Liu, 2009)

Design has been defined in different ways by different fields of inquiry at different points in time, and it needs some careful explication. As far as we know, the term entered English around the 1500s. The Merriam-Webster Dictionary definition of design is: (i) to create, fashion, execute, or construct according to plan, (ii) to conceive and plan out in the mind, (iii) to have as a purpose, (iv) to devise for a specific function or end, (v) to make a drawing, pattern, or sketch, (vi) to conceive or execute a plan. At the most minimal level, designing seems to have making, planning, sketching, creative action, and possibly problem-solving as key features. The latter focus is often foregrounded in practitioner definitions of design. Perhaps one of the most famous one-line definitions comes from Herbert Simon's assertion that to design is to "devise courses of action aimed at changing existing situations into preferred ones" (Simon, 2019: 129). More recently, Ken Friedman and Erik Stolterman offer us a more contemporary view by noting that design is not simply a general abstract way of working:

> Design takes concrete form in the work of the service professions that meet human needs, a broad range of making and planning disciplines. These include industrial design, graphic design, textile design, furniture design, information design, process design, product design, interaction design, transportation design, educational design, systems design, urban design, design leadership, and design management, as well as architecture, engineering, information technology, and computer science.[4]

We could similarly add to this list landscape architecture, which combines aspects of architecture and urban design with those of horticulture, ecology, land art, civil engineering, and geography, and which throughout its disciplinary history has struggled to place itself on the spectrum between "art and instrumentality" (Weller, 2006), as both a creative discipline rooted in design and a technical one rooted in science.

Attempts to come up with a singular authoritative definition of design, which can bridge the historical and the contemporary, the abstract and the applied, nevertheless confront a number of challenges. Alfredo Gutiérrez Borrero (2020) has observed that historically, while Indigenous, Arabic, Chinese, and Indian cultures have had rich traditions of prefiguring and making, they did not historically understand such practices using the term "design." Demarcating such activities as design is very much a product of Euro-modernity. More generally, as design culture entered into circuits of global trade, colonization, and exchange, the term has endlessly expanded into all manner of material and immaterial practices (Latour, 2008; Escobar, 2018). This has made neat demarcation exercises even more difficult. For example, Guy Julier, in his definitive account of the relationship between contemporary design and neoliberalism, has suggested that

> design is far too variegated in its practices, far too widely deployed and far too diverse in how it is understood and used for us to be able to express a singular definition for it. Instead, we have to take into account the different temporalities and territories that it operates in. We have to understand its various and sometimes conflicting purposes. We must recognize the many formats it appears in and the conjunctions of objects within these and between them. No object is an island. No one definition of design is enough. (Julier, 2017: 2)

Design, then, at minimum, we might note, is a historically and culturally variable set of creative and material practices that has been understood in different ways in different cultures. Design fields, moreover, may seek to present themselves as clear and distinct, but the boundaries between fields are often in flux and have changed over time. These complexities that erupt around almost all design discussions mean that many of these conversations drift toward a focus on the contemporary design industry, "as is" or what we might refer to as "business as usual" design. This is certainly one important and unavoidable point of reference since it is the space that is occupied by most commercially focused working designers. But a growing body of designers (Boehnert, 2018; Wizinsky, 2022; Stein and Carr, 2025), critical architects (Barber, 2023; Förster et al., 2022; Carlisle, 2020), planners (Stein, 2019), critical landscape architects (Fleming, 2019; Carlisle and Pevzner, 2013; Pevzner, 2019), and design theorists and critics (Fry, 2009; Tonkinwise, 2003; Fry and Nocek, 2020; Willis, 2018) have all argued over recent times that continuing business as usual is not always the most generative or productive space to think through the ways in which design, in all its many historical and geographical expressions, could relate to decarbonization, sustainability, or struggles for environmental or social justice.

Through facades, packaging, rendering, styling, streamlining, prototyping, and performative promises, the mainstream design industry that took shape in the Euro-American world in the early twentieth century and then amplified itself across much of the world through circuit of capital, trade, and empire has long been tightly allied in some of its most prominent manifestations with fossil capitalism (Malm, 2016), carbon modernity (Iturbe, 2019), and the expansion of consumer culture, commercial real estate, and private interests (Stein, 2019) in the Global North and the subordination of the majority world in the periphery. Often reflecting this reality, many elite centers of architectural, planning, and design education that serve as key props of this industry have long bolstered formalist, instrumental, and aesthetic approaches to making, which cleave making from history, design from politics, surface appearances from exploitation and suffering, and the studio from the world. Indeed, this preference for understanding design as a "gentlemanly creative practice" (Krasny, 2019) that is largely untethered from the mess of history and politics has intensified as "neoliberal common sense" has come to pervade design education. The worlds of commercial design and architecture have demonstrated a remarkable ability to tolerate appalling labor practices and a cultivated innocence for exploring the entanglements of race, class, gender, empire, and other modes of subordination and ecological unraveling. (Julier, 2017; Baxi et al., 2018; Cayer et al., 2016; Deamer, 2020; Fleming, 2019, 2023; Franco, 2024).

Designing the Just Transition starts from what we take as a self-evident proposition that the current institutional, political, economic, and intellectual configuration of these aspects of the design industry, design education, and design thinking are *not* going to save us, but that they are to some considerable extent inscribed in the crisis. Yet, a significant body of work and activism in critical design studies, critical architectural studies, design philosophy, decolonial design, and design activism now suggests there are many more interesting and productive ways of approaching design, which can open up other horizons and other imaginaries. Design and the related fields of architecture and planning are often reduced to their role as a service profession for capitalism. However, this is not the only story to be told.

Disciplinary Histories and Philosophical Anthropologies of Design

The rise of disciplinary histories of specific design fields is a product of the professionalization of design in the twentieth century and its evolution into academic fields and subfields in the Euro-American Academy. Even though many practicing designers have (and continue to) move across and practice architecture, urban planning, interior design, and so on, disciplinary histories of architecture, planning, and design are often engaged in demarcation and legitimization exercises. As such, many classic architectural histories begin by demarcating the history of architecture as located in Mesopotamia and settled societies, focusing on the rise of the distinct category of "the architect" as emerging out of the Renaissance and key moments when something called an architect was distinguished from a builder or master mason (Aureli, 2016; Ferro, 2024). Design history (itself largely a product of the 1970s) has sought to distinguish the emergence of the "professional designer" from the craftsman, often giving attention to the rise of pattern makers in the early factories of the Industrial Revolution in England who exercised increasing influence over the production and labor process through the provision of templates for production in the early industrial revolution. With artisanal goods increasingly giving way to mass-produced goods, we see new demands for product differentiation by design as well as the growth of mechanization and alienation in the labor process (Forty, 1987).

Such disciplinary histories of design have their uses: they can bring fields into focus, allow for common languages and problématiques to emerge, consolidate professional identities, and allow disciplines and subfields to evolve. Such approaches can help push back against the endless presentisms of certain kinds of design thinking and design journalism that assume that today's particular neoliberal configuration of design, architecture, and planning is natural, universal, and inevitable. A historical approach to design reminds us that design fields have been institutionalized in various ways across different histories and geographies. As critical design historians and theorists have emphasized though, formalist and object-based design histories taking their cues from art history may have their uses, but they can also fetishize their object of study, collude with design ideology in obscuring production processes veering into antiquarianism or connoisseurship, and ultimately tell us very little about the range of people and ecologies involved or impacted in processes of making. More generally, approaches that essentialize their design taxonomies can become exercises in distinction, boundary maintenance, and control and draw from long-standing power-laden distinctions and divisions between mental and manual labor that have long and tortured histories (see Fry in Dilnot, Fry and Stewart, 2015; Tonkinwise, 2022).

More sociological approaches to design history, influenced by critical feminist, post-/decolonial, and Marxist scholars and the work of Pierre Bourdieu have underlined that the distinction between "designing" and "craft" has often been deeply gendered, colonial, and classed, allowing for the elevation of some kinds of design practices as "more worthy" of social and artistic status and others as lesser. Lilly Irani and M. Six Silberman have noted that ideologies of "design creativity" in digital design and human–computer interaction design that predominate in Silicon Valley are often premised on problematic distinctions. Notably, distinctions between "the creative designers" (notably entrepreneurs, management, VCs, and designers) who work above the application programming interface and are involved in "programming protocols that allow someone to invoke code, services, or action without concerning themselves with the details of its implementation" and data processing workers who are quite incorrectly treated as noncreative, offering mere programable labor (Irani and Silberman, 2016). Forms of design focused on producing luxury goods and spaces for the wealthy or that are more directly in conversation with the fine arts and museum culture (such as some aspects of graphic design, industrial design, and apparel design) have historically benefited from higher status and rewards from modes of creativity such as textiles, knitting, street art, or self-build housing, which is associated with those of "lower" class, gender, and racial status.

The formal distinction between design and craft may have emerged in the English factory, but as Arindam Dutta has emphasized, these designs often drew inspiration from Indian art and craft practices and relied on Indian artisans to reproduce these designs en masse (Dutta, 2006). Cameron Tonkinwise (2022) has noted craft has been regularly associated in more traditional modes of design theory with the "repetitive" making of premodern or non-European people (often collapsed together) in contrast to modern (European) design, which is deemed to be marked by self-conscious innovation —a distinction that does not hold up under critical scrutiny (see also Akama and Yee, 2019). The Brazilian architect Sérgio Ferro has further suggested that the contemporary professional field of architecture is premised on separating architectural design from the many other forms of skilled architectural and design work that occur around the construction site. What has followed from this, in architectural history, Ferro argues is a persistent attempt to elevate the architect as artist and "author" of the building above the many other hands, skills and forms of *design labor* that are involved in construction (Ferro, 2024). These distinctions between architect and laborer run counter to the original vision of the Bauhaus, the hub of design pedagogy that proved so influential in launching modern architecture, which was specifically founded with the idea of uniting artists and craftsmen. As Walter Gropius wrote in the Bauhaus's founding manifesto, "Let us then create a new guild of craftsmen without the class distinctions that raise an arrogant barrier between craftsman and artist! Together let us desire, conceive, and create the new structure of the future, which will embrace architecture and sculpture and painting in one unity," (Gropius 1919).

To resolve some of these vexing difficulties many recent critical and decolonial currents of design history and theory have been drawn to more anthropological and philosophically inflected modes of design inquiry that often start from a much more expansive philosophical anthropological claim, notably that design is central to the human as *homo faber*, that humans are tool makers, prefigurers, and transformers. The designer Victor Papanek offers probably the most expansive entry point into this discussion in his text, *Design for the Real World*, famously declaring:

> All Men are designers. All that we do, almost all the time, is design, for design is basic to all human activity. The planning or patterning of any act towards a desired foreseeable end constitutes the design process. .... Design is composing an epic poem, executing a mural, painting a masterpiece, writing a concerto. But design is also cleaning and reorganizing a desk drawer, pulling an impacted tooth, baking an apple pie, choosing sides for a backlot baseball game, and educating a child .... Design is the conscious effort to impose meaningful order.
>
> (Papanek, 1972: 3)

Failings of gender aside, if Papanek's almost unbounded view of design potentially raises more issues than it resolves, Tony Fry and Anne-Marie Willis have developed a rigorous body of critical design philosophy premised around the starting point that design is "our ability to prefigure what we create before the act of creation" (Fry, 2009, 2013). The human is a designing being (prefiguring and planning) whose designing designs the human in turn. As Anne Marie Willis elaborates, "we design our world, whilst our world acts back on us and designs us" (Willis, 2018). As we shall see in Chapter 5, this ontological view of design has led to some generative new design histories of prefigurative practice (see Fry, 2013) that open up all manner of productive engagements between design and the rise of histories of anthropocentric landscape and urban ecological transformation that have proliferated in historical ecology and environmental history and post- and decolonial histories of indigenous institutional innovation (e,g., White, Rudy, and Gareau, 2016; Escobar, 2018; Graeber and Wengrow, 2021; Tonkinwise, 2022).

Design drawing from more recent design histories or embedded in a philosophical anthropology of the human offers gains for thinking about what designs for just transition might entail. Notably, such approaches hold the potential of escaping a narrow professional silo-ism and the Eurocentrism that invariably follows from this. Rather than locking us into disenchanted narratives about the failings of professional design and architecture industries in the United States to address justice and decarbonization, such discussions open up broader histories and geographies to think with and against. They allow us to recognize that the distinctions between architecture, design, and planning are not carved into the nature of things but are historically varied and changing. They allow us to acknowledge that the vast majority of goods, buildings, and services that have historically existed in the world (and still a good deal of design work today) continue to be designed by *non-credentialed designers* or credentialed designers, who are often working with or appropriating the creativity, practices, and innovations of lay people (Rudofsky, 1987; Brand, 1995; Manzini, 2015; Wilbert and White, 2011; Berger and Irvin, 2022). They allow us to see that designing goes on in our material worlds long after the professional, architect, designer, or planner has left the scene (see Wilbert and White, 2011; Leadbeater and Miller, 2004; Julier, 2017; Dharia, 2022) and that these acts of maintenance, care, and repair of our designed worlds are all too often deeply classed, raced, and gendered activities (Ukeles, 1969; Berger and Irvin, 2022; Dharia, 2022; Franco, 2024). More specifically such approaches draw our attention to the fact that historically lay people have driven a great deal of adaptation to climate change through building and landscape design and transformations of agricultural, cooling, and hydrological systems and that drawing from this contextual knowledge will be critical as we adapt to life on a warmer planet.

Approaching design in terms of "worldmaking" (Fry, 2009; Willis, 2018; Escobar, 2018) further brings into view the ways in which design, architecture, and planning (and landscape architecture, which we are implicitly including whenever we mention architecture, design, and planning) in their various moments of formation and reformation have long contained all manner of dissenters, activists, visionaries, self-proclaimed prophets, and oddballs that have insisted design should align with radical movements to unearth other approaches and possibilities. There have long been moments when design, architecture, and planning have been open to systemic critiques of what exists and provided spaces for dialogue with radical and revolutionary social movements about the material, visual, spatial, and cultural forms that could support more emancipatory futures. These moments, while sometimes more durable than others, have created spaces at different points in time where different forces and voices have sought to envisage more sustainable material cultures and more just modes of worldmaking and desire shaping.

Philosophical anthropologies of design focused on how we become human through design (Fry, 2013) have greatly expanded engagements with design across social theory (Latour, 2008; Escobar, 2018) and offered interesting speculations on how design fields may well have to be significantly reconfigured in the future, under pressures of ecological crisis or climate change, in ways that disciplinary accounts of the just transition often struggle to contemplate (see Fry, 2009; 2013). However, for all its philosophical inventiveness and accenting the agency of design as a general human capacity, ontological design discussions have had less to say about the ways in which design labor is sociologically institutionalized, stratified, and constrained through the organization of work and through the broader capitalist political economy of the design industry (Julier, 2017; Boehnert, 2018; Deamer, 2020).

Design Politics, Design Labor, and a Political Ecology of Design

> Design serves a social and political function that is not always acknowledged within the design industry. While there are prominent movements in design working towards socially responsive practice, these efforts are hampered by the manner in which power relations are reproduced by design and the lack of analysis of these dynamics. Design is a practice that functions to mediate social relations. Typically, it reproduces the values and priorities of those who determine which design problems are to be addressed. The interests of powerful groups are manifested in design.
>
> Boehnert (2018: 1)

Grappling with the relationships between design, architecture, and its political-sociological, economic, and ecological conditions of existence is clearly critical for understanding both the social and ecological impact of these fields and the constraints and enablements that shape the capacities of design to act in the world. Nevertheless, both Peggy Deamer and Guy Julier have observed that with some honorable exceptions, historically these relationships have rarely been foregrounded in design education or design history, anthropology, or theory. Peggy Deamer has noted that architectural history has often been "oddly devoid" of attempts to foreground and excavate the relationship between design and capital (Deamer, 2013 1). Similarly, Guy Julier has argued that the relationships between design, neoliberalism, and processes of privatization, marketization, and financialization have "received little attention in design history" (Julier, 2017). Fortunately, over the past decade—through Deamer and Julier's own scholarship, as well as further interventions by Joanna Boehnert (2018) and Matthew Wizinsky's (2022) explorations of the relations between graphic design and communicative capitalism, Sam Stein (2019), Namita Dharia (2022) and Ijlal Muzaffar (2024) writing on the relations between architecture, colonialism and the real estate and construction industries, and Jomy Joseph's (2024) engagements with the relations between industrial design and capital (among others) —we have started to see numerous attempts to theorize the political economy of design at site, industry, country, and cross-national levels and situate design labor within the broader division of labor.

Much classic sociology of design and architecture (largely emergent in the United States and the UK) has tended to situate credential design fields within the Weberian sociology of the professions (see Gutman, 1988), accenting the role that design professionals have often played as relatively privileged intermediaries situated between capitalist owners and workers, involved in tasks that often involved controlling, ordering, or deskilling blue collar workers. A range of more recent interventions have argued that although such mid-twentieth-century accounts have their insights, they do not adequately capture how much credentialed design work has been transformed through the spread of neoliberalism, precarity, globalization, and technological changes.

Peggy Deamer and members of the Architecture Lobby, for example, have persistently made the argument over the past decade that it is vital for design and architecture to acknowledge that designers and architects *are workers*, the work they do occurs in all kinds of exploitative contexts, they face many threats common to all workers (deskilling, casualization, proletarianization, the "gig" economy), and it is only through recognizing this fact, developing solidarity with other forms of design labor, and organizing architects and designers *as workers* that improvements can occur across the design industry (Deamer and Bernstein, 2010; Deamer et al., 2015).[5]

Industrial design scholars such as Jomy Joseph have further suggested that if it is still helpful to think of credentialed designers as members of the *professional-managerial class* in very general terms, the social function of credentialed design work in the neoliberal design industry needs to be differentiated further. Notably, design work contains some aspects closer to a *managerial class* "privy to higher-order visions and planning for profit and growth." At the same time a good deal of design labor is closer to a *precarious working class* "trained in particular creative craft traditions" that are continually under threat from technological deskilling and automation. With the rise of service design and user-experience design (UX), many forms of design labor are also involved in gendered, classed, and racialized *emotional labor and care work*, focused on the social reproduction of everyday life. Such work may be "white collar work," but it is often blurring boundaries with frontline service workers. Joseph suggests that specific design fields and subfields can be seen as stratified along this *spectrum* of managerial/precarious labor and care work with much depending on "the amount of social power design disciplines have negotiated for themselves" (Joseph, 2024). We might also add that the proximity of credentialed designers to management, precariat, or service/care workers also is shaped by the differing historical genealogies, status, and power of different design fields, their capacity to achieve occupational closure (e.g., through licensing, certification, or unionization); the organization structure of the design industry and workplace (large or small private business, sole practitioner, co-owned/cooperatively owned firm, public employee, self-employed), degrees of seniority, security, and status in employment settings (partner, manager, associate, junior, intern, independent freelancer/contractor), and the labor relations and employment laws of specific national design cultures. For example, around 90 percent of Swedish architects are unionized, whereas in the United States, unionization of architects is almost nonexistent, with only a single private-sector architecture firm unionized as of 2024.

If credentialed designers, planners, and architects located in the high-income core of twenty-first-century capitalism can increasingly be located in a range of "contradictory class locations" (Wright, 1985), a second set of questions emerge from thinking about the relationship between credentialed designers and the many other kinds of design workers and design adjacent workers that are central to actual execution of any design task. The writings, research, and interventions of scholars and activists centered around the *Who Builds Your Architecture Group* (WBYA) has been critical here arguing that the focus of design conversations should indeed shift from "how buildings are conceived by architects to how they are materialized by a broad network of people including architects, construction workers, and a host of other actors" (Baxi et al., 2018: 5). For the WBYA group this requires tracing the many forms of *design labor* through *global design construction networks* (Baxi et al., 2018) and rendering these classed, raced, gendered, and colonial entanglements visible and legible so that we can understand the global relations of exploitation and uneven exchange that pervade the global design industry and do something about it.

The recent rise of *Production Studies* (Thomas et al., 2015), greatly inspired by the writings of Sérgio Ferro, is, in many respects, a complementary development in its attempt to reclaim the building site and broader processes of industrial manufacturing as key sites of engagement with design labor. As we shall see in Chapter 5, Production Studies has not only drawn attention to the extent to which a great deal of traditional *and* critical design education has generally obscured the relationship between production, labor, and design—but it has sought to recover a range of marginalized design-labor histories and mobilizations from the New Architecture Movement (NAM) and the feminist architecture Matrix Collective (Matrix, 1984) to the participatory architectural traditions championed by figures such as Giancarlo di Carlo and Lucian Kroll to innumerable traditions of self-build and DIY design across the Global North and South. Production studies is informed by a desire to critically interrogate the classed, raced, and gendered accounts of elite architectural and design discourse that is premised on sharp differentiations between "those who make" from "those who design." It is a conversation that has argued design labor always brings together a range of different technical, conceptual, intellectual, and craft skills that are found across the design professions, the trades, and other design workers, including citizen designers. In doing this, much of this work has also attempted to think about possibilities that might exist for building cross-class design-labor alliances between architects, designers, tradespeople, and construction workers.

Developments in labor-feminist and decolonial design discussions have further highlighted the flows of racialized, classed, and gendered design labor, from debt-burdened designers on precarious contracts, young architectural workers forced to intern for free, back-office digital workers, to the armies of subcontract workers, day laborers, and migrant/undocumented construction workers that successful design projects are dependent on but continually render invisible (Allen, 2020; Ansari, 2024; Benjamin, 2019; Saad, 2020; Dharia, 2022; Franco, 2024; Muzaffar, 2024; Stein and Carr, 2025). Bárbara Estreal and Marcelo Ramirez (2024), for example, have argued that North–South dynamics have to be grappled with if we are to have an adequate account of design labor. Notably, they observe that cost-cutting measures to outsource design work by US and European companies from their own in-house or contracted teams to design teams in the Global South has perpetuated extractive colonial relations and economic and intellectual exploitation within the design industry. Northern design firms can attract local design talent in the Global South "with the promise of engaging in cutting edge design work", but Southern design workers then "often find themselves relegated to the manual labor of the design world—executing tasks under tight constraints, with little room for creative autonomy or significant contribution to the conceptual stages of projects. The intellectual and creative output of these designers is taken back to the Global North, enriching the product ecosystems and market dominance of these corporations, while contributing minimally to the local design landscape and economy" (Estreal and Ramirez, 2024).

Perhaps one of the most striking and original interventions that has emerged out of the intersections between design-labor studies, gender, political ecology, and decolonial studies is Namita Dharia's *The Industrial Ephemeral: Labor and Love in Indian Architecture and Construction* (2022). This text provides a remarkable accounting of the transient human and material landscapes produced by the booming urban political economy of construction in India and the complexity of the power relations shaping the many modes of design labor involved in construction. As Dharia outlines, the construction site in India is a world where caste, class, kinship relations, gender, and regional affiliations all play a role in the ordering interactions and agencies of design labor through real estate decisions, architectural professionals, white collar design workers, tradespeople, and construction workers.

It is the intersections of caste, class, and familial relations that shapes the nature of design labor on the construction site. As Dharia notes: "While some laborers climb economic ladders to become small-time contractors and foremen themselves, the positions of site supervisor and construction manager are often reserved for individuals with engineering degrees or familial connections (often these are individuals from small-town, lower-middle-class backgrounds, yet who have greater caste privilege)" (Dharia, 2022: 13). The construction site is a space moreover where elite urban ecologies are constructed for the rich and very different urban ecologies are experienced by the workers "Cement and the chemicals applied to it to manipulate curing enter the lungs and skin as workers wade through them" (Dharia, 2022: 48). Indeed, it is a space of unique dangers "Every single person I spoke with on construction sites had witnessed death on one of them." At the same time though *The Industrial Ephemeral* is a text that while recognizing the construction site, design, and architecture as ordered by caste, gender, and class, and ordered by the dictates of capital accumulation and colonial extraction, refuses the claim that the products of architecture and design simply reduce to these moments either. It is also a world of creativity—from the chalk drawings left by women construction workers on the walls of buildings to the carpenters that Dharia (as a trained architect as well as an anthropologist) has to give building instructions to "who clearly know more than me" (Dharia, 2022: 15). Indeed, Dharia warns that to flatten or ignore the expertise of many different kinds of design labor is not helpful: "This neglect is a form of violence: it erases the contributions of the working class to architecture and allied design fields. Ignoring labor politics also erases the diverse knowledge systems and exchange of knowledge that design relies upon, marginalizing non-dominant-caste or minoritarian knowledge in particular" (Dharia, 2022: 16).

If North–South dynamics or building practices in the Global South provide one lens to view design labor from, the recent research of landscape architect and historian Michelle Arevalos Franco (Franco, 2022, 2024) suggests that we should also be alert to the existence of "global Souths" in the "global North" and the ways in which the "environmentalisms of the rich" can effectively hide the labor-ecologies of the subordinate classes. Notably, focusing on labor relations in the contemporary landscape architecture economy in the United States, Franco argues that landscape architecture is underpinned by migrant and racialized working-class labor, who find themselves constantly negotiating unsustainable and precarious labor regimes, but they are largely invisible to the field. In the United States she observes a great deal of landscape labor, alongside domestic labor—housework and childcare—is done by Latine immigrants or migrants. This work, as Franco notes, "literally provisions care for living landscapes through growing, planting, pruning, tending, and other immediate physical negotiations." This landscape brings its own particular knowledge and expertise of ecology, planting, and landscaping to the work, but like the invisibility of migrant gendered labor of nurses and cleaners and maids and other servants that the Global North draws for social reproduction, this landscape labor, is marked by "obscurity and precarity in the economic order." Much landscape labor in the US is done by workers on guest worker visas in working contexts that are reliant on exploitative agents charging high fees for job access, room, and board (which can ensure workers never get out of debt). They often have to endure abysmal labor conditions characterized by exposure to toxic chemicals and poor health and safety conditions. Wage theft is common, and there is little legal protection for such labor abuses. Nevertheless, she observes that discussions between clients and designers about ecology are often welcomed, but issues of social and labor justice do not enter the picture. As she notes, "Ecology is much more palatable to elite clientele than are issues of economic justice, social equity, and marginalization of human beings" (Franco, 2022: 591). Similarly, within landscape academia, Franco observes, even though progressive discussions of the need for designers to rethink their practices around an "ethics of care" that foregrounds the importance of working with "non-humans as partners" have become popular, Franco notes, approaches that also attend to class relations, center *the expertise of labor* or draw ecology and labor into conversation are much less commonly foregrounded.

Attending to the labors of design in all their complexity is important *for* thinking about what design for the just transition might entail in that it allows us to think of design not as some kind of disembodied creative art produced by a lone mostly male, upper-middle-class genius/hero of the Ayn Randian cliché who is going to save us all. Rather, attention is drawn instead to design as a form of *collective labor and work* that enrolls many hands and impacts many broader political ecologies. It allows us to think about the ways in which *design labor* as is, is "differentially stratified" (Deamer, 2020), shaped, enabled, and foreclosed by particular power relationships around class, caste, race, gender, the unequal and colonial division of labor, the compositions of labor markets, cultural practices, levels of unionization, and political ecological conditions and contexts (Brenner and Schmid, 2014; Dharia, 2022; Muzaffar, 2024; Stein and Carr, 2025).

Attending to design labor as a whole also opens up thorny issues about the often-fraught historical relations that have existed both *within* credentialed design fields and *between* credentialed designers and other kinds of design workers (Saad, 2020). Thinking about the labor of design draws attention to the divisions that have long existed within the design industry between "mental" and "manual" labor, between design labor that has been socially classified as "professions," white collar workers, "tradespeople" "semi-skilled" and "unskilled" design labor (Krasny, 2019; Dharia, 2021; Franco, 2022; Ferro, 2024). As Pier Vittorio Aureli has emphasized, such discussions allow us to see how many more kinds of contemporary design labor that operates within neoliberal design economies are precarious, "expendable", "unpaid" and "wasted", out of site and out of mind (Aureli, 2016). It allows us to recognize, as Aureli also emphasizes, that long after the architect or designer departs the scene there are many more forms of gendered, classed, and racialized design labor that continue to be involved in maintaining, repairing, updating, repurposing, abandoning, or disposing the buildings, landscapes, services, and designed urbanscapes we move through (Aureli, 2016). More positively, it allows us to ask questions about whether there have been in the past or are still in contemporary times ways of *institutionalizing design, architecture, and planning, and building alliances between design workers* that can help us think about how design labor oriented to achieving just transitions could act most effectively in the world.

This turn to the question of labor in design converges with and potentially builds bridges with a broader attempt to bring into view the political ecology of labor that has occurred in many currents of the environmental social sciences over the past decade. The project of ecological design as we shall see in Chapters 2 and 3 has long sought to force acknowledgment that all modes of design are interdependent on and impact all manner of ecological and climate processes, relations, and nonhuman agents. Over the past decade and more, some critically important interventions in environmental history and geography (e.g., Harvey, 1996; Moore, 2015; Barca, 2024; Huber, 2022) as well as feminist science and technology studies (Haraway, 1991; 2016) have further suggested that our understanding of such socio-environmental relations are severely delimited if, in foregrounding "nature" as backdrop to all designing, we remove labor from the story and obscure the ways in which labor works through the web of life to produce the shifting environmental histories and historical geographies of capitalism. Here, Jason Moore's *Capitalism and the Web of Life* has vividly retold the history of ecological imperialism through his examination of the double movement of how "the mosaic of relations that we call capitalism work *through* nature; and how the mosaic of relations we call capitalism works *through* that more limited zone we call capitalism." We have seen a broad range of important interventions seeking to recover and render visible the role of subaltern (Dharia, 2022) and gendered labor (Battistoni, 2017a, 2017b; Barca, 2019, 2024), indigenous labor (Estes, 2019), and migrant (Franco, 2024) and working-class labor in making all manner of urban and rural natures, landscapes, and ecosystems. Such approaches have specifically sought to bring into view *the agency of labor* in the web of life and they are approaches that press the question of who is going to do the work of transition, and what kinds of classed, raced, and gendered *design labor* is going to count as we strive to design and build post-carbon futures?

Making visible the different kinds of labor present in the design and redesign of our worlds does not magically resolve the many conflicts embedded in both design and the ecology/decarbonization project. However, it does squarely draw attention to the many different kinds of creative labor and design intelligences that exist within the many worlds of design that need to be rendered visible to move just transitions forward. As Sérgio Ferro (Ferro, 2024) and MacKenzie Wark (Wark, 2015) argue, it forces us to pay attention to the techniques and technologies that capital uses on the building site or the design workplace to *break apart co-operation and solidarity*. It also pushes design discussions beyond implausible individualist accounts of designers as worldmakers and forces design to think about the broader kinds of social relations and hegemonic alliances that will need to be brokered between design, design workers, and their publics if any plausible path to a just transition is to be accomplished. Such an emphasis draws attention to ways in which the just transition is a material and ideological project, a project that evokes the imaginary as much as a political or policy conundrum.

Why *Imagining* the Just Transition?

> So many researchers and commentators of climate change and societal futures speak of the imperative to change our ways: we ought to do this and that, to have this and that … we must decarbonize, and we must have justice. Too few, to date, have offered compelling alternative social and spatial visions of what this might mean…. Transformative futures demand the production of alternative visions.
>
> Kian Goh, 2021: 185–6

> A problem like climate change is going to need unprecedented collaboration between different kinds of knowledge and labor. Such a challenge may well require of us that we produce new kinds of production relations, new technologies, new kinds of affective culture, new ways of organizing the world.
>
> McKenzie Wark 2013.

Finally, let us address the question of why designing the just transition by necessity involves *imagining* the just transition at *material* and *ideological* registers. This book tries to structure a conversation between design and the critical social sciences but it tries to do so in a fashion avoiding the "design savior" complex that can pervade much design journalism, unhelpful "critique everything/propose nothing" tendencies that can inform those manifestations of critical geography or political ecology that dissolve all into contingent localisms or broader imperial tendencies in social and political theory that dismiss the practical sciences. McKenzie Wark (2013, 2015) has long argued that humanist or social science critiques of *technological or design determinism* can pose important challenges to the design solutionism that pervades our culture. However, she has also long argued that the all-too-comfortable armchair critic of technological culture can oftentimes end up overcorrecting and collapsing into modes of *social determinism* that are incurious about the material or design features of possible alternative futures or the kinds of technical and practical knowledge that needs to be *brought into dialogue with social critique* to make different kinds of worlds possible. Wark's *Molecular Red* (2016) offers a damning critique of "the traditional indifference, disdain, or lordly sense of superiority" of critical theory to the applied and practical sciences in particular. The retreat of the Left into the academy Wark notes, has produced endless modes of "High Theory" that seeks to "adjudicate, legislate or police" the practical sciences. However, these are modes of engagement that all too often dismiss too quickly practical knowledge, knowledge produced through labor's work on nature. In contrast, Wark argues we need critical approaches to our designed worlds that are "critical and constructive, collaborative and comradely" (Wark, 2016: 120).[6] These are approaches attentive to the many forms of labor (scientific, affective, precarious, reproductive) that work on nature to produce the designed world we live in; they are approaches that try and traverse and translate the many knowledges produced through this work with a view to being open to more plural utopian possibilities that might be partially revealed by this work. Similarly, Donna Haraway, Bruno Latour, Kian Goh, Kim Stanley Robinson, Octavia Butler, and David Harvey, in very different ways, have also argued there are very distinct limits to styles of social critique that critique all but ultimately bristle at any mention of the value of speculative design-futuring or show little interest in examining and grappling with how social life is systematically *designed, engineered, materialized, physically made, ecologically embodied, and embedded* (Haraway, 1991: 2016; Latour, 2008; Butler, 2000; Goh, 2021). Of course, it is the case that, as Harvey has argued, perhaps most notably in *Spaces of Hope* (2000), resolving these dilemmas by proposing form or design-based utopias that leaves social relations intact is of limited value. But as Harvey also notes, to move from this observation toward advocacy of an abstract philosophical utopianism of pure process, which has no formal or institutional content, is equally limited in its value. As he notes, one of the invaluable contributions that figures such as Walter Benjamin and Henri Lefebvre have made to historical-geographical materialism and critical theory is their insistence "that we do not merely live in a material world but that our imaginations, our dreams, our conceptions, and our representations mediate that materiality in powerful ways"; hence both of these thinkers were deeply interested in "spectacle, representations, and phantasmagoria" (Harvey, 2004: 18–19). For Harvey, keeping form and process in dialogue is the key here as is a critical defense of the imaginary.[7]

Evoking the virtues of an alliance between the socio-ecological imagination (White et al., 2016) and design of course has its dangers. We take the view in this book that—as Kate Soper has recently underlined—arguments for socially and ecologically sane living cannot just proceed through an obsessive focus on *limits, curtailments, and constraints*—an unfortunate feature of much contemporary degrowth discussions. Rather, for such a project to be successful, it will have to be deeply involved, attending to material possibilities and reconstructive agencies that open up other courses of action, to the politics of future making and desire shaping (Soper, 2020). Design here is interesting as opening up one way into this discussion because, as Anne-Marie Willis underlines, it is "right there in the middle of it all." It is a material practice deeply concerned with making, but as she also notes, it is also tied up in a broader cultural project because "Design channels, directs, focuses, crystallizes inchoate yearnings, anxieties, fears, insecurities. Design for difference, for transformation, for a better life, to be respected, to live meaningfully" (Willis, 2018: 3). Design-futuring then (see Fry, 2009) joins a cluster of other fellow travelers, from cli-fi and sci-fi writers to assorted subordinate traditions of social and political theory that argue futures are always embedded, even in the most mainstream narratives of decarbonization.

In framing the climate crisis as a challenge that is going to require ongoing and iterative responses (Tonkinwise, 2003), *Designing the Just Transition* looks for the resources we can draw together and push back against the tremendous foreclosing narratives of misanthropy, climate determinism, climate nihilism, and the necro-politics of fossil capitalism that so dominate our climate discussions today.

Plan of the Work

In what follows, we outline how the argument will proceed in this book.

In the first chapter, we unearth the history of labor environmentalism and the just transition and trace its development over the past four decades. We explore the complexities, ambiguities, and tensions that have run through this conversation and highlight how it could engage with design.

In Chapter 2, we dig further into the historical legacies that have engaged design, planning, and architecture with questions of ecology, climate, "Nature," and labor. Advocates and critics of the ecology/design story have long been drawn to telling singular stories of how these currents have come together, often narrowly focused on the United States and the counterculture's impact on design. In this chapter, we seek to shake up this narrative by exploring what we argue are *three* foundational moments in early- to mid-twentieth-century design-ecology discussions. Notably, we explore a range of attempts to align design and ecology from the Bauhaus to garden city advocates and regional planners, which largely flourished between 1900 and the 1930s and where planning, public institutions and the role of the state are lively foci of attention. We then look at a series of debates that take off from the counterculture and the *Whole Earth Catalogue* but circulate across further diverse currents of social, socialist, feminist, and black ecology, which are marked much more by Left and Right libertarian sensibilities, movement politics, and strong claims made about the necessary relations between localism, decentralization, and a politics of ecology. Finally, we pivot to considering an important continental Europe/Latin American set of exchanges that take place about the relationship between design, political ecology, and planning that emerges from the design writings of Tomás Maldonado, Gui Bonsiepe, and Victor Papanek. Once again, we find that for Maldonado and Gui Bonsiepe "small" is not necessarily "beautiful"; ecology is not reducible to ecologism; and the local is not the site of all wisdom and insight. Rather they argue conditions of combined and uneven development in Latin America require designers to politically engage with both the dilemmas of the developmental state and the question of democratic planning.

Chapter 3 starts from the energy crisis of 1974 and considers how, after this moment, design, architecture, and planning increasingly come under pressure to address questions related to "energy efficiency," "sustainability," and finally, carbon and climate justice. This chapter explores the reactions these pressures generate within the worlds of design, planning, and architecture, which, as we shall see, range from affront, outright contempt, and resistance (notably from many currents of the US architectural avant-garde) to "design for sustainability" and visions of *Natural Capitalism* (Hawken et al., 1999). We also suggest, though, that growing recognition of both the failings of green neoliberalism to deliver on its own terms and growing critique of business-as-usual design has generated new conversations. Discussions of adaptive reuse, green urbanism, repair and retrofitting, discussions of just green urban planning, and the need for design to engage more explicitly with questions of carbon and climate, have opened up a new terrain of engagement. And whilst all these developments, as we shall see, are imperfect, compromised and subject to all manner of contradictions, we try and move beyond the dead hand of total critique to consider how such interventions could bring together new alignments among those concerned with environment, labor, and design justice.

If many currents of design informed by liberal, Malthusian, romantic, and technocratic environmentalists have studiously avoided engagement with questions of labor, production, and work, what resources do we have to bring labor and working-class environmentalism into richer dialogues with design? In Chapter 4, we argue that there have been some important moments in the history of design, architecture, and planning where workers and labor advocates have proposed expansive visions for how work, production, housing, and broader urban–rural relations could be redesigned for more labor, ecology, and worker-friendly futures. This chapter in particular explores various historical attempts to align struggles for economic and industrial democracy with struggles for a different labor- or worker-centered vision of design, urbanization, and planning, which remain largely dismissed or forgotten in mainstream design history. Here we move from William Morris's advocacy of "a factory as it might be" to Dolores Hayden's material feminist architects, experiments in worker-friendly design, architecture, and planning in Amsterdam, Berlin, and Vienna, to Scandinavian, Brazilian, and British experiments in industrial democracy. We conclude by looking at a range of contemporary developments from attempts to envisage worker-centered human interaction design to proposals for labor and community-friendly designs for circular economies that might usefully inform labor-environmental design politics today.

If designing is a universal human capacity and designing the just transition necessarily involves inviting many more people than credentialed or professional designers to "remake the future," how can we think the relations between credentialed and non-credentialed designers? Chapter 5 approaches this question by considering how some of the most prominent currents of critical design and transition design thinking have sought to place publics and the climate crisis at the center of their analysis. For nearly two decades and more, Tony Fry and Anne-Marie Willis have compellingly argued that politicized processes of making and prefiguring (what they refer to as re-directive practices) are going to be constitutive features of the attempt to build survivable futures on a rapidly warming planet (Fry, 2009; Willis, 2018). Ezio Manzini and his collaborators have argued that a sustainable future society needs to bring the public into the process of cocreating and redesigning systems and services, but also new cultures, experiences, and ways of living that might meet diverse human needs in less impactful ways (Vezzoli and Manzini, 2008; Manzini, 2019, 2015). Such work has stimulated further developments in transition design (Irwin, Kossoff, and Tonkinwise, 2015) and calls for "designs for the pluriverse" (Escobar, 2018, 2020). This chapter explores the many contributions that these currents make to struggles for just transitions. We also probe some tensions or gaps concerning how such currents of critical design studies address (or fail to address) questions of labor, class, scale, and the relationship between design and public institutions.

If the mantra of 1970s-influenced sustainable design was that "small is beautiful" and that bottom-up is best, the Green New Deal era of climate policy with its focus on achieving just decarbonization through public infrastructure and direct public investment, brought the question of design's relationship to the state and public institutions back on the agenda for the first time in decades. Often driven by conversations between critical social scientists, architects, urban planners, and landscape architects, discussions of design for the Green New Deal reignited interest around municipal, regional, and national-scale interventions for achieving just decarbonization and brought new objects into focus, such as housing and schools that have often been missing from ecological design for decades. In Chapter 6, we trace the contributions of the Green New Deal to the design-labor-decarbonization discussion. We explore what the design and environmental history of the original New Deal has to offer the design imaginary, and we consider what remains of this agenda in the light of the unraveling of Biden-era climate policy by the Trump administration.

In Chapter 7, we explore a range of climate futures that design, planning, and architecture could find themselves enrolled in as climate change unfolds. Here, we draw from political theory, cli-fi, speculative design-futuring, and contemporary design practice to sketch out a range of ways in which design and architecture is currently being deployed as facilitator of unjust transitions through neoliberal green theming, decorative green urbanism, and climate Haussmannization and multiple proposals for post-carbon fortress ecologies for the 1 percent. We also explore a range of alternative design futures from utopias inspired by a reinvigorated vision of green social democracy to visions of bottom-up solar punk and eco-maker futures, indigenous design, and pluriverse futures. We also grapple with the kinds of unsettling challenges that designs for a carbon removal society and designs for a world marked by AI, automation, and planetary-scale computing present to all emancipatory post-carbon imaginaries.

Finally, in the concluding chapter to this book, we consider the fate of the design politics of the just transition in dark times, times when a great deal of first-world environmental and climate politics seems to have reached an impasse. In this chapter, we attempt to make sense of this moment exploring the ascendance of China as green tech superpower and swirling currents of climate nihilists, so-called climate realists, and theorists of abundance engulfing the old world, and we take stock of the resources that might help us move beyond the "melancholic paralysis" (Wark, 2015) that seems to pervade all climate conversations. Here, we suggest, often well-meaning attempts by the well-heeled to mobilize around modes of *technocratic ecomodernisms* that treat working people as entirely passive and inert, forms of *degrowth* desperately trying to sell a green politics of less to working populations brutalized by decades of austerity, or *doomist modes of environmental politics* that maintain the only options on the table are scenarios that range from "bad to catastrophic" are destined to fail as political projects. In contrast, we suggest that what an engagement between labor environmentalism and design potentially brings to the table is a relentless focus on the creative human agency, skills, and capacities of working people that could be channeled toward making, building, and designing just, sustainable, and convivial post-carbon futures. An engagement between questions of labor, design, and ecology forces us to think about how we are going to design, make, plan, and build post-carbon futures on *an ongoing basis*, and it makes the claim that we can live well on a warming planet if we can redirect the *productive* and *creative* power of labor to accomplishing these ends. We argue that a key task now for a design-labor politics oriented toward the just transition is that it must be part of building the broadest coalitions that can push back against the politics of disenchantment that now courses through the ecological Left and the now incipient fascism that lurks behind the climate-sabotaging politics of the populist Right. Our bet is that the just transition understood as an invitation to engage in democratic worldmaking offers the best hope of transcending these currents.

1

The Just Transition as a Social, Political, and Design Imaginary

In 2014, *The Guardian* newspaper reported that more than 500 immigrant workers from India had died in Qatar since January 2012, on top of 385 Nepalese construction worker deaths running up to the World Cup. A week later, the award-winning and widely admired British-Iraqi architect Zaha Hadid, who designed the Al-Wakrah stadium in Qatar, responded to a query about the conditions of immigrant labor on construction sites, stating: "I have nothing to do with the workers … I think that's an issue the government—if there's a problem—should pick up. Hopefully, these things will be resolved." Asked by a journalist if she was concerned, Hadid added:

> Yes, but I'm more concerned about the deaths in Iraq as well, so what do I do about that? I'm not taking it lightly but I think it's for the government to look to take care of. It's not my duty as an architect to look at it …. I cannot do anything about it because I have no power to do anything about it. I think it's a problem anywhere in the world. But, as I said, I think there are discrepancies all over the world. (see Riach, 2014)

Technically, Hadid was correct. Architects are not formally or legally responsible for the health and safety of construction works on-site. For some, the media controversy that followed Hadid's statement was nothing more than a "brouhaha" with Lee Rosenbaum asserting that "the ultimate moral responsibility lies with the client, not the architect" (Rosenbaum, 2014). Nevertheless, Hadid's interview circulated around the globe and prompted another extended discussion in architecture and design circles about how we might start to think about the labor of design and the broader human rights, environmental, and climate issues that confront the design industry. We would say *another* because scarcely a year before, the design industry had been rocked by the Rana Plaza building collapse. An eight-story factory in Bangladesh, which manufactured apparel for some of the leading sustainable design companies, collapsed, leaving over 1,100 workers dead. Both incidents prompted some design critics to argue that it was time that architecture and design got their house in order. Many others in the design world shrugged their shoulders and questioned whether design could contribute to labor or environmental questions.

In this chapter, we attempt to do three things. Firstly, we provide a condensed discussion of the history of labor environmentalism and the just transition and elaborate on why its advocates in the labor movement and beyond have suggested it is central for making progress on decarbonization and for envisaging a labor-friendly sustainable transition. Second, we map out in some preliminary way that design and the just transition could productively inform each other. And finally, we consider how more recent sympathetic and not-so-sympathetic critics have problematized, interrogated, and expanded accounts of the just transition, revealing that it is indeed, as the political theorist John Barry and his collaborators have argued, an *agonistic* field of political conflict (Barry and Ellis, 2014; McIlroy et al., 2022).

Labor Environmentalism, the Oil Chemical and Atomic Workers Union, and Tony Mazzocchi

The post-war US labor movement, all too often tied to the coat strings of the Democratic Party at home and more often than not allied with the Cold War foreign policy ambitions of the US abroad, is rarely viewed as an important space for radical conceptual or strategic innovation in international labor politics or a space for environmental concerns. Recent work emerging from environmental labor studies has unsettled these assumptions in interesting and valuable ways. If what we might call the standard history of the rise of post-war environmental movements in the United States has tended to start with the publication of Rachel Carson's *Silent Spring* in 1962, the mobilizations of a mostly white and middle-class environmental movement across the late 1960s and early 1970s giving rise to legislation marking the beginning of the environmental regulatory state, historians of labor-environmentalism have brought other vantage points and actors into this discussion.

Andrew Van Alstyne, for example, has documented how it was the United Autoworkers (UAW) that hosted one of the first gatherings to bring together environmental activists, community organizers, and trade unionists at their Black Lake Conference in 1970, a gathering that helped build momentum for the passing of the Clean Air Act later that year.[1] UAW contributed financially and organizationally to the first Earth Day in 1970, and indeed, the UAW went on to host one of the first conferences on environmental justice in 1976 entitled "Working for Economic and Environmental Justice and Jobs" (Van Alstyne, 2015). Such activities presage indeed a range of emerging debates occurring in the US labor movement in the late 1960s and 1970s concerning how organized labor might engage with the rise of the environmental movement and environmental concerns (see Stevis, 2013; Newell and Mulvaney, 2013; Stevis and Felli, 2015; Healy and Barry, 2017; Sweeney and Treat, 2018). Beyond the UAW, though, perhaps the most consequential space for reflection on labor-environmental alliances emerged from the Oil Chemical and Atomic Workers Union (OCAW), prompted in particular by the thinking of OCAW union official and labor leader Tony Mazzocchi.

By the late 1960s, both Tony Mazzocchi and his sometimes interlocutor, the scientist, ecologist, and labor activist Barry Commoner, had concluded that it had become vitally important for the future of organized labor and the environmental movement to move beyond the "hippies-versus-hardhats" tensions that had come to pervade environmental–labor relations of the era. Both of these thinkers believed that organized labor had to build relationships with environmental movements and develop common interests if there was to be any hope of building a "new working-class economic crusade" for justice, jobs, and the environment (Leopold, 2007). As Mazzocchi's biographer Les Leopold has documented, Mazzocchi's entry point into these issues came through his desires to build a multiracial and anti-discriminatory "social unionism" that would not only move beyond business unionism by servicing the working class "in the workplace and in the community," address gender and racial discrimination but directly address the growing evidence that industrial corporations in the US were knowingly exposing nuclear, chemical, and energy workers to toxic and even lethal substances (see Leopold, 2007: 16–17). Mazzocchi's broader "conceptual breakthrough," as Leopold outlines, emerged through his engagement with Rachel Carson's *Silent Spring* and his attempt to highlight the limits of liberal environmentalism.

While *Silent Spring* (1962) drew attention to the impact of chemical toxins on humans in the home and animal health in the broader ecosystem, for Mazzocchi, the limit of this inquiry was that it stopped "at the factory gate" (Leopold, 2007). Carson may well have played a critical role in alerting the general public to the problem of pollution. However, *Silent Spring* presented the organization of production generating toxins *as a private matter*. As Leopold elaborates:

> In warning consumers about the dangers of pesticide residues on food, she started the chain not in the factory where the pesticides were manufactured but with those who sprayed the chemicals … Carson's prize-winning narrative never mentions the black hole of production, where thousands are sickened or killed by multiple exposures … Carson's blindness, Mazzocchi recognized, was fundamentally rooted in class. (Leopold, 2007: 226)

To achieve these ends, Mazzocchi, Commoner, and their emerging allies in academia, such as Jeanne Stellman, embarked on a range of attempts to build alliances between environmentalists and labor across the 1970s and 1980s. *Environmentalists for Full Employment* brought environmentalists and unions together and published what Stevis, Morena, and Kraus have described as "the first green industrial policy that addresses work and workers" (Stevis et al., 2019; Stevis, 2023). Mobilizations around occupational health and safety arguably generated some of the most tangible and practical accomplishments of this period. Here, both Mazzocchi and Commoner (along with their allies) did much to lay the ground for the environmental health movement and worker-oriented environmental health professionals. The passing of the Occupational Health and Safety Act of 1970 in the United States is one critical milestone, even if Mazzocchi argued that its legalistic framework did not fully acknowledge the need for *worker control* of health and safety issues.

If connecting the dots between environmental health and the workplace can be seen as the first basis for labor–environmental struggles, it is important to underline that both Mazzocchi and Commoner, across the 1970s and 1980s, went on to develop a much more expansive vision of labor-environmentalism seeking to connect environmental, labor, and community struggles around pollution and environmental health with broader struggles for economic democracy and the expansion of the welfare state. For example, one of the earliest conceptualizations of the just transition that emerged out of engagements between OCAW and the Sierra Club took the form of demands that the federal government support not only a "superfund for communities" impacted by toxins but a "superfund for workers" that would compensate labor impacted by environmental regulations and offer opportunities for retraining (Mazzocchi, 1993). Mazzocchi argued that demands for occupational health and safety should be the beginning of further demands for greater worker control over the *pace* and *content* of production. He supported arguments for industrial conversions, such as using defense workers' skills to transition to building public buses and subways (Leopold, 2007: 159). However, he also argued for legislation that would "...reduce boring repetitive, dirty and dangerous work to a minimum," reduce work time and retirement ages, provide sabbatical for workers, and fundamentally expand the possibilities to live more by working less (Leopold, 2007: xv; 414–45). Commoner's *Making Peace with the Planet* (1990) elaborated further arguments for the application of labor-friendly industrial ecologies at scale to clean up the output of particular plants. Notably, he argued that new modes of social and design innovation would also be required to eliminate the waste endemic to industrial capitalism. Commoner suggested, in addition, that a just transition required the embrace of forms of democratic planning that could facilitate the development of eco-industrial strategies for sustainable futures (see Commoner, 1990).

Mazzocchi's basic approach to building a *labor-focused environmentalism* can be seen as marking a distinct break from liberal-individualist and austerity/Malthusian modes of environmentalism, currents which were very dominant in the largely white, upper-class environmental politics of the 1970s–1990s. Rather than starting with "the problem" of a generically and power-free construction of "people" and their "problematic appetites" and then proceeding to moralize about their "mindless consumption," Mazzocchi's analysis starts from the premise that "pollution always starts in the workplace, and then moves into the community and the natural environment" (Leopold, 2007: 246).[2] What follows from this line of thought is the proposition that "you can't be concerned about the general environment unless you're concerned about the industrial environment, because the two are inseparable" (Mazzocchi cited in Leopold, 2007: 272). Workers are often doubly exposed to pollutants in the workplace and their communities; labor environmentalism should aim not just to protect workers from toxins but eliminate them entirely (Leopold, 2007: xiv). But as Les Leopold and Dimitris Stevis (2023) have observed, we can also find in Mazzocchi's advocacy an account of the just transition that is much broader than simply pushing for a narrow productivist, workerist, or class reductionist agenda.

Mazzocchi took the view that (i) struggles to expand workers' power over health and safety should also engage with questions surrounding *the pace and quality of work*, but also *investment decisions* (Leopold, 2007: xiv). He argued that some plants and industries are incompatible with a healthy workplace and a healthy planet and that those who work in such dangerous conditions deserve full pay and benefits for life (Leopold, 2007: 416). (ii) Mazzocchi further argued that the objective of labor-environmentalism should not be to provide more opportunities for toil but to "live life to its fullest by working less, not more" (Leopold, 2007: xv). Labor environmentalists should fight for dramatically reduced work time since all workers deserve more free time, regular sabbaticals, and the right to a rich and fulfilling life outside work (Leopold, 2007: 414–18). In this respect, Mazzocchi and *OCAW*'s account of the just transition can be seen as an attempt to *synthesize* classic old Left/social democratic/productivist trade union labor demands with social unionism and New Left post-productivist or even post-work demands.[3] Yet in contrast to the anti-statism and anti-industrial bent of much of the US New Left and influential currents of the US environmental movement, as Dimitris Stevis has underlined, we can also see in *OCAW*'s attempt to establish a party of Labor in the United States in the 1990s, a view that the just transition should be synonymous with *green industrial policy* and advocacy of a "stronger social and environmental welfare state" (Stevis, 2023: 13). As Stevis notes, the scale of OCAW's vision of the just transition was substantive:

> Even as a stand-alone policy, this would have been a major expansion of social welfare in the USA and most other countries ... OCAW placed just transition within a broader agenda that included green industrial policy. Moreover, just transition was part of the even broader platform of the Labor Party that the OCAW tried to create throughout the 1990s. The party's 1996 platform called for an end to bigotry, universal health care and education, family and community time through the reduction of working time, the regulation of corporations and the reduction of their power, and worker participation in the design and organization of work. (Stevis 2023: 13)

We might say then that Mazzocchi and OCAW's account of the just transition is an attempt to think about the ways in which a labor-environmentalist politics might facilitate positive environmental impacts and address workplace democracy while also offering a broader program that could make real material benefits to people's lives. Finally, Mazzocchi suggested—somewhat prophetically—that workers do not have fixed interests, and unless they are organized around class lines, issues of race, immigration, and nationalism can easily be used to build xenophobic and authoritarian right-wing working-class movements (Leopold, 2007: 441). Progressives and environmentalists should understand that democratic and green social unionism is the only thing that stands in the way of this development.

Co-opting and Diversifying the Just Transition: Gender, Race, and Southern Perspectives

If the just transition can be seen as a foundational concept for labor-environmentalism, it has also ebbed and flowed through many spaces well beyond such circles over the past three decades, operating as a radical demand for mobilizing social movements as well as becoming one more co-opted vocabulary to describe business as usual (Stevis et al., 2019). As we shall see in Chapter 6, it is a term that has been used to frame climate and environmental policy in many green, labor/left, and social democratic organizations, advocacy groups, and political parties, often converging with discussions of green new deals (Healy and Barry, 2017; Stevis and Felli, 2015; Sweeney and Treat, 2018; Stevis, 2023). After 2015, the concept of the just transition moved into the UN climate process through the COP process of international climate summits via advocacy and lobbying of the *International Trade Union Confederation* (ITUC) and the *International Labor Organization's Workers* Office (ACTRAV), as well as through collaborations between the ITUC, the International Labor Organization (ILO), and the UN Environment Program (UNEP) (Stevis and Felli, 2015). The incorporation of the language of just transitions into the preamble to the Paris Agreement in 2015 was followed by further centering of the just transition at the 2018 COP 24 in Katowice (branded "the just transition summit"), giving rise to the *Solidarity and Just Transition Silesia Declaration*. In the wake of these developments, talk of just transitions became (for a brief period) a lingua franca of international climate politics and diplomacy between 2015 and 2024. Perhaps more comically or tragically, commitments to "just transitions" also started to appear in all manner of corporate social responsibility documents between 2015 and 2022, including many produced by energy companies after the term was endorsed by an increasing array of climate diplomats and politicians. As the economic geographer Brett Christopher has observed, BP, Total, and Shell between 2016 and 2022 all made public declarations that they would shrink their oil and gas exploration. However, with energy prices surging in the light of Russia's invasion of Ukraine in 2022, all these companies swiftly abandoned such commitments, and we have seen all these companies make a massive pivot back to hydrocarbon production since this moment. In June 2023, Shell dropped its plans to cut production and announced "it would invest $40 billion in new oil and gas production between 2023 and 2035" (Christopher, 2024). As of 2025, Shell's website nevertheless still proclaims, "We aim to contribute to a just transition by working with governments and society to make a positive economic and social impact, while minimizing negative effects, throughout our energy transition journey."[4]

If co-optation and the deployment of just transition talk for greenwashing purposes is certainly one key aspect of the career trajectory of the concept perhaps a more constructive aspect of the discussions that have unfolded around the term in climate justice, indigenous, and socialist feminist circles is that it has also opened up productive tensions as well as insights and the potential for new alliances that can expand our understanding of the labors of transition.

For example, discussions of just transitions among resurgent currents of socialist feminism have been particularly productive in the past decade in helping us understand the labors of transition that problematize and break from class reductionism (Battistoni, 2017b, 2022; Barca, 2024; Luke, 2023; Daggett, 2018; Fraser, 2022). Socialist ecofeminists have long argued that environmentally destructive and patriarchal capital accumulation proceeds through the continual appropriation and gendering of labor in the formal economy, accompanied by capturing the unpaid labor of "women, nature, and colonies" (Mies, 1986: 77). They have argued that gendered, racialized, and classed processes of social reproduction are central to the maintenance of the current order and will emerge as key points of tension for all transition proposals, just or otherwise.

Such observations have been explored in a number of ways by resurgent feminist scholarship and activism focused on potential socialist feminist contributions to just transitions.

For example, Alyssa Battistoni (Battistoni, 2017a, 2017b, 2022) has observed that the "hard hat focus" of a great deal of just transition discussions, preoccupied with questions of how we can transition traditional blue-collar workers to green-collar jobs, is important, but it can also function to limit our political imagination if it does not link these concerns with the growing sectors of the working class in the United States and elsewhere that are situated in the service economy and what might be called *pink collar work*. Pink collar refers to teaching, nursing, care work, service work, and so on. It is these sectors, rather than manufacturing, that constitute the critical growing sectors of the United States and other affluent world economies. Much of this pink-collar work, Battistoni notes, is exploitative and underpaid, disproportionately occupied by women, immigrants, and people of color. However, in many cases, it is also labor that is central to human flourishing. It is a sector that has demonstrated renewed labor militancy over recent times and potentially a low-carbon economy oriented toward human flourishing could well give rise to an expansion of this work (under different terms and conditions, to be sure). What follows from this insight, then, is the proposition that while finding ways to address the concerns, skills, and agency of energy workers, the trades, and manufacturing workers is important (Huber, 2022), such a strategy alone is likely to be insufficient if it does not also involve efforts to organize the "nurses and teachers, care workers and service workers who are already doing the work that will be foundational to a low-carbon society oriented toward the flourishing of all" (Battistoni, 2017a; Fraser, 2022; Winant, 2019).

Intersectional labor approaches to the just transition, attending to how race/class/gender and indigenous struggles around energy and food justice invariably mesh with broader struggles over environmental justice, incarceration/criminal justice reform, public health, urban and rural land use planning, and so on have further diversified accounts of the histories of just transitions and enriched accounts of how just transition approaches can think about the classed, gendered, and racialization nature of labor. In the North American context, J. Mijn Cha's important recent intervention *A Just Transition for All* (2024) has noted that Mazzocchi reached out to the emerging environmental justice movement in the mid-1990s, producing the Just Transition Alliance in 1997—a coalition of labor unions and environmental justice organizations (Cha, 2024: 31). In many respects, it could be argued that following this moment, in the United States, it has been environmental justice movements such as The Just Transition Alliance and the Climate Justice Alliance as well as political mobilizations by frontline and fence-line communities of color to environmental toxins, pollution, and health risks that has informed a good deal of the political horizons of just transition discussions in the US over recent times (see Akuno and Nangwaya, 2017; Cha, 2024; Lennon, 2025). Such interventions have been underpinned and enriched by historical scholarship exploring the intimate relations between the building of fossil capitalism, the slave trade, and the dispossession of land and life of indigenous peoples across the Americas (Estes, 2019b; Sultana, 2022) as well as more contemporary interventions that have explored historical exclusions of working-class people of color and women from trade unions and the challenges that exist for racialized and gendered workers who find themselves working in more precarious employment sectors (Luke, 2023). Scholarship exploring the different ways in which class, race, and gender impact transition policies have highlighted how not all green jobs are equal, notably, there are "green jobs" in recycling and waste management that can be as dirty and dangerous as old, gray economy jobs. Following conventional economic models, eco-industrial development can generate new global toxic waste streams and forms of ecologically uneven exchange where cleanup occurs in some spaces and places at the expense of others (Little, 2022, 2023). Such work has also highlighted how public/private community renewable energy programs can generate complex and fraught relations between communities of color, environmental NGOs, and green capitalists, as Myles Lennon (2017, 2025) has outlined. Race-, gender-, and class-blind attempts to move green urban design and planning forward can, in certain contexts, further green gentrification (Anguelovski and Connolly, 2021). Technocratic approaches to just transitions, which present the social democratic state as the primary and unproblematic vehicle for decarbonization, could easily evolve platforms and programs at odds with indigenous, First Nation, or pastoral peoples and communities seeking autonomy from the settler colonial state (Whyte, 2017; Estes, 2019; Estes et al., 2021).

Not all of these issues are necessarily intractable. The indigenous Environmental Network (2018) has initiated a generous indigenous-based just transition campaign, developing indigenous principles of just transition that stand in full solidarity with workers, communities of color, and frontline/fence-line communities fighting similar battles. The vision of the just transition issued by the New Zealand Council of Trade Unions in 2017 begins its statement of principles by acknowledging that in Aotearoa, the just transition "requires the active involvement of iwi and hapū"—the indigenous communities. Policy, alliances, and power matter. However, these examples are salutary. They focus attention not only on the daunting political and socio-technical complexity of the project of transition but real fault lines in the just transition discussion and the conceptual limits of class reduction modes of labor-environmentalism to address these issues.

As for North/South relations, scholars such as Eve Annecke, Mark Swilling, Vishwar Satgar, Jacklyn Cock, and Michelle Williams (see Swilling and Annecke, 2012; Satgar, 2018) have documented how the political difficulties generated by carbon and other resource dependencies in resource export-driven countries have ensured that progressive commitments made by Southern unions to just transitions can unravel quickly when push comes to shove. Southern unions that prioritize the importance of obtaining national sovereignty over energy supplies from private and international owners over rapid decarbonization can end up operating with very different understandings of the just transition from Northern environmentalists. A narrow Northern-workerist vision of the just transition can be used by incumbent sectors, such as well-remunerated fossil fuel workers, to slow down and delay deep decarbonization (Mertins-Kirkwood, 2016). As Max Ajl (2021) and Farhana Sultana (2022) have reminded us, a project focused on building "green social democracy" in a Global North that still uses the Global South as a resource and sink will not do. Here, southern agendas and perspectives on just transitions focused on climate debt, land reform, resisting imperialism, and affirming the rights to development and self-determination are going to look very different from northern approaches, and they may be in tension.

Design, Architecture, and Planning as Opportunity and Challenge for the Just Transition?

So, how can design, planning, and architecture aid the just transition? As we have suggested in our introduction, design, planning, and architecture are potentially of critical significance for thinking about just transitions since they are a set of fields that are deeply attentive to the composition of our material culture and the broader infrastructures, systems, and services that currently maintain high-carbon societies and will have to be redirected (Fry, 2009) if we are to have a hope of building just post-carbon and sustainable futures. We have also suggested, though, that the design fields—as presently configured—and the industries and economies that design and architecture are situated within are significant contributors to both labor exploitation, poor working conditions, and the climate and broader ecological crisis. As we marked out in the introduction, we have seen stirrings of a labor consciousness in design and architecture over the past decade. There has been greater recognition that the global building and construction industry, for example, has some of the most exploitative conditions for labor in the world, with more fatal accidents than any other sector, that design-labor on construction sites are regularly exposed to toxic materials, and that the global building industry is marked by widespread toleration of child labor, wage theft, illegal contracts (including indentured servitude), and the suppression of trade unions (Deamer and Bernstein, 2010; Aureli, 2016; Cayer et al., 2016; Julier, 2017; Deamer, 2020; Dharia, 2022; Thomas et al., 2015). We have also seen great attention paid to how the design industry from manufacturing to the textiles, the tech industry to landscaping is embedded in broader "design production networks (Baxi et al., 2018; Franco, 2024) that manifest similar problematic features. However, the recent turn to labor in design and architecture (with some honorable exceptions) has been surprisingly disconnected from more long-standing discussions in design concerned with addressing ecological, environmental, and climate issues. It is only in fairly recent times, particularly responding to the explicit engagement around questions of design, architecture, and planning made by the IPCC over the last decade where we have seen potential for substantive realignments and alliance building between labor-focused design advocacy groups and climate and labor-environmental issues. How might the analysis of the IPCC help guide us here?

The IPCC, in its most recent Sixth Assessment Report (IPCC, 2022), outlines in extensive detail the challenge climate change poses for the design industry. According to IPCC AR6, direct and indirect building sector emissions are approximately 21–23 percent of global total GHG emissions, with buildings responsible for 28–31 percent of total energy-related CO_2 emissions. Energy use in residential buildings contributed 50 percent, nonresidential buildings contributed 32 percent, and embodied emissions contributed 18 percent to global building CO_2 emissions.

Emissions from buildings are steadily increasing: between 1990 and 2019, "global CO_2 emissions from buildings increased by 50%" (IPCC, 2022).

The impact of buildings is not simply given. Rather, it is driven by an interplay of socioeconomic trends, technology, and design choices, which range from increased floor space, energy, and electricity demand to the increase of energy services in buildings and embodied emissions, as well as the characteristics of buildings and the degree to which sufficiency measures are adopted. As such, the IPCC argues that the building sector has significant potential to cut up to 61 percent of building emissions by 2050. They suggest this could be done using existing technologies and design strategies, which would include designs for sufficiency, efficiency, and renewables, designs for retrofitting and repair, and designs for passive cooling, bioclimatic design, low- and net-zero-energy buildings, among other strategies (IPCC, 2022).

Beyond buildings, the interactions of urban systems, politics, and climate systems are key drivers of carbon-intensive energy use and greenhouse gas emissions, sites of risk and loss, and critical sites of potential climate mitigation and adaptation. The IPCC notes that urban settlements generate "about 70% of global CO2-eq emissions, and severe weather events exacerbated by anthropogenic emissions already significantly impact urban dwellers" (IPCC, 2022: 988). All things remaining equal, these impacts are likely to increase with urban areas presently home to 4.5 billion people, an expected further 2.5 billion expected to live in urban areas by 2050, the IPCC estimates that over a billion people will be living in low-lying cities and settlements at risk from "coast-specific climate hazards" (IPCC, 2022).

Urbanization processes intersecting with social inequalities, preexisting vulnerabilities, climate change, and other compounding events (air pollution, land subsidence, inadequate housing) can generate various hazards and vulnerabilities. Some of these impacts we should anticipate could include growing heat stress on human health (urban heat island effect can add two degrees C to local warming (IPCC, 2022: 901), sea level rise and increased droughts and water stress, increases in the frequency and intensity of extreme weather events, greater vulnerabilities to urban floodings, and urban food insecurity. However, they also note: "How settlements and key infrastructure are planned, designed and maintained determines patterns of exposure, social and physical vulnerability and capacity for resilience." As such, many opportunities exist for urban politics, design, and planning to help mitigate and adapt to these impacts and find synergies between adaptation measures, mitigation, and well-being. This could include improvements in urban livability that encouraged low GHG outcomes and offer greater protection against the elements, more thoughtful land use planning for climate adaptation to building protective and adaptive urban infrastructure, more comprehensive social infrastructure from health care, schooling, and housing to social services to minimize vulnerabilities, the expansion of coast flood protection and early warning systems to improve urban preparedness (IPCC, 2022: 909–10). Strategies to cool public and private spaces through "street trees, green roofs, green walls, and other urban vegetation," as well as urban parks and other measures to increase shade, can reduce extreme heat (IPCC, 2022: 949–51). The design of cities and urban areas in an era of decarbonization will also have to proceed using low-carbon materials and draw from a diversity of technical and local knowledge to accomplish its aims (IPCC, 2022: 989).

If building and urban form are critical sites for thinking about climate mitigation and adaptation, with clear implications for the future of architecture, landscape architecture, interior architecture, and adaptive reuse, industrial waste and pollution produced by product design, use, and disposal are also critical issues. Industrial production and consumption are huge emission sites—around 24 percent of direct GHG emissions (IPCC, 2022: 1172). The production of key building materials such as concrete, asphalt, gypsum, metals, brick, glass, and plastic is not only fossil fuel–intensive and contains a heavy carbon footprint but is also hazardous. Waste is also a huge problem. The design industry is currently deeply embedded in linear economies marked by a "take-make-use-dispose" model characterized by massive consumption of natural materials, where something like 90 percent of material inputs are disposed of after a single use (McDonough and Braungart, 2002) with significant waste streams from affluent countries such as the United States now exported to the Global South or ending up in the ocean (Lavender-Law, 2020). Here, it is notable that the IPCC argues that moving toward net zero emissions and containing waste streams is "possible but challenging." This will involve the use of low- and ultimately zero-carbon materials and transformations of industrial production processes, so that they are powered by low-/zero-carbon energy inputs. It will also involve achieving energy efficiencies, embracing material efficiencies, and circular economy solutions, all of which imply a fundamental rethinking of product, service, and communications design.

If the IPCC reports provide us with fairly unequivocal and rigorous, if typically apolitical, accounts of the role that design, architecture, and urban planning are playing in exacerbating the climate crisis as well as highlighting the role they could play in decarbonization, as we shall see in Chapters 3, 5, and 7, growing engagements between political ecology, architecture, design, and urban geography are now deploying a range of strategies and methodologies to highlight the socio-ecological and labor impacts that broader processes of planetary urbanization (Brenner and Schmid, 2014) are having on the transformation and disaggregation of spaces and places around the planet.

As Neil Brenner, Nikos Katsikis, Martin Arboleda, Julie Klinger, Peter Little and Thea Riofrancos have documented, the design industry—mostly centered in global cities in the affluent world—is increasingly reliant on the extraction of materials and labor from an ever-expanding set of "global hinterlands" from supply zones to sacrifice zones, logistical corridors to impact zones, which stretch from the Amazon rainforest to the ocean floor (Klinger, 2018; Arboleda, 2020; Katsikis and Brenner, 2021; Little, 2023). Design can play a key role in obscuring the impacts of these design production networks on labor, ecosystems, agricultural and indigenous communities, and nonhumans through resource extraction and waste disposal, and hiding the value transfers that occur between North and South (Boehnert, 2018; Moe, 2017). One critical question that follows from this work is the role that critical design and planning, allied with political institutional and social movements, could play in visualizing and ultimately containing these impacts (see Chapter 4). A second is how such transitions can be managed across time and space.

The writings of the civil engineers and macro-energy-system researchers Emily Grubert and Sarah Hastings-Simon (2022) here have provided a compelling account of the central role that careful design and planning are going to have to play in managing energy system change. As they detail, decarbonization is a project of building new infrastructure but also retiring the old (Grubert and Hastings-Simon, 2022 1). Moreover, given that very rapid decarbonization will "likely take decades," we are going to encounter a "mid-transition" where, to some degree, two systems, fossil fuel and post–fossil fuel, coexist. This means we are going to need "explicit and coordinated plans not only for zero carbon phase-in, but for fossil carbon phase-out." Processes of phase-in/phase-out "must be aligned in time," while addressing "the multiscalar challenges associated with continuing to provide crucial services while emphasizing sustainability and justice." As they warn, a lack of attention to the need for coordinated planning to achieve decarbonization "represents a major risk to a successful, just energy transition" (Grubert and Hastings-Simon, 2012: 2). Such challenges include multiple risks of "*maladapations*," as "the potential for weak medium-term policies to increase both the cost of transition and the likelihood of locking in carbon intensive policies" since "even rare missteps are likely to have significant and potentially system design relevant impacts on perception, political support, and implementation." As they summarize:

> Without explicit planning, the transition is likely to face major challenges like local economic busts, highly inequitable access to high-quality energy and infrastructure systems, and poor coordination for system-level characteristics like reliability, accessibility, and affordability. Already, evidence from the uncoordinated US coal transition shows increased potential for negative outcomes like economic hardship (e.g., from lost tax revenues and jobs), unfunded obligations (e.g., pensions, remediation commitments, maintenance and monitoring), identity and governance disruptions, and loss of resilience.
>
> (Grubert and Hastings-Simon, 2022: 1–2)

Challenges and Tensions in the Just Transition Framework

The just transition certainly offers no magic elixir for the problems and dilemmas thrown up by attempts to embark on low-carbon or sustainable transitions. Moreover, given that it has become a term used by everyone from radical eco-socialists to the Pope, from environmental justice advocates to Shell Oil, it is more than reasonable to wonder what value it brings to such discussions. Concerns that it could be reduced to the status of an empty signifier, or worse, are real and cannot be dismissed. At the same time, though, it does need to be acknowledged that all forms of decarbonization are already confronting the issues, dilemmas, and deficits raised by discussions and debates generated by the just transition. There are trade-offs in political life. These issues are not going to magically disappear by calling this process a different name (e.g., "transformation") or simply proclaiming that a better academic theory will resolve such issues.

Let us conclude this chapter by bringing some of the most fraught and contentious issues that surround just transition into the open. We might summarize current points of friction around four issues: (1) concerns about how much the union movement will aid the struggle for decarbonization and how much workers can actually gain from discourses of "transition"; (2) the tension between speed and consultation; (3) contestations around the optimal energy mix that will facilitate a just transition; and (4) contestations around the model of political economy that could further just transitions.

Doubts about the value of talk of just transitions for unions are long-standing and have generated some important recurring concerns worth examining in more depth. Resistance to any talk of just transitions has come from workers and union leaders in fossil fuel–intensive industries that have long regarded the just transition as a fundamental threat to their livelihood, unions, and broader communities (Obach, 2004). Tony Mazzocchi's attempt to construct labor-environmental alliances in the 1970s received a great deal of resistance from the construction trade unions who, as Mazzocchi observed in a moment of exasperation, would "pave over the Atlantic Ocean if given the chance" (in Leopold, 2007: 310). The history of *reactionary unionism* is a long and storied affair, particularly in the United States, and should direct us away from any "workerist romanticism" (Coe, 2021). Just transition discourse, at its most instrumental and propagandistic, can often assume that job skills are fungible. As J. Mijin Cha and her colleagues have observed, geographically (and culturally) embedded workers who have obtained very specific skills working in various carbon-intensive industries are not necessarily going to benefit from promises of "green jobs," which may involve extensive retraining, could be far more geographically dispersed, requiring travel and dislocation, and may indeed offer worse terms and conditions (Cha et al., 2021; Cha, 2024; Stein et al., 2025). Older workers, in particular, may not have the appropriate skills or means to move, nor the financial resources to deal with an increase in their cost of living that may come with relocation. Here, unions not unreasonably point out that the promise that displaced workers will be supported by either various modes of compensation or retraining "coal to coding" has a long and disappointing history (Heffron, 2021; Cha, 2024; Stein and Carr, 2025). Some energy unions argue that the decarbonized energy technologies that most align with fossil fuel workers skill sets and offer the potential for well-paid union jobs are located in nuclear power, carbon capture, offshore wind and storage and advanced geothermal—energy sectors that receive uneven support from environmental groups at best (Huber and Stafford, 2023).

Labor-focused transition discussions have to confront the added reality that manufacturing labor is on the decline across the globe, unionized workers constitute an increasingly shrinking share of the labor force of all high-income countries, and the class and occupation structure of advanced capitalist societies has been radically transformed over the past four decades through technological changes, neoliberalism, globalization, and the like. We might say here then that if just transitions and labor environmentalisms in such countries do not directly address the concerns of service workers, care workers, and knowledge workers, who now constitute the vast majority of the working class in high-come countries as well as engage with the downwardly mobile professional middle classes, it is a call that is unlikely to gain much political traction in broader society.

We have also seen ongoing tensions emerge between transition advocates who emphasize that a decarbonization strategy must prioritize and emphasize "speed and scale" (Steffen, 2017; Newell and Simms, 2021; Harry et al., 2024) versus accounts of just transitions that argue "directly impacted stakeholders" have to be at the center of a just transition (Cha, 2024: 2). Alex Steffen has argued, fairly unequivocally, that speed *is* justice, and "predatory delay" in acting on decarbonization will just embed more misery over the long run (Steffen, 2017). Further worries have been expressed that unions can use discourses of just transitions as discourses of climate delay (Harry et al., 2024). Others argue that if the voices of frontline and fence-line environmental justice communities, indigenous communities, and biodiversity advocates are just dismissed, historical injustices will be compounded in the future energy transition, marginalized and indigenous communities of color will again find themselves as "sacrifice zones," and opportunities to think about how we might do things differently this time will once again be suppressed (Mulvaney, 2019, 2020; Hernandez et al., 2019; Cha, 2024; Riofrancos, 2021, 2025a).

Further complexities to these issues emerge from the recognition that the local economies, employment opportunities, and tax bases of many frontline, fence-line, rural, and indigenous communities are reliant on extractive and fossil fuel industries. The realities of such ongoing material dependencies coupled with the desire of many indigenous communities to prioritize tribal sovereignty over their land and meet the pressing immediate needs of their community members can ensure that positions on fossil fuel/decarbonization within such communities can be complex and not uniformly aligned with climate justice demands to "leave it in the ground" (NoiseCat, 2025).

The appropriate energy pathways that are going to produce a just energy transition have been ongoing sources of considerable tension in all energy transition discussions, particularly in debates between environmental NGOs, environmental justice advocates, trade unionists, and energy analysts. For example, distributed energy systems, largely focused on renewables plus batteries, plus energy efficiency underpinned by community ownership models framed around the themes of energy democracy or 100 percent renewables, have long had strong supporters in environmentalist and environmental justice circles (e.g., see Baker, 2021; Jacobson, 2023). In contrast, energy analysts such as Jesse Jenkins and Seaver Wang have argued that all low-carbon technologies come with pros and cons (Wang, 2021), and at least in the US context, if distributed energy sources have a role to play in the new energy system, the proposition that we are going to replace the grid we have with either 100 percent renewables or a patchwork of decentralized grids is not grappling with the engineering challenge posed by decarbonization, the need for a renewables-heavy grid to have clean firm backup, and the extent to which delivering reliable and affordable energy also raises questions of justice (Jenkins et al., 2021; Wang, 2021).

Trade Unions for Energy Democracy (TUED) have suggested that rather than providing a pathway to energy democracy, the actual reality of many community power projects and the development of off-grid alternatives in the EU's neoliberal energy markets has reinforced pathways to unjust energy transitions. Notably, it has given rise to a range of "green gated energy communities" that have increased the capacity of (mostly) higher-income groups to partially exit the grid while pushing the inflated costs of maintenance and energy production of national grids onto the broader public (Sweeney et al., 2020). Similar concerns have been raised with Pakistan's current energy path, where the rapid adoption of off-grid solar by higher-income households seeking to avoid grid failures and unreliable energy may well have led to a solar boom, but it has also led to escalating costs for lower-income households who remain dependent on the national grid for their energy needs (Malik, 2023). The concern here is that without national planning and strategic coordination of centralized and decentralized energy sources, focused on equity and universal provision of energy resources, an unplanned market-driven distributed solar boom could end up producing a two-tier energy system with clean power for the wealthy and dirty power for the rest. The only solution to this, TUED argues in Europe, is public ownership agenda that can integrate a range of low-carbon energy sources into a public pathway model for a decarbonized grid (Sweeney et al., 2020). Matt Huber and Fred Stafford have argued in the United States, the shoring up of public ownership under a centralized grid model will be essential for securing just decarbonization and grid reliability. Moreover they have pointed out that there are distinct fractures emerging between environmental NGOs that favor renewables-only pathways to decarbonization and many energy unions that frequently lean to more of an "all of the above" approaches to decarbonization favoring a range of low-carbon energy sources such as utility-scale wind and solar, nuclear power, advanced geothermal technologies, and carbon capture and storage (Huber, 2022; Huber and Stafford, 2023).

There is significant ongoing tension in all just transition discussions surrounding the broader political economy that best supports just transition outcomes. Whether the best pathways to pursue decarbonization are going to be achieved through the pursuit of degrowth, green growth, or growth-agnostic strategies remains a deeply contentious matter in all transition discussions (see, for example, Huber, 2022; Barca, 2024). Degrowth scholars Irina Velicu and Stefania Barca have argued that the vision of the just transition deployed by many international trade union bureaucrats in the ILO and the ITUC is often completely devoid of any clear conceptualization of justice, largely informed by a business-friendly green growth ecomodernist agenda and orientated to ensure that trade union bureaucrats get a seat at the green capitalist "transition table" (Velicu and Barca, 2020). Following this, Barca has argued that a post-capitalist, feminist, and labor-friendly vision of degrowth attentive to planetary boundaries has to be at the center of a just transition (Barca, 2024). Many just transition advocates have further argued that North–South reparations, recognition of the need for the Global North to decarbonize and dematerialize as fast as possible to create breathing space for the Global South to develop and ultimately an agenda of "contract and convergence" must inform a just transition informed by the need for global justice which addresses the legacies of colonialism. Many Northern trade union currents—some of whom are sympathetic to *some* post-growth discussions of alternative economic indicators, support for global South development and addressing quality of work/life issues—have countered that it is vitally important to avoid the language of degrowth because it is associated by their members with austerity, job losses, or economic stagnation/decline (Trappmann et al., 2025). Such concerns have also been raised by Southern environmental justice voices, some of whom have suggested that outside Northern NGO circles, "the term degrowth is not appealing, and does not match people's demands" (Rodríguez-Labajos et al., 2019). Despite its growth in Left academia, as Buch-Hansen notes, degrowth "has neither support from a comprehensive coalition of social forces nor any consent to its agenda among the broader population" (Buch-Hansen, 2018: 157). Others have argued that if degrowth is strong on rhetoric, outside of wartime conditions, we have no historical examples of democratic electorates voting for a "planned descaling of production and consumption," and if they did, the implementation of such a project is likely to be far more fraught than its advocates are able to acknowledge (Buck, 2023).

In many decarbonization discussions now, the alternative path to degrowth is often flagged as an eco-statist vision of the just transition that seeks to decouple emissions from growth through a post-neoliberal green growth or a growth strategy focused on green industrial policy, fast deployment of low-carbon energy systems, green investments in green public goods (such as low-carbon housing, transportation, etc.), and delivering post-materialist improvements in quality of life (short working weeks, etc.) (e.g., Mazzucato, 2015, 2021). As we shall see in Chapter 6, such a theory of social change informed a good deal of the Green New Deal conversations that took place between 2018 and 2022. Degrowth advocates have long countered that we have little evidence that decoupling can occur quickly enough to avoid dangerous levels of climate change and, more broadly, that the social democratic eco-statist pathway is floundering everywhere—with the defeat and rollback of Biden-era climate and energy policy being the most recent example. At the same time, we also have very little empirical evidence that Left political forces will be able to stem the hemorrhaging of working-class political support to the populist Right by agendas that seek to frame the just transition as a project of "living within limits."

The Power Geometries of the Just Transition

In this chapter, we have made the argument that the perspectives of labor environmentalism and the just transition offer distinct and important frameworks for thinking about the design politics of decarbonization and ecological transitions more broadly. As we have seen, in contrast to discourses of environmentalism that have promoted an individualistic understanding of the environmental crisis premised on universal responsibility, consumption practices, and the need for "collective sacrifice," labor environmentalists have long argued these are the wrong starting points for a coherent political analysis of the problem we face and fairly politically useless for building solidaristic broad coalitions for change that actually addresses people's material needs and concerns alongside climate and environmental concerns (Mazzocchi, 1993; Räthzel et al., 2021; Huber, 2022; Stevis, 2023; Barca, 2024). Rather, as we have seen labor-environmental approaches seek to connect consumption practices to the "the hidden abode of production," as well as connect these processes of consumption and production to processes of extraction and waste disposal (Klinger, 2018; Arboleda, 2020; Katsikis and Brenner, 2021; Little, 2023; Riofrancos, 2025a). More generally, and in contrast to a lot of design thinking that, when forced to think about how design is situated in broader socio-ecological aggregates and relationships, can fall into largely diagrammatic and apolitical modes of systems thinking that can tidy up the world or modes of naturalistic reductionism that fall into ecologism, the often-fraught discussions around just transitions suggest we need to seriously confront the conflicts that exist both within and across modes of racialized, gendered, and class labor that stand in the way of achieving a just transition. The just transition is an *agonistic*, gnarly concept (Barry and Ellis, 2014), which brings the mess of history and politics back into design and ecological thought, and it requires grappling with what the geographer Doreen Massey has referred to as numerous "power geometries" (Massey, 2005) that have to be negotiated in any conceivable project of transition.

Massey's conceptualization of power geometries is an invitation to think in more fractal ways than systems theory about how the complexities of socio-ecological are shaped by (for example) classed, raced, and gendered labor, but it also invites us to consider how these relations are further refracted across space/time, geography/history, scalar registers, and by forms of uneven and combined development that are to be found within and across the Global North and Global South (Massey, 2005). The just transition is usefully approached through this imaginary because as we have seen, for a project engaged in thinking about designing the just transition, it prompts enquiries about the political subject of transition and *whose labor* and *what communities* should be acknowledged and prioritized in the politics of the just transition? The just transition raises questions of *spatial and sectoral justice*—should the just transition prioritize industrial workers in deindustrialized regions, or should it attempt to think in broader terms about how much the rise of postindustrial service economies has significantly transformed classed, raced, and gendered labor relations across the Global North and the Global South. The just transition raises further questions about *the scales of justice.* Notably, at what scale (local, municipal, regional, national/federal, international) should publics be engaged with and what is the scale that should ultimately be sovereign when we argue that workers or publics should be consulted in processes of transition? Such spatial and scalar questions also intersect with a whole series of vexing *temporal* questions, which touch on any number of issues from how a just transition operationalizes the recognition that there are divergent historical responsibilities for climate change to whether a just transition should also acknowledge that there are divergent development timelines for acting on decarbonization.

For design, the just transition further draws attention to the extent to which the design, construction, tech, culture, and building industries (configured as they currently are), are key drivers of climate and ecological breakdown. It also draws attention to the extent to which a great deal of "design-labor" that is now enrolled in these carbon-intensive neoliberal design economies is increasingly isolated, expendable, disposable, and precarious (Aureli, 2016). Questions from this emerge as to whether it makes political sense for designers, planners, and architects to continue to politically distance themselves from other design adjacent workers or whether it would make more sense now to find ways to organize and build relations of solidarity across design-labor constituencies. It also poses questions about the insights and the limits of past moments of design, planning, and architecture that have sought to take questions of environment, ecology, climate, and labor seriously. It is this issue that we will explore further in the next four chapters.

2

Environmental, Ecological, and "Nature-Inspired" Design Histories

The field of inquiry that has variously been referred to as "ecological," "sustainable," and "green" design has sparked a wealth of ideas, propositions, prototypes, schemes, and dreams for not only imagining but materializing other worlds and other social ecologies over many decades. If some of these propositions have ultimately been discarded, found to be counterproductive, half-baked, or impossible to scale (see variously: geodesic domes, inflatable buildings, composting toilets), other enthusiasms (for wind turbines, solar cells, energy storage technologies, electric vehicles) have seemingly failed but then seeded the basis for some of the most significant industries of the twenty-first century. Many ecologically inspired design schemes have been partially implemented with optimistic political assumptions (garden cities, green urbanism, the green "sharing" economy) only to generate mixed or sometimes disappointing results or unintended consequences (dormitory suburbs, green gentrification, the surveillance economy, performative consumer environmentalism). Many more projects still linger in the uncertain twilight of hyperbole, possibility, and partial realization (industrial ecologies, circular economies) or are subject to ongoing research inquiry (emotionally durable design, low-carbon steel and concrete, sustainable aviation and shipping, carbon capture and sequestration technologies) where it isn't easy to make conclusive judgments about their future trajectories.

If a great deal of writing about the relationship between design/ecology and environmental questions has pivoted from uncritical claims about the redemptive power of design to "save us all" to damning contrarian diatribes, our aim in this chapter and the chapters to come is to tell a more complex story. In this chapter, we focus on three moments in the development of conversations about ecological design.

We begin by focusing on two definitional moments in Anglo-American design conversations: the 1900 to late-1930s period and the counterculture 1960s to early 1970s era. The chapter then concludes by exploring a third moment, notably a continental Europe–Latin American exchange about the relationship between design, *political ecology*, and planning and the state that emerges from the design writings of Tomás Maldonado, Gui Bonsiepe, and Victor Papanek.

As we will demonstrate, these attempts to align ecology and design in each of these moments are informed by concerns about a range of differently framed environmental problems—from concerns about industrial pollution and sanitation, to colonial and eugenicist desires to embed racial hierarchies in a concrete design plan. We can find socially and ecologically informed account of *the region* inspiring advocacy for rational technocratic planning and anarcho-communist utopian schemes to remake society. Countercultural enthusiasms for the politics of ecology give us Malthusian population bombers and advocates of black liberation, advocates of DIY politics, and democratic socialist champions of the developmental state. We probe the limitations and flaws of these moments from a labor-environmental perspective. As we shall see, biomimetic reductionisms and environmental determinists abound. But our aim in this chapter is not simply to capture the diverse kinds of design politics that have been triggered by design-ecology discussions but also to keep an eye open for usable histories (see Martin, 2018), loose threads and incomplete dreams, partially compelling propositions that got sidelined or derailed, and half-plausible plans that in different circumstances may have gone another way and might be worth resuming.

Nature as Map, Muse, Mirror, and Milieu

The attempt to locate a terrain of engagement between design, planning, ecology, and politics is often viewed as a unique legacy of the mid-1960s to mid-to-late 1970s, as is the explosion of interest in "alternative technology," "eco-design," "green architecture," and so on. This is certainly a pivotal moment. Nevertheless, the proposition that "nature" can operate as a map, muse, or mirror or define the milieu for design is arguably as old as any design practice. Julia Watson has argued that it is a constant theme that one can find running through vernacular architecture and indigenous design practices (Watson, 2019). In European design history, the idea that designers should seek to discern an "order of nature" (whether understood through cosmology, geometry, myth, poetics, or science), which will provide the map for the ultimate balance and "correct" relationship between town and country, landscape and volk, building and body, is a proposition at least as old as Vitruvius, who offers us the only treatise on architecture to survive from antiquity, and probably a great deal older. Leonardo da Vinci's remarkable sketches of what he called "ornithopters" were deeply influenced by his studies of bird flight. Similarly, the notion that either climate or broader environmental conditions has not only influenced design choices and outcomes but may mirror the achievements of different human societies or allow us to rank different kinds of people in order of merit, character, or capabilities is equally long-standing (Cheng et al., 2020). As Irene Cheng has observed, the *L'Habitation Humaine* exhibit at the 1889 Paris World Exposition produced by architect Charles Garnier and historian Auguste Ammann presented a history of building following a "natural" evolutionary sequence from primitive to advanced races with Chinese, Aztec, Inuit, and African contributions assigned to a natural prehistory (Cheng et al., 2020).

The period from the 1880s to the 1930s is a moment where design, urban planning, and the science of ecology all begin to take shape in recognizably modern forms as professional fields of inquiry. We see many modes of design advocacy deploying ecological motifs and concerns to propose design interventions, the scale of which we have yet to see before or since.

Figure 1 Phaeodaria, Plate 61 by Ernest Haeckel Kunstformen der Natur (1904), plate 61.

Lydia Kallipoliti has observed that design and ecology were often fused at some very early moments in developing the science of ecology. From Ernst Haeckel's genealogical tree to Johann Wolfgang Goethe's Urpflanze, Kallipoliti argues we see "taxonomical thinking" emerge as a design endeavor (Kallipoliti, 2024). In some of the earliest attempts to think and promote the science of ecology in diagrammatic form, we see the visual, the descriptive, and the normative often come together in complex amalgams that are difficult to untangle. As she observes, Haeckel's demarcation of the field of ecology which was concerned with the study of the relationships among living organisms and the biotic and abiotic environment they inhabit led to a range of taxonomies, maps, and visualizations that sought not only to classify but also to elaborate on and normatively speculate about the order and roots of the natural and social world. As attempts unfold to "situate design in an ecological paradigm" or explore interdependencies between the natural and the social, more broadly, we start to see the rise of a particular socio-ecological milieu, which, in turn, is generative of a bewildering range of political imaginaries. These endeavors are not simply concerned with "what the world is made of" but also seek to illustrate "what the world should be" (Kallipoliti, 2024).

From Imperial Design Ecologies to the Bauhaus

Peder Anker's research into the imperial origins of ecological science (Anker, 2001) and his investigation of the subsequent complex legacy of how ecological concerns informed the thinking of the Bauhaus (Anker, 2010) capture some of the complexities at stake. Anker's *Imperial Ecology* (2001), focusing on the development of the science of ecology during the last years of the British Empire, highlights how the evolution of ecological discourse in the natural sciences quickly becomes entangled in various colonial attempts to cement "natural" racial hierarchies and extractive schemes to make such social ecologies concrete through design. As Anker notes, the scientific ecologist and later Prime Minister of South Africa, Jan Smuts, made major contributions to the empirical and conceptual development of ecological science, arguing that ontological holism was essential for developing ecology as a discipline. Smuts subsequently used such a conceptual framework to draw race, design, social Darwinism, and ecology together in a particularly potent and problematic way to structure the guidelines for apartheid across southern Africa. Holism in ecology could quite happily be used to justify the design of a society framed around racial hierarchy and concepts of "evolutionary fitness" (Anker, 2001). The search for "nature in balance" (a persistent fixture in "classical" ecology pre-1970s) drove all manner of coercive conservation projects that identified "ignorant natives" as the sources of "disruption" and opened up extraordinary flows of fauna and flora from the colonies to key scientific centers of botanical imperialism such as Kew Gardens.

Anker's (2010) reading of the ecological interests of the Bauhaus provides some interesting further insights as to how progressive and regressive viewpoints can come together in design in strange amalgams at this point in time. As he notes, many key figures in the Bauhaus, such as Walter Gropius and László Moholy-Nagy, combined rhetorically leftist-leaning politics and genuine concerns with the labor and environmental issues of their day with the view that the biological sciences might offer to design "practical solutions to social and environmental ills" (Anker, 2010: 3). Moholy-Nagy saw "nature as a construction model," was drawn to vitalist traditions of biology, and looked for "prototypes in nature to determine functionality" (Anker, 2010: 14). Walter Gropius often espoused a broader program still for design education, which proclaimed design as the space that could unite the arts, technology, and the sciences. Julian Huxley, the British ecologist and occasional Bauhaus interlocutor, in contrast, saw genetics and evolutionary biology shading toward eugenics as a pathway to design a new society (a set of assumptions lampooned in his brother Aldous Huxley's novel *Brave New World*). The Bauhaus-inspired science fiction writer, futurologist, socialist, and meta-intellectual H. G. Wells, in *Things to Come*, stepped further into the mix by offering a Malthusian-socialist vision of a future marked by planning, egalitarianism, and shades of eugenics. Alexander Korda's film adaptation of Wells's novel visualized such a future, deploying Bauhaus set design by Moholy-Nagy (Anker, 2010).

Anker suggests that Bauhaus's concerns with functionalism and design interventions that maximized light and air and promoted modes of collective living that could realize the "rational, emotional, and environmental needs" of "the full human being" possessed the potential to ground design in a humanist environmentalism. Gropius did spend a good deal of his career engaged with and attempting to build affordable, high-quality, and finely designed mass public housing for working people, even if the results were somewhat mixed. The Bauhaus was quite successful in synthesizing across a number of design and art disciplines, bridging some of the ever-present architect–artist versus craftsman–laborer tensions, and generated a design curriculum that is still emulated in countless architecture schools. Nevertheless, it could be argued that the Bauhaus at the end of the day never really connected their interests in labor, nature, and design in a coherent program for production, construction, or design education. If we were to acknowledge that the Bauhaus's enormous impact on art, architecture, and virtually all modes of design in the twentieth century holds few parallels, it could be observed that many of the Bauhaus's engagements with environmental questions, labor questions, or ecology never really rose above conceptual or aesthetic inquiry. Ecology largely remained a point of inspiration for thinking about how to enhance aesthetic qualities or functional efficiencies in design rather than directing design toward a transformative program.

Sanitary Design Ecologies

If the dissemination of concepts and insights derived from the science of ecology provides one driver of design–environment conversations in the early twentieth century, public health and the desire to eliminate or contain disease and pandemics provide a second. Sara Jensen Carr has observed that when urban planning and public health were professionalized and institutionalized in the late nineteenth century, they were fields that were quickly forced into engagement to address epidemics of yellow fever, typhoid, and cholera in the industrial city. The result was a range of practical and pragmatic coalitions between these fields and government officials, which, through street paving, the construction of underground sewerage systems, the promotion of personal hygiene, and municipal waste removal, significantly improved urban health and comfort (Carr, 2021: 62). Within architecture and planning, as Jeanne Haffner notes, such impulses helped to feed progressive era aspirations among "socially minded planners and architects" that they would be able to build "not just new forms of housing but novel environments that, in turn, would transform the lives of those who resided in them." It also led to a particular preoccupation with nature/landscape as a "site of sanitation" in the modernist movement, which when allied with ecology could easily move beyond the aspiration to maximize the specific well-being of residents and expand further to the proposition that design, architecture, and planning could be tools for improving the "moral behavior" and "evolutionary fitness" of the population (Carr, 2021). Carr notes that the key objects of such professional concerns were frequently the "problematic behavior" of the poor and working classes. As approaches to public health moved from a broad ecological focus on multiple influences in the environment to a focus on pathology. Carr argues "we see a linear etiology between single origin and outcome at the individual level" (Carr, 2021: 59) and a growing focus on the notion that design and architecture might shape individual behavior.

Carr demonstrates how much these notions running alongside concerns about sanitation critically informed the numerous city-wide projects of Le Corbusier. which, from the Plan Voisin to Radiant City, are particularly preoccupied with structures and plans that can protect inhabitants from the dirt, dangers, and disease of the street and urban life and rather create sanitized and safe modern contained spaces to avoid the mess of the city. Many ironies follow from this, though, in Le Corbusier's urban ecological design schemes, which remain with us today. Notably, with his "towers in the park," we are lifted out of nature to sanitized spaces where we can both escape "contaminated nature," allow spaces for managed nature to flourish, and create individualized automobile-centric corridors to move quickly through space. Le Corbusier's thinking here draws together sanitation and moral control, speed and ergonomic efficiency, providing a template for all manner of auto-centric urban renewal projects across the 1950s and beyond, with mixed results at best.

Olmsted, Howard, and the Garden City Movement

The extent to which a search for ecological holism in design leads to reactionary outcomes certainly remains contested. Gideon Kossoff, for example, has argued we need to distinguish between conservative and radical currents of ecology and holism in the history of design (Kossoff, 2015). If, at the beginning of the twentieth century, we can identify all manner of social Darwinists, nationalists, and conservative voices lining up to claim that evolution "reveals" Nature red in tooth and claw, Kossoff also reminds us that we can also find some of the foundational figures of radical geography—such as Peter Kropotkin and Élisée Reclus—arguing that if the implications of evolutionary theory are understood correctly this reveals mutualisms, symbioses, and self-organizing processes as far more important than competition. And as such, modelling society on this reading of "nature" can as easily justify an anarcho-communist future society defined by mutual aid, self-organization, and commoning as much as it justifies capitalist competition and imperialism. Similar sensibilities can also be found in the evolution of landscape architecture and town planning.

For example, Frederick Law Olmsted (1822–1903) and Ebenezer Howard (1850–1928) were both hugely influenced by late-nineteenth-century utopian socialist, anarchist, and municipal socialist discussions, giving rise to a shared sensibility that the design of healthy democratic urban spaces had to be carefully planned at appropriate scales. Reacting to the unplanned mess of Victorian slums, Frederick Law Olmsted, in his early years, was a follower of utopian socialist Charles Fourier and viewed industrial capitalism as a "transition stage" between "barbarism" and municipal socialism (see Hayden, 1982: 11). He further maintained that the massive expansion of urban parkland and access to public gardens in the city could not only "counteract the squalor of tenements" (Haffner, 2021) but provide a further means to improve the public health and leisure opportunities of the working classes. His park designs with Calvert Vaux which developed a distinct urban ecological aesthetic highlighting the natural drama of rock ledges, lakes, fens, streams, and forests within a city content also aspired to creating democratic spaces that could be shared by both wealthy and working-class publics. Olmsted's influence within the field of landscape architecture—which he named and founded as its own discipline, distinct from either architecture or landscape gardening—is still unparalleled, and the practical legacy of his vision of proto-ecological urbanism endures to this day in many of the remarkable parks and gardens that are his legacy from Central Park and Prospect Park in New York to the

Emerald Necklace in Boston and the many other park systems his office developed in the decades that followed.

On a technical level, Olmsted's parks were cutting-edge integrations of natural systems into urban environments, with the introduction of innovations from forestry and civil engineering, and the development of novel methods for creating convincing moments of "nature" in the city. However, as Sonja Dümpelmann has demonstrated, a closer look at Olmsted, particularly attending to questions of race and class, reveals a more complex legacy.

If Olmsted's early work is marked by an enthusiasm for constructing spaces inflected by utopian socialist and civic republic sensibilities, these youthful commitments rather give way to a more conservative vision of landscape design as social reform later in his career. Working-class and communities of color were displaced by many of his projects, which increasingly tilted to serving wealthier clients. For example, residents of Seneca Village, a community of free Black Americans (as well as some Irish and German immigrants), that was home to the largest group of African American property owners in pre–Civil War New York, were forced by eminent domain to sell their land to the city to make way for Central Park. While Olmstead was a sharp critic of the conditions of slavery during his travels through the Southern states just before the Civil War as a reporter for the *New York Times*, on the whole, as Dümpelmann notes, he was "never more than a moderate abolitionist" and his "attitudes toward Black and indigenous populations were benevolent yet paternalistic" (Dümpelmann, 2022). While Olmsted's contribution in advocating for the preservation of natural wonders like Niagara Falls and Yosemite saved them from extractive exploitation and helped lay the groundwork for the larger National Parks movement, his projects also contributed to the erasure of indigenous peoples from the places he fought to preserve.

Ebenezer Howard's vision of garden city futures, meanwhile, is widely misunderstood in part due to associations with actual garden cities built in the twentieth century and Jane Jacobs's (1961) polemical dismissal of Howard as a design-centric advocate of dull dormitory suburban towns. But actual engagement with Howard's writings and proposals reveals a thinker engaged in a political and design project to reconstitute both urban form and ownership patterns at multiple scales with bigger ambitions than Olmstead *or* Jacobs.

Howard's plan for garden cities outlined in his book *Tomorrow: A Peaceful Path to Real Reform* (1898/2010) proposes a series of human-scaled urban settlements that would surround and deconcentrate the industrial metropolis. Howard's proposition indeed is somewhat futuristic. Far from being isolated, he argued garden cities should be connected to the metropolis and one another by efficient and low-polluting electric railway links which would allow workers access to fast and convenient inter-municipal travel for leisure or work. Each garden city would be rationally planned in terms of urban form and economic outputs. Notably, Howard argued that garden cities had to, first and foremost, provide affordable and spacious housing for workers with their own private gardens as well as spaces for collective gardening. The garden city would be walkable but also have trams and have urban centers designed around many cultural institutions, public buildings, spacious boulevards, and public parks. They would

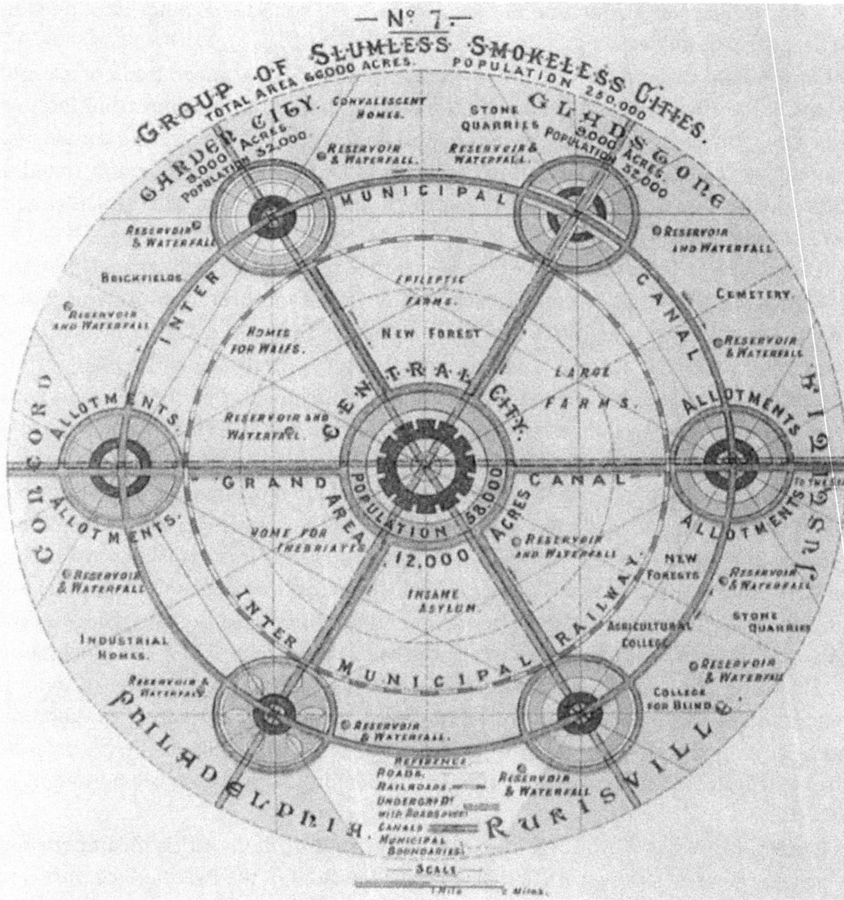

Figure 2 Ebenezer Howard the Garden City Concept from Howard, Tomorrow: A Peaceful Path to Real Reform (1898/2010).

be surrounded by zoned areas for manufacturing and then greenbelts, allowing for agriculture or wildlife to flourish.

Perhaps one of the most interesting parts of Howard's proposal though is how he advocates for an economic basis to the garden city that draws inspiration from mutualist, associational, and cooperative socialist traditions. Howard believed a "collective commonwealth" of garden cities would be sustained by a mixed economy, which would include both private enterprise, workers cooperatives, and municipal ownership of land and utilities, within a system of robust municipal democracy. With land and key assets held in common at the municipal level, it was argued that rents on property and business could be used to fund a local welfare state and pensions.

Howard's cooperative commonwealth can be seen as one of the first modern attempts to think about multi-scalar social and ecological planning that avoids bureaucratic collectivism or Victorian capitalism. But if *Tomorrow: A Peaceful Path to Real Reform* outlines a detailed vision of how garden cities could function, then as Jane Jacobs observed in *The Death and Life of Great American Cities* (1961), Howard did not have much to say about the future of the central metropolis. At the height of the misery of the industry city, Howard's general view was that de-concentration of the metropolitan core would reduce the core to a more manageable size and livable form. Jacobs argued that this move ensured that Howard essentially gave up grappling with the question of how large, bustling great urban forms could remain vibrant and dynamic. This is an oversight that is repeated in a good deal of garden city and regional planning advocacy after Howard. Moreover, Howard never fully elaborated how the relatively self-sufficient underpinning economic vision for garden cities relates to the central city nor did he, in practice, manage to obtain funding for implementing his own mutualist vision of the garden city. The realities of the Garden City developments that did occur in the UK, such as Letchworth and Milton Keynes, did not approximate Howard's original vision. With the innovative social content of Howard's plan removed, the UK ended up with many "garden cities" that effectively became provincial suburban towns.

Nevertheless, Howard's garden city vision did inspire an international set of garden city movements and got town planning advocates enthralled by the prospects of building garden villages, garden districts, garden suburbs, cités-jardins, and ciudades jardins around the world from Europe to Japan and Brazil. And as we shall see in Chapter 4, some of these projects in Berlin, Amsterdam, and Vienna resulted in the construction of carefully planned social housing for workers and their families surrounded by green and healthy living environments. More generally, these ventures also triggered broader conversations, particularly in the United States, in the possibilities that might exist for ecological, social, and labor concerns to be addressed in larger-scale attempts to do regional planning.

Regional Planning and Abolitionist Democracy: Mumford, DuBois, and the RPAA

Lewis Mumford founded the *Regional Planning Association of America* with the conservationist and planner Benton MacKaye and the architect Clarence Stein in 1923. They soon drew into their fold, among others, labor organizer and housing policy experts Catherine Bauer and Edith Wood, (who both went on to shape New Deal era housing policy), the economist and "big power" energy enthusiast Stuart Chase, Charles Whittaker (architect and editor of the *Journal of the American Institute of Architects*), and the architect Frederick Ackerman.

The founding members of the *RPAA* found some degree of overlap in their enthusiasms for the writings of Ebenezer Howard. But it was the Scottish biologist and sociologist Patrick Geddes who provided one of the earliest influences on the group. Geddes, in *City Development* (1904/2018) and *Cities in Evolution* (1915/2022), argued planning should begin with a survey which situates urban form in its broader social, ecological, geological, climatological, economic, and geographical region. He went on to suggest that we should conceptualize cities and regions in terms of a flow of energy, matter, and human and nonhuman organisms. A regional view could additionally be used as a further template for engaging in top-down master planning but also to facilitate public engagement in bottom-up civic participation. As Colin Ward has observed, a central theme of Geddes's work is "the idea that the average citizen has something positive to contribute" to improving their environment. Geddes was convinced that "each generation has the right to build their aspirations into the fabric of their town" (Ward, 1991: 110).

Lewis Mumford's regional planning vision can be seen as an attempt to incorporate some elements of Howard's and Geddes's visions into a new synthesis. Following Howard and Geddes across numerous writings, Mumford argued that cities should be imagined in organic and holistic terms, carefully located and tailored to the environment and resource endowments of their region. What could emerge from this, according to Mumford, was a planned "republic of regions" based on compact and integrated communities, participatory democracy, and new decentralized technologies. Subsequent "members" of the *RPAA* (which was always a loose affair) did not always align directly with such utopian agendas. However, it is a group certainly open to thinking about what a labor and environmentally friendly planning might look like and drawing inspiration at different points in time from the British Garden City Movement, "Red" Vienna's experiments in social housing, Soviet planning, feminism, the technocracy movement, New Deal liberalism, and the civic republican traditions of New England and its town hall meeting.

In general, and whatever their differences, we can find a common social democratic view running through all these *RPAA* members. They believed that regional planning tools, public ownership, zoning, and industrial regeneration focused on maximizing the democratic and ecological health of city-regions could offer the best hope of developing "balanced" urban–rural futures. Different members of the group had different design foci within this framework, ranging from the construction of affordable housing (Bauer, Wood, and Whittaker), industrial regeneration of rural areas (MacKaye), regionalism (Mumford), national planning (Ackerman), energy (MacKaye, Ackerman), labor, and public leisure (MacKaye, Bauer). Nevertheless, they sought to articulate policies and proposals to provide a new labor- and worker-friendly vision of a socially and ecologically engaged planning mode.

For example, MacKaye's advocacy for the Appalachian Trail provides one of the earliest and most concrete examples of the kind of project the *RPAA* sought to develop public support for, which could connect the economies of urban and rural areas and offer real opportunities for working people to experience the natural world, while also allowing for conservation. MacKaye first proposed a two-thousand-mile public walking trail running from the borders of Canada in the north to Georgia in the south in 1921. A trail running the full distance was completed in 1938, and while maintaining the scale of MacKaye's vision, some of the more radical edges of the proposal were lost in execution. MacKaye's proposal had been deeply informed by his socialist conviction that public ownership and workers' rights to wilderness spaces for leisure were key to any viable conservation policy that would balance humanity's and nature's needs. Whereas the wealthy could visit the National Parks out West to get their fill of outdoor recreation—or "those in the East who can afford time and funds for an extended trip in a Pullman car," as MacKaye put it—he advocated for more egalitarian opportunities for leisure: "the camping grounds to be of the most use to the people should be as near as possible to the center of population. And this is in the East" (MacKaye, 1921). In its original intent, he envisaged that the trail would initiate "a system of planning embracing the technical skills of the forester, the engineer, the architect, and the agriculturalist" with the "active involvement of the people" (Spann, 1996: 29). As such, he hoped the trail would include public facilities such as shelter camps with co-operative housing projects forming small communities, and food and farm camps, which taken together would provide both rest opportunities and rural employment for working people. He also hoped the trail construction would lead to an energized labor movement that would make greater demands for more leisure as they engaged in collective bargaining with their employers (Mattson, 2020). Whereas the trail "might attract outdoor enthusiasts for an enjoyable hike in the woods, the broader goal was to relocate large populations into cooperative, nucleated communities located along the length of the mountain range" (Nelson, 2019). The trail gave rise to a more ambitious vision of the potential of rural revitalization through planning and public engagement, seeking to operate at the scale of an entire region, redirecting patterns of settlement and leisure for the whole of the eastern US.

In his later years, MacKaye sought to develop a much more ambitious vision of a national development strategy, notably "a People's Appalachian empire … stretching from New England to deep in the South," which would reclaim land for productive use and revitalize communities and new visions of Michigan reclaimed by anchor farm and timber colonies devoted to sustainable forestry (Spann, 1996).

The architects Charles Whittaker, Clarence Stein, and Frank Ackerman focused more on urban concerns—housing and overcrowding—but all notably shared a collective desire to view the architect's role less as a singular artist and more as one of many design professionals who could work together and with the public to achieve "a great and universal human service" (Spann, 1996: 8). Whittaker wrote an article advocating the building of 10 million homes. All admired the Garden City movement in Britain, the speed at which postwar housing developed in the UK, and embraced the project to design and build planned communities. This led to experiments in building affordable garden suburbs in Sunnyside, Queens, and Radburn, New Jersey, marking the first exercises in community planning in the United States. More broadly, Whittaker sought to create "A League of Professions," bringing professional men and women into a force dedicated to changing society and operating as "a powerful counterforce to business and the profit motive" (Spann, 1996: 8).

If Whittaker maintained a benign technocratic vision that a coalition of architects, scientists, engineers, and others could "liberate the professions from the domination of selfish interests" (Spann, 1996: 9), Frank Ackerman held to a more radical vision still of "making Architecture and Planning a democratic movement aligned with an awakened citizenry" (Spann, 1996: 8). To advance this vision, he argued that city planning should be taught in public schools "to train them to evolve practical solutions from life experience" (Spann, 1996: 11). He further advocated that concerted attempts should be made to establish more cooperative relations between the design professions and the building crafts to build housing, particularly planned communities. Ultimately, Ackerman advocated for "effective national planning," which he believed could "guide the whole of American society" (Spann, 1996: 16). This would involve thinking beyond state boundaries, embracing city and regional planning, and planning for rural areas. However, Ackerman believed that effective urban–rural planning in the United States ultimately required "public control over investment" (Spann, 1996: 17) and a move from an individualist ethos to a more cooperative worldview.

The *RPAA* had an ambiguous relationship to the question of scale and the future forms that could bring democracy and design together. Whilst Mumford saw decentralization of the urban core through the establishment of appropriately scaled garden cities and revitalized rural economies as necessary to achieve a balance between city and region and a more participatory democratic culture, other members of the *RPAA* understood the project of city-regional planning in rather more expansive and technocratic terms. For example, Stuart Chase and Benton MacKaye increasingly saw regional planning as the means to achieve rural revitalization and the doubling of living standards (Spann, 2016: 69). A key strategy would involve making production facilities more efficient through "Giant Public Power." Spann notes that giant public power involved building "one great rational system of electrical power production and distribution" that would bring cheap energy even to isolated rural areas. MacKaye and Chase argued that this could be achieved by locating giant electricity-generating stations at coal mines, through the much greater use of hydropower, and through the development of a new grid marked by high-voltage electricity lines that would run "to every part of the country" (Spann, 1996: 69). They were joined in this Giant Power campaign in the 1920s by prominent political figures like the New York governor Alfred E. Smith and Pennsylvania governor (and prominent forester and conservationist) Gifford Pinchot. Both MacKaye and Chase saw the anticipation of such a program in Roosevelt's proposed Tennessee Valley Authority. Benton MacKaye enthusiastically supported the TVA and saw it as a key opportunity to establish the central importance of regional planning for the New Deal. He became a regional planner for the TVA and advocated building regional centers or garden cities around metropolitan centers. He also sought to address the dilemma of how citizens' control over policymaking could reconcile with the need to draw from technical expertise in planning. MacKaye's proposal for resolving this amounted to a vision of regional governance that would undercut the power of state governments and facilitate democratic governance by an enlightened regional technocracy. But such a proposal for planned regions held little political appeal for either federal or state officials in the United States, and did not move forward, aside from the two power authorities on the Tennessee and Columbia Rivers (the TVA and Bonneville Power Administration), where the flood control, irrigation, and hydropower benefits were too good to ignore and which had long been championed by powerful senators and coalitions. Mumford similarly engaged in city master planning and advocated in 1932 that Roosevelt should embark on a massive national housing program that would provide low-interest loans for the construction of "community units -villages, urban neighborhoods, whole towns" (Spann, 2016: 165). In contrast to MacKaye's vision of regionalism, Mumford's vision of the future involved advocating for an entirely new social and ecological order that would involve the fusion of science and humanism (Spann, 2016: 162).

The ecological vision of regional planning espoused by the *RPAA* in many parts appears remarkably progressive, if not radically transformative, for the United States. However, gaps exist in this intervention. Mumford's polemics against "giantism" and his hope that self-sufficient city-regions might transcend urban sprawl remain ambiguous. For example, it is unclear how the governance needs of large metropolitan cities would fit into the city-region. More strikingly, in an era of United States history marked by legalized segregation, the enforcement of Jim Crow policies in the South to prevent Black engagement at the ballot box, lynching, and systematic discrimination in housing and land ownership, the *RPAA* said little substantively about how regional planning might directly address the struggles of racially unequal societies where non-white citizens experienced subordination through the exercise of extraordinary levels of state-sanctioned violence. Here, it is interesting to contrast the *RPAA* with the concerns that animate the writings of W. E. B. DuBois on the history and politics of urban–rural matters. As Reinhold Martin has observed, Mumford knew DuBois well, and he had engaged with the NAACP at the fringes. He was a principled anti-fascist and publicly spoke up against the horrors of racial supremacy and fascism brewing in Europe across the 1930s. Mumford's engagement with "the racial question" at home, though, rarely moves beyond a general liberal humanitarian call for fair treatment of all humanity (Martin, 2020). The disconnect here with DuBois's near-contemporaneous scholarship is striking. DuBois in *Black Reconstruction in America* (1935/2014) offers a comprehensive historical account of the particular ways in which the promise of Black and indigenous dispossession from land by white settlers acts as a key means through which white capital bought off the loyalties of the white working class. Yet, Mumford's grand meta-theorizing and sweeping Hellenic and Europhile histories of urban civilizations developed in *The Culture of Cities* among other places (Mumford, 1938) give no space to this observation. Members of the RPAA lived through the great migration of Southern Blacks to the urban North and the publication of DuBois's *The Philadelphia Negro* (1899/2017), but the RPAA, for all its concern with labor issues, had very little to say about the plight of Black and indigenous labor, the racial dynamics of the US city or the extent to which regionalism might require a head-on confrontation with the use of regional and organicist myths by white supremacists.

Here we might also observe some further gaps in the general vision of the RPAA. MacKaye's vision of rural regeneration through "agricultural colonization" and community settlement of "wild lands" that would give rise to Jeffersonian "planned farm communities" does not address the issue of Black and indigenous dispossession. Both Mumford and Benton MacKaye idolized the New England town (and history of town meetings), its human-scale forms, and its participatory culture. Oddly enough, though, RPAA members such as MacKaye were not beyond talk of the virtues of the Appalachian range as a "traditional society" inhabited by "the purest form of original American Anglo-Saxon stock" (cited in Spann, 1996: 73). DuBois was initially a supporter of FDR's New Deal rollout and the economic development that it could have brought to Black workers, but quickly soured on the implementation, which left unequal and racially discriminatory employment and housing policies in place, even for federal projects like the TVA (as we discuss in Chapter 6). The onset of the Second World War brought entreaties for DuBois to muffle his criticism in the name of public morale and the war effort, along with a quieting of calls to end segregation in the army or pass anti-lynching legislation. Ultimately, the RPAA's vision of public power and regionalism never fully grappled with DuBois's claim that unless the United States built an "abolitionist democracy" (DuBois, 1935/2014) that fully enfranchised African Americans and other subordinated and racialized and marginalized peoples, it would never even begin to approximate its democratic claims.

The Counterculture Explosion Part I: Bucky, Brand, and the Whole Earthers

The postwar period in the United States, marked by the Cold War abroad and McCarthyism at home, largely shrinks the space for progressive conversations generated in architecture, planning, and design by Howard, Geddes, Mumford, and the RPAA. With overwhelming social priorities now focused on reorienting industry and design to address pent-up consumer demand and with states turning to technocratic planning (among other tools) to rebuild infrastructure and housing as quickly and cheaply as possible, prewar homilies about the virtues of craft, artisan productions, the potential of garden cities, or idealistic visions of ecologically sensitive regional planning seemed out of step. The RPAA's hope for intelligently and publicly planned communities that would be government-led and seek to give rise to compact and walkable cities, surrounding garden cities, and broader city regional forms that could "correct the imbalance" (Spann, 2016: 51) between town and country may have continued to be concerns of European planning (and re-emerge much later again through the New Urbanist debates of the 1990s. However, they are mostly pushed to the sidelines in the United States. In contrast, highway construction, "urban renewal," and government-enabled but mostly private developer–led rapid suburbanization and sprawl became much more important in shaping urban form in the United States.

Within architecture and from an environmental dimension, as Daniel Barber has documented, the embrace of high modernism and the international style saw the rise of the sealed facade and the fully conditioned interior, which increasingly cut mainstream architects off from the environmental conditions that surrounded them (Barber, 2016: 2). The design fields largely focused on producing products and dreams that would consolidate suburbia. Nevertheless, sometimes operating under the mainstream and sometimes alongside were design experiments that would open up different pathways later.

The 1950s may have given rise to autotopia rather than Mumford's ecotopia; still, it is a period marked by interesting anomalies and contradictions, where other socio-technical developments laid the grounds for other possibilities. As Barber (2016) notes, within pockets of architecture, there were important experiments in solar house construction in the United States, particularly in the southwest—in California and New Mexico—using solar thermal and passive cooling systems, admittedly with varied results. In the Global South more broadly, we saw developments in climate-attentive modes of architecture that deployed innovative shading and cooling through screens, fins, louvers and brise-soleils, and other architectural means to control the way the sun enters the building, rather than reliance on mechanical ventilation and air conditioning, underpinning incipient currents of climate design methods (Barber, 2023).

Peder Anker has observed that Cold War military competition and the space race drove research and development into solar and wind energy, new materials, scenaric planning, cybernetics, systems thinking, and experimental structures that could be deployed quickly and easily in a military confrontation. Many of these technologies found new purposes and uses in very different ways as the 1960s counterculture explosion revived the design–ecology relationship. The architect, designer, and futurist Buckminster "Bucky" Fuller looms large here as a critical figure in the military-to-counterculture ecology interface.

Fuller is perhaps most known today for his introduction of the geodesic dome into architecture, his various "dymaxion technologies" from the prefabricated dymaxion house to the streamlined and futuristic dymaxion car, to his unfolded polyhedral dymaxion map of the globe, and his proposals for airborne floating and underwater cities. However, as his eclectic body of design work expanded to scenario building, world game planning, and system theory, so did his ocular pronouncements on the fate of "spaceship Earth," culminating in the claim that an age of "planetary crisis" required a synthesis of design science and system theory that would ultimately allow us to replace the irrationalities of politics with a "design science revolution" that would open up a more rational comprehensive approach to solving humanity's problems (see Anker, 2010). Such bold claims resonated with a generation of young radical designers, engineers, architects, and scientists disgusted with the Vietnam War and mainstream politics and looking to some possible synthesis of design, technology, and ecology as an escape. This project was arguably made most manifest through the work of the self-professed Fuller disciple, sometime ecologist, and full-time cultural zeitgeist definer Stewart Brand and his *Whole Earth Catalog*.

The *Whole Earth Catalog* and the magazine *Co-Evolution Quarterly* provide a starting point for thinking about how ecology and design established a new terrain of engagement from the late 1960s onward. As Andrew Kirk has suggested, "Whole Earth provided a publishing vehicle and an audience for a new generation of 'outlaw' designers who were exploring the possibilities of ecologically sensitive housing, industrial design, energy production, and farming" (Kirk, 2007: 9). Rather than providing a body of original ideas, projects, and proposals, as Simon Sadler has insightfully observed, we might see *Whole Earth* as operating like a giant information aggregator—almost like a precursor of today's search engines, offering access to tools and books to change the world (Sadler, 2008). Stewart Brand and his colleagues modeled the catalog on the Sears and L. L. Bean mail-order catalogs (Turner, 2010), with some further points of aesthetic inspiration drawn from the underground media and zines of the 1960s, and curated an exploding sixties scene intending to furnish the reader with "tools" that could ensure they would "find his own inspiration, shape his own environment, and share his adventure with whoever is interested" (Brand, 1968).

The catalog was indexed into broad sections: Whole Systems, Shelter and Land Use, Industry and Craft, Communications, Community, Nomadics, and Learning, which documented all manner of emerging ecological technologies from solar and wind to geodesic domes and dymaxion technologies (from cars to houses), aquaculture and architecture, to offering further product reviews of tools, manuals, and kits that might be of use to commune dwellers. *Whole Earth* did not sell any products, but it offered readers the contact information of vendors who wanted to get their hands on this kit. *Whole Earth*, though, was not simply a product reviews catalog; rather, it explored all material and cultural practices from folk art, tepee construction methods for beginners, back-to-the-land farming equipment for rural communards, to psychedelics, DIY education, systems theory, and emerging computation technologies. This latter category also contained significant further intellectual content in the book reviews section, highlighting the writings and work of place- and context-sensitive designers such as Ian McHarg, Bernard Rudofsky, and Christopher Alexander, urban theorists such as Lewis Mumford, and emerging free market libertarian heroes such as Ayn Rand and Milton Friedman.

Buckminster Fuller's influence over Brand can be seen in many places. For example, the *Whole Earth Catalog* shared with Fuller an enthusiasm for dymaxion technologies, prefabricated structures, geodesic domes, cybernetics, and systems theory. However, perhaps Fuller's most substantial legacy to *Whole Earth* lies with his notion that a post-ideological, makerly politics of design might allow counterculture ecologists, designers, engineers, and makers to sidestep the emerging political contestations unfolding around the New Left, the Civil Rights movements, as well as more militant voices still arguing for Black Power or anti-imperialism. Here, it is

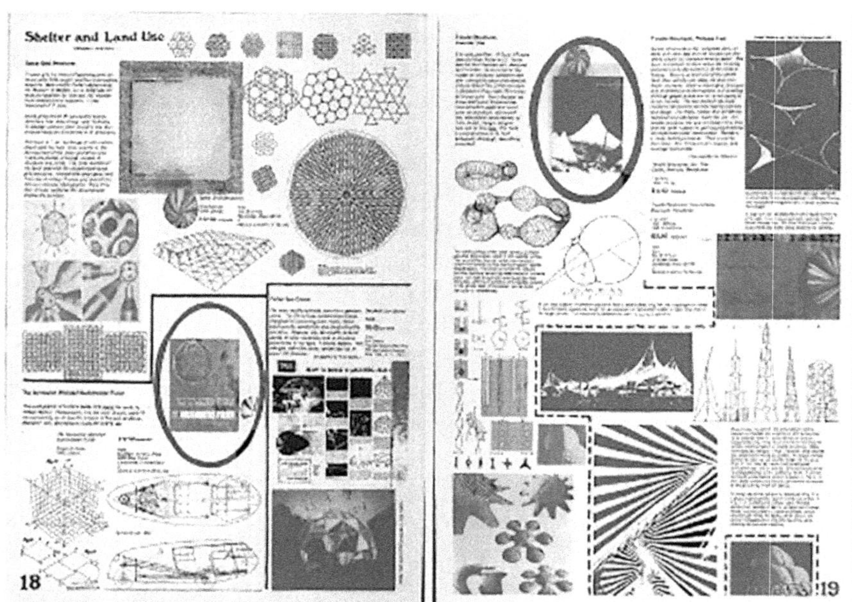

Figure 3 Assembling the last issue of the *Whole Earth Catalog*. Menlo Park, Calif., May 28, 1971. Stewart Brand, 32, publisher on the phone. Photographer Richard Drew; AP Copyright. Reproduced with Permission. A page from the *Whole Earth Catalog*. Copyright Stewart Brand. Reproduced with permission.

interesting to compare the shift in the politics of design that occurs between the high point of the *Regional Planning Association of America* and the *Whole Earth Catalog*. If both these moments contain utopian design impulses, for the RPAA, as we have seen, it is the potential of master planning allied with civic engagement, and designers and architects allied with labor and public agencies that is seen as the way forward. In contrast, *Whole Earth* offers a much more individualist and consumerist path to the age of ecology. It is bourgeois bohemians with private incomes, counterculture dropouts, radical designers, engineers, and assorted other makers that are now offering the path to liberation.

The Counterculture Explosion Part II: Liberatory Technologies of Solidarity—Bookchin and the Radical Technology Movement

If the *Whole Earth Catalog* offers one influential path to think about the design–ecology relationship, where design, in many respects, became a means through which politics could be transcended, this is not the only vision of a design-politics that we can locate during this critical period. The late 1960s and early 1970s also saw the explosion of a range of groups and organizations across the East and West Coast that sought to cultivate, develop, and propagate ecological design and alternative technologies.

The New Alchemy Institute in Woods Hole, Massachusetts, led by scientists John and Nancy Jack Todd, established itself as a research and education center developing projects in "living machines" (notably constructed ecosystems that would use solar power and living organizing to treat wastewater and remediate polluted water environments by mimicking natural ecological processes without chemicals); bio-shelters (solar greenhouses that sought to develop self-contained ecosystems combining plants, animals, soil, and insects to create closed-loop life cycles for food production, waste purification, and energy efficiency); organic agriculture and aquaculture, and "self-sustaining" energy-efficient buildings. *The Farallons Institute* in Berkeley, California, brought together architects, engineers, and biologists in the Bay Area under the guidance of Bill and Helga Olkowski and the architect Sim Van der Ryn to develop "urban scale ecological technology." The end product was the building of the *Integral House* a demonstration project built and maintained by *Farallons* members that sought to build an energy-efficient urban house made up of sustainable materials that could also integrate energy conservation with food growing and waste disposal. Van der Ryn was subsequently appointed "state architect" in 1978 by Governor Jerry Brown in California and founded the Office of Appropriate Technology. *The Institute of Local Self-Reliance* in Washington, DC—heavily influenced by the somewhat idiosyncratic figure of Karl Hess—sought to incorporate ecological technologies such as neighborhood gardens and urban aquaculture into broader strategies for community development, local banking, and local financial systems in the Adams Morgan neighborhood of Washington, DC. Perhaps most interesting, though, from the standpoint of leftist social theory is *The Institute for Social Ecology* (ISE) in Vermont.

The ISE was founded by Murray Bookchin and Dan Chodorkoff in 1974. Bookchin had developed an extensive body of reflections on ecology–technology–design–society relations dating back to the early 1960s (see Bookchin, 1971), and the ISE sought, in many ways, to develop a political and programmatic vision of the role of ecological technologies in building an emancipatory politics. If the ISE experiments in community technology and eco-design largely followed the lead of *The New Alchemy Institute*, *The Institute of Local Self-Reliance*, and *The Farallons Institute*, and often collaborated with such institutes, convening shared meetings in Vermont to advance discussions of eco-design and technology, it is in the theoretical and conceptual interventions where Bookchin and his colleagues made their primary contribution to "the movement."

Bookchin shared many of Stewart Brand, John Todd, and Karl Hess's enthusiasms for the potential of distributed ecological technologies, automation, and cybernetics. However, he also argued in a series of rather more politically pointed essays and texts across the sixties and seventies (collected in Bookchin, 1971, 1980, 1982) that the ecological pathologies of capitalism were not going to be resolved by strategies for social change reliant on getting "tools" into the hands of countercultural radicals or relying on well-meaning green CEOs to bring about a new society. More generally, he suggested that the proposition that design and technological interventions could be judged as optimal just because they are "small," "intermediate," "appropriate," "hard," or "soft" was misleading:

> The historical problem of technics lies not in its size or scale, its "softness" or "hardness," much less the productivity or efficiency that earned it the naïve reverence of earlier generations; the problem lies in how we can contain (that is, absorb) technics within an emancipatory society. "Small" is neither beautiful nor ugly; it is merely small. Some of the most dehumanizing and centralized social systems were fashioned out of tiny technologies. Still, bureaucracies, monarchies, and military forces turned these systems into brutalizing cudgels to subdue humankind and, later, to try to subdue nature. (Bookchin, 1982: 240–1)

So what should be the standard that informs our design and technology choices for an ecological society, and how would this be organized? Drawing inspiration from Hellenic thought, Hegel, Mumford, and critical theory, Bookchin argued that attention must be given to humanist and ecological considerations in making our technological and design choices and to the broader social, political, and institutional matrix—the social ecology that design and technology reside within. This requires paying attention to the ecological impacts of design, planning, and technological development. It additionally requires us to ask further questions about the patterns of ownership and control and the extent to which the use and deployment of such technologies might cultivate autonomy, a healthy sense of interdependence, or merely reinforce existing patterns of social domination. In this respect, he argued, there was indeed nothing necessarily progressive about the rise of an eco-technology industry:

> Solar power, wind power, methane, and geothermal power are merely power insofar as the devices for using them are needlessly complex, bureaucratically

controlled, corporately owned, or institutionally centralized. Admittedly, they are less dangerous to the physical health of human beings than power derived from nuclear and fossil fuels. Still, they harm humanity's spiritual, moral, and social health if treated merely as technologies that do not involve new relations between people and nature and within society. (Bookchin, 1980: 226)

Bookchin further argued that the proposition that ecology alone, either as science or cosmic ontology, geometry, or complex systems theory, could operate as a singular pre-political guide to action was misguided—engagement with critical social theory was also required.

Bookchin's interventions on eco-design questions then are notable for their general distrust of the US alternative technology movement's tendency to replace political analysis with cool tools and systems theory. His interventions are further notable for their broader hostility to all forms of Malthusian, scarcity-orientated, and romantic environmentalisms of an anti-humanist and technophobic flavor and their general view that it is only through advocating for and seeking to bring into being a post-scarcity and urban-focused social ecology that this movement had any chance of building a broader base of support beyond the counterculture (see White, 2008). For example, it is argued, in his 1965 essay "Towards a Liberatory Technology," that whatever the endpoint of ecological limits of capitalism, we must acknowledge that "bourgeois society ... if it achieved nothing else, revolutionized the means of production on a scale unprecedented in history." This technological revolution, culminating in cybernation and the rise of new eco-technologies, has created "the objective quantitative basis for a world without class rule, exploitation, toil or material want" (Bookchin, 1971: 33–4). The essay goes on to foreground a range of ways in which a further extension of existing tendencies in manufacturing toward automation and the expanded deployment of computer technologies via cybernation could open up vast potential for delimiting toil and improving the energy efficiency of all manner of manufacturing, mining, and agricultural modes of production. It is further speculated in the essay that an ecological society could derive its energy from "solar, wind, hydroelectric and geothermal energy ... heat pumps, vegetable fuels, solar ponds, thermoelectric converters, and eventually, controlled thermonuclear reactions" (1971: 119). As Bookchin acknowledges in other writings, the development of such an ecological society would further require a transformation of production and consumption, a "new industrial revolution, one which will replace a patently obsolete, highly centralized wasteful technology designed to produce shabby, short-lived junk commodities in immense quantities by long-lived, high-quality, useful goods that satisfy rational human needs" (1974: xx). To be sure, at some point in the future, a free ecological society would "be transformed from a wealth of things to a wealth of culture and individual creativity" (1982: 69). However, in the meanwhile, people will have to have their fundamental needs met—and a degree of abundance will have to be spread to all. Bookchin ultimately argued that this new technological and design settlement could only be achieved by cultivating and expanding newly democratized municipalities practicing economic and political democracy.

Bookchin's eco-humanist, ecotopian, and post-scarcity interventions in the alternative technology debates in the 1960s and 1970s were not always well received by

Figure 4 Institute for Social Ecology, Cate Farm, Vermont, 1975. Photographer. Mark Ivens. Copyright Institute for Social Ecology. Used with permission.

a movement that was heavily preoccupied with environmental limits and the coming ecological catastrophe, and he often clashed with the growing market orientation of the US alternative technology scene (see Bookchin, 1980). His ideas, though in a design context, were taken up enthusiastically by the Radical Technology movement in the UK. Centered around *Undercurrents*, a "radical science and people's technology" magazine, the Radical Technology Group shared many common reference points with Whole Earthers, soft technologies, and others. Nevertheless, they more explicitly believed a future for design should situate itself not in new age nostrums but "the rag-bag anarcho-utopian-visionary-revolutionary tradition," which they suggested included everyone from John Ball, Gerald Winstanley, William Godwin, Thomas Paine to Robert Owen and Charles Fourier, the diggers, the levelers, and the Paris Communards to Howard, Geddes, and Kropotkin. It is Bookchin's essay *Towards a Liberatory Technology* that is described in *The Radical Technology Handbook* as "perhaps the single most influential piece of writing on radical technology" (Boyle et al., 1976: 279).

The term "radical technology" was purposely chosen by the editors of *Undercurrents* to distinguish itself from alternative technology, which in its fixations with wind generators, methane digesters, autonomous houses, and so on, was "too much concerned with pure gadgetry, especially of the environmental variety, for use by the affluent to soothe their consciences and amaze their friends at safe distances from the cities." As Boyle and Harper outline, by using the term radical technology:

> We wanted to express an ideal of a technological organization that was part of a total movement towards a new kind of society, but at the same time to assert the belief that technology itself matters, not just who controls it—that is, in other words, not only the relations of production—but the means themselves must be changed to permit the achievement of a just, stable and fulfilling society. (Harper and Boyle, 1976: 267)

The Radical Technology Handbook was organized not unlike a condensed *Whole Earth* review. It was divided into seven sections: Food, Energy, Shelter, Materials, Communications, and "other perspectives" to review the multitude of ways in which society and technology might be reorganized around more "humane, rational and ecological lines." The text, though, breaks with *Whole Earth*'s ban on explicit discussions of politics. Instead, the emphasis is placed on the importance of discussions of "hardware" (machines and technical methods) and "software" (social and political structures). The handbook highlights explorations of the usual 1960s/1970s interests in organic gardening, hydroponics, tree farming, and autonomous and self-built housing, and it considers the potential for energy generated by solar, wind, and wave. However, unlike *Whole Earth*, elaborations of the possibility of such technologies, *The Radical Technology Handbook* is further interlaced with advocacy of workers' control and workers' co-ops, discussions of street farming, the economics of autonomy, and visions of a green-anarcho-socialist city of the future provided by the anarchist artist and illustrator Clifford Harper.

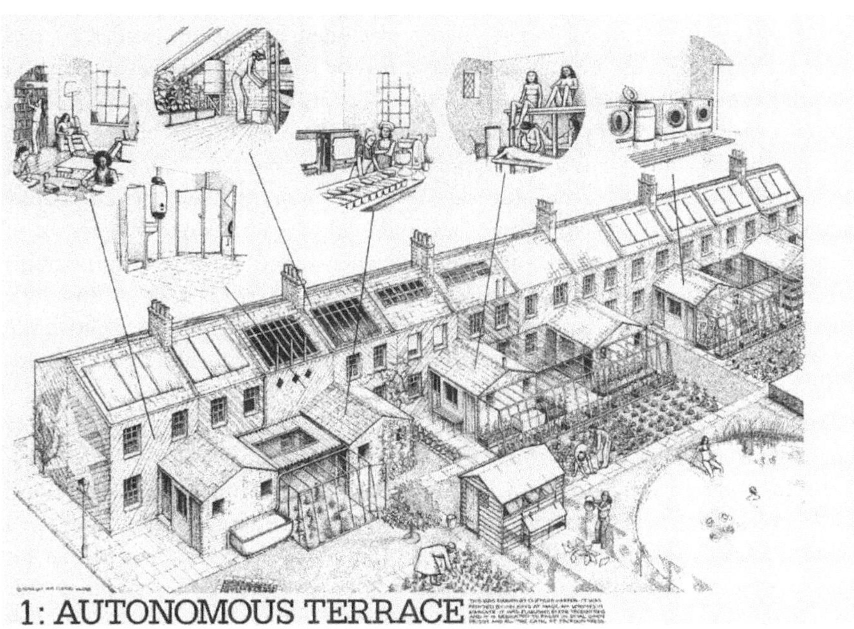

Figure 5 Clifford Harper, The Autonomous Terrace. The Radical Technology Handbook. (1976) Copyright Clifford Harper. Reproduced with permission.

Gender Blind Spots in the Alternative Technology Movement

After the revolution, who's going to pick up the garbage on Monday morning?
Mierle Laderman Ukeles 1969 *Maintenance Art Manifesto 1969!*

If we go back to low technology alternatives, that is labor-intensive rather than energy-intensive solutions, whose labor will we depend on?
Judy Smith *Something Old, Something New, Something Borrowed, Something Due: Women and Appropriate Technology* 1983, 69.

Fred Turner has observed that one key feature of the *Whole Earth Catalog* is that it was ultimately focused on addressing an audience that was "masculine, entrepreneurial, well-educated, and white." Marked by a celebration of "systems theory and the power of technology to foster social change," it was a project that, for all its professed radicalism, would turn away from questions of gender, race, and class and toward a rhetoric of individual and small-group empowerment (Turner, 2010: 97). The limitations of this vision did not go without challenge.

Women were extensively involved in the broader alternative technology movement of the 1960s and 1970s in the United States, both in terms of driving projects forward and offering pertinent critiques of the movement. Nancy Jack Todd was the cofounder of *The New Alchemy Institute*, one of the key centers of alternative technology in the North East and author or coauthor of a range of influential publications advocating for the virtues of ecological design, living machines, bio-shelters, and eco-cities which she argued should be further underpinned by feminist principles of care and solidarity as well as ecological values (Todd, 1977, 2012 Todd and Todd, 1994). Lois Britton, a mathematician by background, was a key figure in assembling, producing, and distributing the *Whole Earth Catalog* (Kirk and Winton, 2007). The artist, environmentalist, feminist, and advocate of participatory art and design, Mierle Laderman Ukeles, had explicitly drawn the counterculture's attention to the question of unequal gendered labor in all manner of maintenance work (including what she called "earthwork") in her *Maintenance Art Manifesto 1969!* The Women and Technology Project, led by molecular biologist and zoologist, feminist activist, and AT advocate Judy Smith, in particular, offered some pointed reflections on the gendered limitations of the movement.

Smith's booklet *Something Old, Something New, Something Borrowed, Something Due: Women and Appropriate Technology* was first published in 1978 by *The National Center for Appropriate Technology*. Here, Smith argued, "With certain exceptions, the AT movement is a very male-dominated movement reflecting very traditional attitudes towards women and women's work" (Smith, 1978). Smith argued that for many alternative technologists, "women merely receive provided services, rather than participate actively in the creation of alternative futures" (Smith, 1978: 3). She additionally noted "Certain high-level technologies," particularly reproductive and contraceptive technologies, technologies that lessen the burden of manual labor, and labor-saving technologies around the household, "may be necessary for women's autonomy, and if so, an emphasis on labor-intensive decentralized technology could destroy women's chance to move out of their traditional roles" (Smith, 1978: 13). Notably, Smith argued that if appropriate technologies did not demonstrate to low-income women in particular that "AT can lower their survival needs," attending to issues of poor-quality housing and fuel poverty, it will be an irrelevance. Following up on these thoughts, some five years later at a conference on *The Future of Technology and Women* held at San Diego State University, Smith noted an ongoing and distinct gendered division of labor within the AT movement:

> At solar energy conferences, all the technical talks are still being given by men; however, all the education and public relations talks are given by women; sex role separation continues even within these "advanced" movements. The historical correlations are fascinating; men continue to do the "hard work" whilst women do the "soft" work, despite the fact that we are professing a "new technology." ... Time and again we find women in appropriate technology dealing with the construction of solar greenhouses, while the "important" projects, such as large-scale energy projects and solar collectors, are addressed by men. Technologies for preparing food and keeping the home are discussed by women and seldom, if ever, by men. (Smith, 1983: 66)

Smith pointed to further unresolved issues in the proposition that a future ecological society would involve some degree of re-ruralization with a vast expansion of gardening, food cultivation, aquaculture, and food preservation. Underpinning this advocacy was the commonly held notion that a society living within ecological limits would be *labor intensive* rather than *energy intensive*. This raised further questions about who exactly was going to do this work in ecotopia and how the domestic division of labor would avoid "back to the land" fall most heavily on the backs of women. The assumption that a new domestic division of labor would magically emerge in the new society to resolve such issues was treated as an object of skepticism for many women in the AT movement who had difficulty getting their male partners to wash the dishes in 1960s communes. As Smith observed, many of the people pronouncing the virtues of "voluntary simplicity" were men, but the men espousing those ideas were "men, making decisions based on the same old values" (Smith, 1983: 66). Smith went on to observe that bringing a feminist lens to appropriate technology debates raised questions about the potential tensions between energy-saving cooking or food technologies advocated by some AT men and the impact of such changes on women's time and labor. As she argued: "Appropriate technology people, downplaying the importance of women's time say, 'It's not a big problem if women have to spend more hours baking bread in a wood-burning stove, or if they don't have as much leisure time because they have to raise a garden and do all the food processing too'" (Smith, 1983: 67). "Technologies," Smith observes, "can either expand or restrict women's roles" and technology discussions need to take this on board. There are modes of high technology, such as contraception, that have been of great benefit to women and modes of low technology that are backbreaking and unhelpful. Ultimately, the issues were clear-cut for Smith:

> If we wish to live in a sustainable, ecologically sane, non-sexist society where women are equal and have power, women as a group must be involved in the decisions that affect their lives, decisions that are now being made without them. (Smith, 1983: 69)

Black Ecology/White Ecology/Apartheid Ecology and Alternative Technology

Race or the struggle for civil rights was rarely explicitly foregrounded as a topic of discussion in the key AT publications of the 1960s and early 1970s and reflects the general avoidance of race by the 1960s mainstream environmental movement.

In 1970, writing in *The Black Scholar* in the same year as the first Earth Day which marked the beginning of the "environmental decade" that would see the creation of the first federal protections for air, water, and endangered species in the United States, Nathan Hare offered a pointed critique of the class and race limitations of the white environmental currents of the era. Hare argued that ecology was potentially of "momentous relevance" for the liberation of Black people. However, the inability of "white ecology" to even begin to address the specific issues faced by working-class urban African Americans—such as air pollution, slum housing, overcrowding, inadequate climate control in urban housing, dietary deficiencies, poor food access, and poor health outcomes, as well as the overrepresentation of black workers in dirty and dangerous jobs—had ultimately left much of the ecology agenda an irrelevance for black people. Hare suggested from this, "No solution to the ecological crisis can come without a fundamental change in the economics of America, particularly with reference to blacks" (Hare, 1970: 7) and, indeed, systematic decolonization abroad.

Hare's critique was followed up some five years later in the May 1975 edition of *Black World* by Terry Jones. In *Apartheid Ecology*, Jones argues along similar lines to Hare that "the concept of ecology in everyday American life is potentially one of the most relevant forces imaginable in the ultimate liberation of Black America," even though at present "it blatantly overlooks Black Americans and their environmental interests" (Jones, 1975: 5–6). As Jones notes, black migration from the rural South to the Northern cities after the First World War and the subsequent white flight to the suburbs gave rise to fundamentally segregated and unequal conditions of housing. Racism, segregation, and capitalist urbanism ensured that "decaying buildings, polluted air from factories, and crowded living conditions make up the ecological environment of inner-city Blacks," while "suburban whites have an ecological environment that tends to be more pleasant, though not entirely problem-free" (Jones, 1975: 8). Crowded conditions, poor services, and a broader lack of recreational facilities are only one outcome of apartheid ecology; employment, poorer schools, poorer public health, and mental health outcomes are some of the "environmental problems" generated by housing shortages, ruthless landlords, excessive rents, and poor repairs. Jones notes that such conditions of *apartheid ecology* have been actively facilitated by the federal government through highway construction, subsidized mortgages for whites, and redlining for blacks, all encouraging white flight from cities and the segregation of the suburbs.

Hare and Jones were writing seven to ten years before the first protests broke out in Warren County, North Carolina, which would later be recognized as the kicking of the Environmental Justice movement. It was not until 1978 that Robert D. Bullard began his study of landfills in the city of Houston and their disproportionate placement in Black neighborhoods, which would go on to show a pattern of locating toxic waste sites in minority neighborhoods that would later be documented across the whole South (Bullard, 1990) and across the whole of the United States (UCC, 1987).

For Jones, redress lay not in DIY programs or tinkering with technology, but in explicitly political pathways: "In order to effectively alleviate the apartheid housing situation that has developed, state, local and federal governments will have to intervene in a more decisive and positive manner" (Jones, 1975: 16). As Jones noted, "In the final analysis, if we are really worried about ecology, we will take steps to eliminate the overcrowded, inhuman living conditions where a segment of the population is degraded, brutalized, and compartmentalized. We must develop methods which begin to ensure ghetto residents the same privileges and rights as the suburban dweller" (Jones, 1975: 17).

The social basis of apartheid ecology is squarely located in concerns rarely prioritized by alternative technologists or other environmentalists in the 1960s and 1970s: urban segregation, the broader environmental challenges of cities, and the crisis of housing for racialized urban populations. Jones was echoing demands that had been voiced by both civil rights activists as well as Black liberation organizations throughout the 1960s, most prominently the Black Panther Party. The Ten-Point Program of the Black Panther Party in 1966 focused on housing, healthcare, education, and technology alongside more militant demands for self-determination, reparations, an end to police brutality, and an end to wars of aggression. It stated as its fourth demand "We want decent housing, fit for the shelter of human beings," and as its tenth demand "We want land, bread, housing, education, clothing, justice, peace, and people's community control of modern technology." Yet, the Panthers had a different approach to DIY politics than Jones. Notably, rather than waiting for answers to these demands from government, the Panthers took matters into their own hands and pursued what we might call today radical "service design" at the community scale. Seeking "survival pending revolution," the Panthers developed various community survival programs across the late 1960s and early 1970s: from free breakfasts for school children to free community-based health care, free shoes to free pest control, free housing cooperatives to a free plumbing and maintenance program. It was these community survival programs that the Panthers sought to showcase, beyond their larger revolutionary agenda of militant self-defense and system change, when Brand's *Co-Evolutionary Quarterly* handed over the entire Fall 1974 edition to the Panthers to guest edit.

As Simon Sadler (2015) has documented, the Panthers took this opportunity to highlight their community survival programs to Brand's audience of hippie eco-technologists and assorted other makers. In their section on "land banking," the Panthers explicitly called out the mainstream environmental movement for ignoring the inner cities and the immediate economic concerns of inner-city residents:

> The inner-city resident (due to the uncontrolled growth and over-development of the inner-city) must focus on much more immediate aspects of survival first,

> such as employment, housing, education (and facilities), food and clothing and other basic necessities. He has little time to contemplate the beauty or necessity of "open space" or more parks. Thus, the ghetto dweller of the inner city sees the environmental movement as a luxury he cannot afford.

So, what might bridge the gap between the nascent environmental movement and movements for racial justice? Here one suggestion made by the Panthers in this issue is that the demand for community control over real estate and community design of parks and open space could potentially open space for "the effective merger" of the aspirations of the environmental movement with "struggles involving the Black and poor of the inner-cities" (Black Panther Party, 1974). In a demand that resonates in many cities just as clearly fifty years later, the Panthers not only specifically called for and implemented community survival programs to address the provision of basic needs but also for public takeover of vacant lots, which could be "acquired from their absentee landlords and developed into much needed mini-parks, tot-lots, community gardens, cultural or recreation centers" providing job development possibilities as well (Co-Evolution Quarterly, 1974). It could be observed here that such a call emerges decades before what would become core environmental justice demands start to permeate the mainstream (still predominantly white) environmental movement.

It is interesting to note though, that in the interventions of Hare, Jones, and the Black Panthers, of the various critiques we can find of the color blindness of the environmental movement of the 1960s and 1970s, it is the lack of attention given to the urban built environment and *housing* in particular which connects all three interventions. Of course, it would be incorrect to suggest that the alternative technology movement in the US paid little attention to housing as a topic. If one reviews the alternative technology and eco-design manifestos, catalogs, and practices of the 1960s and 1970s, as we have seen, they are very much concerned with elaborating the social and ecological potential of geodesic domes, structures made of prefabricated or found materials, solar houses, retrofitting houses and integrating such proposals with food growing, waste disposal, such as The Integral House of the Farallones Institute. However, reflecting the middle- and upper-class social basis of much of the hippie movement, the postwar commitment to suburban private development, and the sheer abundance of space in the United States, a great deal of these design interventions largely defaulted to back-to-the-land housing experiments, escape pods, or even second homes for middle- and upper-class white downscalers as the radical and communal energies of the counterculture dissipated. A great deal of the impulse experiments in the 1970s and 1980s to build "bioshelters," "ocean arks," "autonomy houses," and "Earthships" may well have generated innovative design experiments. However, such projects rarely, if ever, evolved toward proposals to plan and build affordable, healthy housing for working people en masse. Indeed, as we shall see in the next chapter, it was not until the 1990s with the rise of new urbanism and green urbanist currents in planning, the development of the Passive House standard in Germany, notably in Darmstadt in 1990–1, and the incorporation of energy efficiency and green housing standards into social/public housing integration in the late 1990s to early 2000s in Germany, Austria, and cities like Vienna that we would start to see genuine moves to build social ecologies for all focused on the urban environment.

Papanek, Schumacher, Bonsiepe, and Maldonado: From Design for the Real World to Project Cybersyn

Let us conclude this chapter by mapping two influential but rather different views on the future of design–environment–society discussions that take shape at the end of the counterculture-ecology-design moment and generate ongoing debates within design that remain unresolved to this day.

If the Black Panther Party, Bookchin, and advocates of Black Ecology are some of the few currents in the US countercultural ecology-design discussions of the 1960s and 1970s to directly acknowledge the problem of colonialism, imperial war and global inequality, following this, divergent clusters of design thinking emerge around the question of how a radical vision of design might situate itself within broader center–periphery conflicts or actually address the pressing needs of the majority world.

One such answer to these issues was proposed by the industrial designer and design academic, Victor Papanek, who wrote his enormously successful polemic *Design for the Real World* in 1971. The text, influenced by the environmental and anti-authoritarian spirit of the 1960s, offered a blistering indictment of the failings of corporate-driven industrial design and a plea that design systematically rethink its whole agenda to fundamentally challenge corporate priorities and instead place ecological and humanitarian concerns at its core. Papanek unflinchingly asserted:

> As long as design concerns itself with confecting trivial "toys for adults, killing machines with gleaming tailfins, and "sexed-up" shrouds for typewriters, toasters, telephones, and computers, it has lost all reason to exist.

What was needed was not a design industry that produced more shabby, poor-quality, useless, and toxic rubbish for a consumer society but rather a design industry and profession that reoriented completely toward making ecologically sane, repairable goods, designs that met real human needs, as well as designs that addressed the needs of marginalized groups such as design for disability and appropriate design for the "third world" (Papanek, 1972).

Design for the Real World went on to lay out a compendium of alternative possible pathways for design to follow in the years ahead: from the more prosaic call for the need for relatively simpler packaging to more visionary propositions that designers should reconsider the merits of electric cars, battery-powered bicycles for adults and sailing ships with computer-steered rigging as more ecologically benign transport technologies to arguments that we should question the virtues of "space guzzling homes." The text also highlighted work completed by Papanek, his students, collaborators, and assorted fellow travelers, which included cutlery for the severely handicapped, tricycles designed for children with cerebral palsy, and most famous, the tin can radio "for the Third World" that used paraffin wax and a wick as a power source (designed by Papanek and George Seeger).

Design for the Real World certainly had an enormous impact. It is unquestionably a brilliant, readable, and quotable polemic and an intervention that, in many respects, foreshadows design for disassembly, biomimicry, and sustainable design, design for disability/inclusive design, and humanitarian design. Moreover, in quick succession, we see the publication of a series of complementary design adjacent interventions, notably E. F. Schumacher's *Small is Beautiful* (1973) and Ivan Illich's *Tools for Conviviality* (1974). Both followed Papanek, offering reinforcing critiques of the failings of industrialism, consumer society, high modernism, and "development" in the "overdeveloped world" and the "underdeveloped world." All these interventions are informed by the view that it is not simply capital or the state plan that is the problem but *gigantism* more broadly. All are inspired by the new politics of ecology in the Global North but also non-Western ontologies and the notion that a great deal can be learned from local craft practices, indigenous and peasant knowledge, which already offer or could allow for the further development of appropriate or intermediate technologies for "the third world."

As Papanek later acknowledged, Schumacher amplified his own formulation that "nothing big works." Schumacher's own key intervention of this period *Small Is Beautiful* is a text written by an economist rather than a designer. However, his intervention, like Papanek, is strikingly disillusioned by both United States and Soviet-led visions of development as modernization, with their shared assumption that the "third world" should embark on a developmental path marked by large factories, industrial agriculture, large-scale electrical power projects, and ultimately a consumer-based, urban-focused economy. Schumacher, more explicitly than Papanek, suggested that such a project largely benefited local elites but did little to support the real needs of rural populations. What was needed was a robust critique of both modernization theory and practice and a rejection of the key force propelling all these developments, notably the endless quest for economic growth. Again, the alternative for Schumacher would support policy, political, and design interventions that would work with the grain of local endowments, local skills, and cultural strengths and were appropriate for maximizing well-being. Such intermediate technology involved neither a romance of the primitive nor a fetishization of the West but rather a focus on technologies that were affordable, made from local materials and resources, and easy to repair and allowed groups to meet their own needs. As Illich added, what was needed for the Global South was not high modernist designs that generalized addiction to the consumer society but a design culture that could distinguish between "tools of domination" from "tools for conviviality" and cultivate the latter.

If the broad contours of an appropriate design and technology movement can start to be discerned from the interventions of Papanek, Illich, and Schumacher and given further momentum by the publication of the Limits to Growth report in 1973, which declared "endless growth on a finite planet" was impossible, it is important to note that not all currents within radical design or further afield were convinced by this post-development or post-growth agenda.

As Papanek, Illich, and Schumacher's influence started to find organizational expression in the intermediate and appropriate design and technology movements of the 1970s, influencing design and technology advocacy in the United Nations and particularly European and British NGO groups focused on advocacy for "the Third World," we can also see significant pushback coming from a range of different quarters. A rather different vision of the role that a political ecology might have for design, planning, and architecture, particularly in the Global South, emerges out of a series of conversations traversing Latin America and Continental Europe, heavily shaped by the interventions of Tomás Maldonado and his pupil, Gui Bonsiepe.

Tomás Maldonado was a crucial figure in the Argentinian avant-garde painting scene, an industrial designer, a member of the Argentine Communist Party, and one of the leading design theorists of the legendary Ulm School of Design in Germany. It was at Ulm where he introduced systems thinking, cybernetics planning, and semiotics to an innovative design curriculum in the mid-1960s. He published his key text, *Design, Nature, and Revolution: Toward a Critical Ecology*, in Italian in 1970 and in English in 1972. While sharing with Papanek, Illich, and Schumacher a desire to center the new politics of ecology for design, he nevertheless offers a series of suggestions as to how design might align with *democratic planning* and a *humane industrialism* while not giving up on modernity, utopia or a potential productive role of the state.

Design, Nature, and Revolution starts from the proposition that a historical and materialist entry point into ecological design must reject ecological romanticism or all modes of nature–culture dualism and acknowledge "we have lived in an environment partly constructed by ourselves," indeed, the "making of our environment and the making of ourselves has been a single process." Maldonado suggests from this that the ecological crisis raises questions about how we might reestablish a conversation between design, democratic planning, and utopia. Those most concerned with "the environmental impacts of consumer society" often imply a direct accusation against planning as a source of responsibility for this crisis. While Maldonado acknowledges that this is not without grounds, given the technocratic and elitist fashion in which planning has been practiced in the Soviet bloc and Western Europe, he argues "the polemic against consumer civilization shows a tendency to sacrifice hope in planning too completely and too readily" (Maldonado, 1972: 13). Maldonado notes that the matter of utopia, dismissed for too long in design, has returned of late among a range of architects and planners in the fashion of "formulating ideal models of future cities" (Maldonado, 1972: 27). Here, Maldonado has in mind a range of mid-to-late-1960s architectonic and urban utopian proposals from Ron Herron and Bryan Harvey's Walking City, Dennis Crompton's Computer City, the Plug-in-City of Chalk, Cook and Compton, the la Ville cybernétique of Nicolas Schöffer, the "Metabolist structure" of Arata Isozaki, the Ville Mobile-Ville Spatiale of Yona Friedman, the Helicoidal City of Noriaki Kurokawa and the Isotopon of Rudolph Doernach. While these proposals are commended for being "very innovative and stimulating," they repeat some of the classic failures of spatial utopianism of old, notably:

> On the one hand, they refuse to undertake any action that implies a planning compromise with the environmental needs and pressures of the present. On the other hand, they also refuse to hypothesize any decision-making trajectory that might make these megastructures realizable in the future, and that would enable

> them to foresee the environmental needs and pressures of which they would eventually become part. (Maldonado, 1972: 27)

The political failure here is wrapped up in an approach to design that relies on images and prophetic words alone, and the fallacy that "design and planning will resolve the problem that politics has left unresolved." Here, Maldonado argues that we see the move toward a typical technocratic utopian imaginary that believes design can ultimately abolish politics, rendering it obsolete. While Maldonado does not mention him, the figure of Buckminster Fuller and Stewart Brand clearly hovers over this whole conversation.

Maldonado argues that what is needed is, yes, a recognition that the conditions of human existence can be transformed for the better by design, architecture, and planning, but what is further required is a combination of the technical or design imagination and what C. Wright Mills called the "sociological imagination." We need more than technocratic design utopianism or design nihilism, but rather an "operative praxis of design" (Maldonado, 1972: 27). This would involve drawing together critical social and political consciousness and design consciousness. Here, Maldonado argues Ernst Bloch's defense of utopia as the principle of hope has some value, but what is missing from Bloch and other revolutionaries' fetishizing action *is a combination of critical and technical consciousness*. Both are needed and potentially come toward the demand for applying environmental design and planning "on a massive scale."

Design, Nature, and Revolution never quite elaborates on the socio-ecological design forms that could institutionalize a "praxeology of design." Instead, and guided by Gramsci, Maldonado suggests we might see "the designer" as an organic intellectual "fulfilling his task in the bosom of civil society." *Design, Nature, and Revolution* ends with the revolutionary call that designers should reject their assigned role as salesmen for the dominant class and instead ally with those revolutionary forces trying to "destroy the old power" (Maldonado, 1972: 130). But how to move forward?

If Maldonado in *Design, Nature, and Revolution* offers us a suggestive but somewhat elusive agenda for design, it is in many respects his pupil, the German-South American polymath Gui Bonsiepe, who draws out the implications of Maldonado's arguments for designs for the periphery more explicitly. Bonsiepe worked with Maldonado at the Ulm school in Germany, departing at the end of the 1960s to settle in South America. He spent much of his subsequent career in Argentina, Brazil, and Chile, where he worked with Salvador Allende's democratically elected socialist administration on the Cybersyn project in the critical 1971–3 period before Allende's overthrow in the military coup staged by Augusto Pinochet, with further periods spent later working on interface design in Silicon Valley. He reviewed Papanek's *Design for the Real World* in 1973, which led to an extended exchange between the two figures (see Penin, 2021).

Bonsiepe suggested that while Papanek had written a brilliant polemic against the design industry and the design professions more generally, *Design for the Real World* was a text that was relatively light on theory or any engagement with political economy. With the key argument of the text placed largely on the evils of "consumer society," the relationship between the political economy of consumption and production largely goes missing. As Bonsiepe further notes, Papanek's general anti-institutional preferences for DIY design solutions ensure that *Design for the Real World* has little to say about how to create change *within large-scale institutions* or how one could begin to think about the humanization of design *that is reliant on manufacturing*. Indeed, Papanek is accused of leaning into certain paternalistic frames that can give the impression that "high tech" is for the First World and "low tech" for the Third World (Penin, 2021).

Bonsiepe's review of Papanek is somewhat ungenerous, as Papanek's response ably demonstrates.[1] However, perhaps what is interesting here is how Bonsiepe's writings go on to elaborate an alternative agenda for "design for the periphery," which has some striking differences from the emphasis of the appropriate technology movement. As Lara Penin observes:

> Both Papanek and Bonsiepe share the view that design could be a force for social and ecological emancipation. However, a critical point of difference emerges between Papanek's bottom-up, anti-institutional and D.I.Y vision of an appropriate design culture for the Third World, which essentially sees the future of the global South as dependent on rural agricultural communities and Bonsiepe's democratic socialist view that the only way that peripheral nations of the world system will be able to resist the imperial power of the core is through a project oriented to sovereign national development, that commits to building publicly oriented industrialization, a design culture that consolidates manufacturing and a broader culture that addresses the needs of the urban proletariat. (Penin, 2021)

Bonsiepe initially attempted to implement such a vision of state-led democratic design and planning through his work with Salvador Allende's administration in Chile. Project Cybersyn attempted to use the most up-to-date computation technologies of the early 1970s to facilitate a real-time model of the Chilean economy that would facilitate democratic planning. While ecological or environmental concerns were not at the center of the project, the project's focus on constructing information flows that would improve resource use, increase technology efficiencies, and more easily identify bottlenecks and points of crisis in the system could have not only improved economic management but also environmental management. The project was hampered by the limitations of the technologies of the time and came to an abrupt stop following Pinochet's coup in 1973. But Bonsiepe continued to advocate for much of his subsequent career for industry policy and an active state as vital for periphery countries to build up their own industrial base as a precondition for social and ecological emancipation. As we shall see in future chapters, it is the relationship between market and state, entrepreneurial approaches and planning, between expert intervention and DIY politics that is going to remain a critical point of contention as ecology and design conversations unfold in future decades.

Political Ecologies of Design

In this chapter, we have tried to capture the convergence of concerns with ecology, nature, design, and planning at three key moments in the nineteenth and twentieth centuries. What emerges from this account is a complex historical picture of architects,

Figure 6 Operations room of Project Cybersyn: Design by a team led by Gui Boniespe. Copyright Gui Boniespe. Reproduced with permission.

designers, planners, and critics who are situated in time and space, working off shifting and changing accounts of what constitutes "environmental problems" and deploying very different ideas about how "Nature" can be defined, engage with or "anchor" design.

We have seen varied ideas emerge about the critical sciences or cluster of sciences that should define "ecological or environmental literacy," and ultimately, very different ideas about how various proposed design ecologies might be socially and politically institutionalized.

This environment/design discussion has on occasion gravitated to the hope that patterning society on "the grain of nature" revealed to us by geometry, mathematics, evolutionary theory, the science of ecology, system theory, or holism will save us. This search for a master/meta-discourse that can integrate all knowledge into a seamless whole is a key preoccupation of all manner of design thinkers in the twentieth century. It is a proposition founded on the undoubtedly correct notion that design acting as a progressive force within society is necessarily an interdisciplinary affair which must draw from thinking across the natural, social, material, and applied sciences to grasp its social and ecological impacts and possibilities.

Yet, we have also seen that the search for a "grand synthesis" between design and ecology has often been an unstable project. It has often involved fixing what are viewed as the core insights of a particular science at one point in time and then struggling to attend to the ways in which debates in a particular science may have moved on. The uptake of ecology by the counterculture is a case in point. The science is read by activist designers as foregrounding equilibrium, harmony, and order in ecosystems in the 1960s and 1970s. These ideas are extrapolated to design systems and codified, but then modest attention is paid by design advocates as to how debates in scientific ecology may have moved on to grapple with the role of disequilibria, instability, chaotic fluctuations, and resilience, which become key preoccupations of ecological research in the following decades (see White, 2008: 110–13).

If the idea that "nature" in the large can provide us with a pattern or pathway to redemption by transcending the always messy and fraught politics of design emerges again and again, we have also seen other approaches pursued in this chapter. Notably, we have also seen a whole range of design and design adjacent currents and thinkers well aware of the dangers of naturalistic reductionism and environmental determinism, and from this, attempting to formulate *social ecologies* and *political ecologies* to address pressing social and ecological needs and concerns. From Howard to DuBois, from Mumford to Bookchin, from the Black Ecologists to Sim Van Der Ryn and the feminist critics of alternative technology, from Papanek and from Maldonado to Bonsiepe, we can also locate sophisticated attempts to bring design, planning, and architecture into more dynamic and provisional alignments with the social and natural sciences, to draw together, as Maldonado outlines, ecology with the sociological imagination and the design imagination.

It should be acknowledged that among critical design currents, we have also seen some fundamental tensions surface on a recurring basis. If all radical currents of design would probably agree with Lewis Mumford that we should try to distinguish "authoritarian" from "democratic" *technics* (Mumford, 1964), how this cut should be made remains deeply fraught. We have seen a recurring thread of design-ecology explorations that have championed local knowledge, DIY culture, vernacular buildings, and small-scale craft interventions that work off the implicit assumption that the project of design should ultimately be providing tools that maximize autonomy, conviviality, and self-management. This current has often come into conflict with other radical designers, planners, and architects (some of whom are situated in or come from the periphery) who argue that a progressive environmental design culture has to find ways to recuperate development, industrialization, or modernization. From this latter perspective, a more emancipatory ecologically aligned design future that addresses the needs of working people is going to require urban and regional planning, large-scale interventions to build new power grids and infrastructures, public housing, parks, and yes, labor-saving devices such as washing machines, tractors, and forms of automation and cybernation. All this is going to require not a wholesale abandoning of urban modernity or an advanced division of labor but a careful sifting of its gains and its limitations.

Of course, it could be argued that not all these agendas are fundamentally opposed. It is plausible to envisage a design-politics that champions durability, repair, open source making, and tools that augment labor and enhance our capacities for conviviality, running alongside design, planning, and architectural inventions that work at much larger scales to facilitate the optimal built environment and infrastructures that could secure more just and equitable post-carbon futures. However, these tensions do remain, and as we shall see in the chapters ahead, these tensions and fissures recur with surprising regularity as design-environment discussions move beyond the counterculture.

3

Green Design as *Threat* and *Promise* for Architecture, Design, and Planning

Coming out of the 1960s, the alternative technology movement produced some real innovations in green design and architecture. But it was the energy crisis of 1973 that propelled green design out of a counterculture/Earth Day niche toward the top table of geopolitics. OPEC's oil embargo, following the 1973 Yom Kippur War, sparked an acute energy crisis in the United States and its oil-importing allies. It spiked U.S. oil prices by 300 percent, triggered gas station chaos with long driver queues, unleashed gasoline rationing that curbed weekend travel, and left cars stranded on highways while schools closed for lack of heating oil that winter. Seemingly overnight, big gas-guzzling cars were out, fuel efficiency and energy efficiency were in, and even President Richard Nixon came on the TV asking families to turn down the thermostat. As energy awareness permeated the architecture and design fields and society at large, we saw a series of shifts in the attitude of these professions as new environmental performance standards became standard practice, but also some unease. In this chapter, we trace a few key moments in the rise of what has alternately been called ecological design, eco-design, "sustainable" design, "design for sustainability," or just "green design". We consider how changing concerns result in changing methods, opening new spaces of collaboration between disciplines as well as movements toward disciplinary resistance.

Firstly, we attempt to capture the ways in which the green technology and green design conversation starting in the 1970s are marked by waves of market, community and state-led innovations responding to the OPEC oil shock, but also moments of inertia and resistance. Innovative basic research proceeded apace through the National Labs, engineering schools, and in some other parts of the public and private sector around energy efficiency, energy conservation, and renewable energy. We explore how many former green design radicals attempt to move forward in the 1990s and 2000s by making peace with the market, resulting in important technical innovations amid a hostile political climate. Encouraged by attempts to develop a greener version of neoliberalism emerging in a range of UN and EU frameworks, we see a whole range of design interventions and agendas develop, from *Natural Capitalism* to green products to voluntary green building certification systems such as LEED, premised on the assumption that a market-led sustainable design transition is imminent and will be led by green entrepreneurs, green consumers, and design-enabled behavior modification. We consider here how this moment gives rise to some important design propositions and interventions that do move the needle, particularly in European design contexts. We also highlight how the search for a "win-win synthesis" between sustainability and progressive neoliberalism has not entirely delivered as expected.

Second, we look at how maverick voices and environmentally orientated design programs are tolerated to greater and lesser degrees by the design academy. However, we also outline how some of the most powerful institutions and forces within design culture, academe, and practice, particularly in the *architectural avant-garde*, view environmental agendas as posing *active threats* to the future of their discipline.

Finally, we see how over time, the academy partially embraces conversations about energy impacts, and then about climate change. This shift comes with new methods, new alignments, and new scopes of concern, beyond the singular building or singular persona of the architect. Notably though, many starchitects display remarkable longevity, and the old reactionaries have not entirely departed the scene (see Schumacher, 2024). Nevertheless, discussions of circular economies, discussions around adaptive reuse, retrofitting, repair, and unbuilding, and engagements with environmental justice, planning, landscape architecture, and green urbanism offer promising ways forward that could in theory align the forces of design, planning, and architecture with the just transition.

Green Energy Technology and the Oil Shock's Reverberations

> Computers won out. Everything else failed. Communes failed, dope failed, and politics failed. The main thing that survived and thrived was the computer.
>
> Stewart Brand 1994 cited in. (Watson, 1998: 236)

As we have seen in the last chapter, solar and energy efficiency technologies evolved across the 1950s and 1960s, with US government–funded research organizations such as NASA and Bell Labs, and national laboratories like Sandia and Los Alamos having had engagements with renewable energy technologies dating back to the 1950s. In design circles, experiments in solar thermal (Barber, 2016) as well as counterculture eco-design discussions during the same time period had already suggested ways to reduce the fossil fuel demands of buildings through design and technology. But the 1973 OPEC oil crisis was a precipitating event that focused attention much more squarely on clean energy technologies and energy efficiency concerns, initially as a set of engineering innovations and policy changes at the scale of the larger energy system that over time made their way into architecture.

Following the oil shock and energy crisis of '73/'74, virtually all Western governments instituted environmental ministries or departments with a focus on energy conservation and energy efficiency, with an aim of reducing oil dependence, and such developments opened up a space for different discourses around energy, technology, and design to emerge. The energy crisis not only generated growing concerns about how US energy security might be endangered by geopolitical crises, but also an increasing awareness of the more advanced state of scalable research into solar in Germany and Japan, as well as an incipient commercial wind industry emerging in Denmark. The Energy Reorganization Act of 1974 set into place US solar energy development and prioritized commercializing solar energy in the United States; the Energy Policy and Conservation Act of 1975 established fuel economy standards for US cars; and the creation of the Solar Energy Research Institute in 1977 by the incoming Carter administration (which would later become the National Renewable Energy Laboratory, or NREL) introduced a centralized research hub charged with accelerating the widespread adoption of solar energy. The research and policy incentives paved the way for the first large-scale solar and wind power projects in the 1980s and the launch of energy efficiency initiatives throughout the 1980s and 1990s.

Within environmental technology conversations, few interventions stemming from the 1970s energy crisis were as influential as Amory Lovins's *World Energy Strategies* in 1973 and his popularization of this argument in his 1976 article in *Foreign Affairs*, "Energy Strategy: The Road Not Taken?" In both pieces, Lovins argues for the virtues of a "soft energy path" that promotes energy efficiency, renewable energy sources, and decentralized energy generation, in contrast to the dominant "hard energy path" of large-scale, centralized fossil fuel and nuclear power—a view closely aligned with that of a generation of environmentalists who came up amid antinuclear activism. But Lovin's synthesis of market pragmatism with ecological systems theory also makes for a decisive break with the free-wheeling Left-libertarian environmental-design spirit of the 1960s. His two critical interventions emerge in a context bookended and, in a sense, turbo-charged by the publication of *The Limits to Growth* (1972), Schumacher's *Small is Beautiful* (1973), and Ivan Illich's *Tools for Conviviality* (1973). Yet, unlike the egalitarian ecotopian horizons of Schumacher or Illich's communitarian-anarchist focus on "tools" that would expand autonomy in a decentralized society, Lovins offered the Carter Administration a relentlessly pragmatic and business-friendly plan to "get the US off oil and nuclear in 40 years." Lovins could ground his arguments in the reality that the United States had emerged in the 1970s as the leading global innovator in solar and wind technologies. He further suggested that an energy transition opened up multiple market-driven efficiencies and cost savings. A "negawatt revolution" pursuing massive energy efficiencies would demonstrate that it was cheaper to save energy than produce it. He argued that the promise of rooftop solar and community-scale renewables, which was largely suppressed by fossil fuel interests and by entrenched oligarchic and conservative public utilities. In contrast, a "soft energy path" could reduce reliance on fossil fuels and nuclear energy and address US geopolitical concerns, while shaking up the moribund energy utility sector through utility reform. Lovins maintained this would ensure the government could set broad rules and then allow private actors to do the heavy work. Far from offering a path to ecoanarchism as Bookchin hoped, Lovins argued that the rapid expansion of decentralized eco-technologies with the pursuit of robust energy efficiency programs would be able to revitalize the market and offer cost savings to all, while bringing bureaucracy, in the form of the public utilities, to heel (Lovins, 1973, 1976).

Lovins's influential intervention here, then, can be seen as marking a broader political moment that both reflects and anticipates the shift of Democratic Party elites away from New Deal liberalism and towards market orientated environmentalism. In marked contrast to the advocacy of "big public power" and centralized regional planning approaches favored by the RPAA and the Progressive movement that we mapped in Chapter 2, in many respects, Lovins's project fused the fascination with decentralized energy systems championed by the counterculture with a more market-friendly and deregulatory approach to the energy sector, an approach that was, in some part, embraced by the Carter Administration.

If the rise and influence of Lovins's thinking on "soft energy paths" in the Carter Administration could be seen as the high point of some of the key arguments circulating around energy circles in the 1960s and 1970s, it could also be seen as the end phase of this moment. In part, such developments can be linked to the growing resistance of incumbent fossil industries and large landowners to the early 1970s environmental legislation, a late 1970s "rate-payers revolt" to the rising cost of energy, as well as the beginning of a broader populist backlash on the political Right to the growth of the environmental regulatory state. Ronald Reagan successfully rode the coattails of such anti-environmental sentiments in his presidential campaign in 1980. Mostly famously, he removed the solar panels that Carter had had installed on the White House during the previous administration and appointed James Watt as Interior Secretary, who favored expanding drilling, mining, and development on public lands. More broadly, the "stagflation" that followed the 1973 energy crisis essentially ended the hegemonic Keynesian-social democratic New Deal postwar consensus, generating all manner of short- and long-term reverberations. Perhaps most consequential for the trajectory of green technologies though was the ways in which the Reagan administration effectively undermined the United States's lead in solar, wind, and other renewable energy technologies by slashing federal funding, rolling back tax incentives, and curtailing support for research and development. The end result was the transfer of leadership and technological prowess in these sectors to countries such as Denmark, Germany, and Japan, and (as we shall explore further in Chapter 8) ultimately China.

Historian of technology Carroll Pursell has argued that by the mid-1980s, most of the countercultural institutions built around alternative technology in the United States similarly had "either disappeared or lost their momentum." While many of the technologies—from solar and wind, to mass transit and alternative agriculture—endured, they did so "without an ideological context that could give them political meaning" (Pursell, 1993: 629). Pursell has suggested that the decline of the radical flank of the alternative technology movement in the United States can, in part, also be explained by the "Reagan revolution" and a lack of ability of these institutions to transform the incentive structure that shaped research and development and federal funding for energy, despite support at the local and state levels in the United States. More generally though the somewhat informal cluster of alternative technology institutions that had experimented with decentralized community-based energy alternatives in the 1970s, for the most part, did not have the capacity to scale their experimental projects or the funds necessary to move from innovative demonstration projects to usable hardware. What Adrian Smith calls "the enrollment dilemma" left many radical advocates of "convivial" and "liberatory" technologies, who hoped eco-technologies might open up communitarian or even post-capitalist possibilities aligned with the visions of Schumacher (1974), Bookchin (1971), or Ivan Illich (1973) dependent on funding from largely unreliable sources such as sympathetic individuals, charities or grants from research councils or local and state governments, which ultimately dried up.

In contrast, and even with the limited environmental interests of the Reagan and Bush administrations, research on renewable energy and energy efficiency continued through the US Department of Energy, government agencies such as NASA, the national laboratories, the aerospace industry, and leading engineering- and energy-focused research institutions such as the University of Massachusetts Amherst, MIT, Berkeley, Carnegie Mellon, and Stanford. All started to develop capacities to test and scale up research and deployment on renewable technologies like large wind turbines or utility-scale solar that could contribute grid-scale energy in ways that were never available to or seen as desirable by many countercultural currents, for whom disconnecting from the existing systems of capitalist extraction was the goal.

In any case, radical energy technologists drifted to academia, NGOs, or local government to work on energy and conservation issues[1] (Smith, 2005). An exception to this general trajectory can be found in the case of the Danish wind industry, which had its roots in activist cooperative movements, folk schools, and other craft and agricultural technology support networks. The Danish folk school at Tvind, for example, built their first wind turbine in 1978, reacting to the same OPEC oil crisis, and pioneering the technical innovations that would come to characterize virtually all contemporary wind turbine designs. For a period, as Smith notes, such grassroots currents were able to gain access to test facilities and electricity grids and develop from this an internationally significant wind industry. But again, the scaling of the industry took it very far from the politicized aspirations of its earlier formation (Smith, 2005). Pursell points to the failure of the alternative technology movement to build a political coalition that could challenge the entrenched dominant forces of agribusiness, energy, manufacturing, and the military–industrial complex. Indeed, all these forces "showed a remarkable ability to expropriate emerging technologies for their own benefit." (Pursell, 1993: 633).

If in the UK and the European context the public sector provides some space for those interested in ecologically focused technologies and designs, energy conservation, and energy efficiency to hang on below the surface, it is notable how in the United States, we see a number of the remaining countercultural currents of eco-design that survive the Reagan–Bush era radically reposition themselves. In California, it is increasingly the information technology revolution that attracts an army of post-industrial utopians such as Stewart Brand and Kevin Kelly, and sees them trade Ernest Callenbach's *Ecotopia* for Alvin Toffler's *Future Shock* and the *Whole Earth Review* for *Wired Magazine*. Ecology funnelled through what Barbrook and Cameron named "the Californian ideology" (Barbrook and Cameron, 1996) arguably provides the basis for the later development of ecomodernism (see Brand, 2009; Asafu-Adjaye et al., 2015). The *Rocky Mountain Institute*, which Amory Lovins started up in 1982 with his wife Hunter Lovins, can be seen following a similar trajectory. From its origins as an eco-design and technology institute, it became known for producing highly skilled energy analysts and engineers trained in green building performance, with expertise in energy and daylight analysis, passive solar design, and advanced building envelope technologies for saving energy, who would go on to influential positions throughout the clean energy and green building sectors. Over the coming decades, it grew into a McKinsey-style green business–friendly consultancy offering route maps to *Natural Capitalism* (Hawken, Lovins and Lovins, 1999), sustainability consulting for large companies, and rebranding itself into the anodyne-sounding *RMI*. The consolidation of a new business-friendly "sustainable development" as a new *lingua franca* of international politics, following the publication of the UN's Brundtland Report in 1987, the adoption of such agendas by the European Union, the UNEP, the signing of the Kyoto Protocol, and the rise of European Green movements and political parties, created new spaces outside the United States for discussions of what becomes increasingly referred to as "design for sustainability."

While the geographical center of gravity of these discussions in significant parts shifts from the United States to international organizations and the European Union, as well as niche footholds in critical universities and industries, the unspoken backdrop to all these discussions increasingly becomes what Mark Fisher calls "capitalist realism" (Fisher 2022).[2]

Green Products and "Natural Capitalism": Eco-Design Goes to the Market

"Design for sustainability" can be seen as opening up a legitimizing space within the broader corporate-dominated design world, leading to two significant streams of research. We see one cluster of sustainable design energies focused on product innovation, with subfields such as green design, eco-design, emotionally durable design, and design for sustainable behavior, as well as discussions of biomimicry, recyclability, and circularity. A second cluster of research focuses on exploring various approaches to systems design forms, from product-service systems design to design for social innovation (see Ceschin and Gaziulusoy, 2016; 2019). Since we will focus on the development of service design in Chapter 5, we will focus here our engagement with sustainable product design approaches.

The recent scholarship of Fabrizio Ceschin, İdil Gaziulusoy, and Joanna Boehnert has provided us with some of the most comprehensive and useful maps of the path that sustainable design followed in Europe as it made inroads into the European design schools and the broader professional fields of industrial/product/service and graphic design from the 1990s onward (Ceschin and Gaziulusoy, 2019; Boehnert, 2018). Ceschin and Gaziulusoy note that a crucial moment here in the professional acceptance of "design for sustainability" is its uptake in the UK by the Design Council and the further cultivation and expansion of European Union funding for academia- and industry-funded collaborative networks, particularly in critical European design and engineering schools in the broad design/sustainability field. Of course, this "greening of design" should not be overstated. Joanna Boehnert's *Design, Ecology, Politics* (2018) underlines that most professional design work remained locked into producing goods and services for market-driven business-as-usual ventures. The sustainability–design interface, nevertheless, was buoyed in the 1990s and 2000s by EU funding, by fledgling wind, wave, and solar industry growth in Denmark and Spain, and perhaps most notably by the German Energiewende, which from 2000 provided expansive public funding for the wind and solar industries through feed-in tariffs, which guaranteed payments at a subsidized fixed price for small-scale producers of renewable energy, greatly expanding solar adoption by homeowners.

Ceschin and Gaziulusoy identify "green design" as one of the earliest responses by the design professions to the environmental crisis in the early 1990s—principally in product design. As they define it, green design on the whole is concerned with reducing the environmental impact of individual products, which might involve attending to the kinds of materials used in a product, replacing "virgin materials" with recycled ones and/or attending to issues of toxicity and replacing hazardous materials (Ceschin and Gaziulusoy, 2019: 8). The very manner in which the problem of free market dynamics is left out of the design brief here delimits the field's impact. As they note: "early implementations failed to replace high-impact products and instead created a new market for more environmentally friendly products," promoting green consumerism. As such and given these limitations, in the second half of the 1990s we see a shift in focus to "*eco-design,*" which is focused "on the entire life cycle of a product" across all its phases, from production through use to disposal (Ceschin and Gaziulusoy, 2016: 8). The EU adopted its Ecodesign Directive in 2005 that set out energy efficiency requirements for energy-using products, with the focus on taking action during a product's design phase, "since it appears that the pollution caused during a product's life cycle is determined at that stage, and most of the costs involved are committed then" (2005/32/EC[3]). Life cycle assessment (LCA) offered important advances by enabling a degree of quantification of environmental impacts, making possible "meaningful comparisons between different product concepts that deliver the same functional unit" (Ceschin and Gaziulusoy, 2016: 16). Nevertheless, LCA also has its limits, with the field being prone to ecological and material reductionism, as Ceschin and Gaziulusoy observe, with many of the social dimensions driving environmental destruction left undertheorized. As they note: "Although early implementations of eco-design resulted in huge environmental gains, once the inefficiencies and 'bad design' were removed from products, the gains started to become marginal and increasingly costly." In short, design research focused on improving the ecological efficiency of single products does not resolve the broader problem of rebound effects, where the ever-increasing consumption of products at the system level ultimately outpaces efficiency improvements at the unit level (Ceschin and Gaziulusoy, 2019: 20). At the end of the day, the improved ability to quantify environmental impact through product manufacturing LCAs could not fix the problem of early disposal of products, since product lifespan was tied not only to physical wear but to a question of consumer psychology.

In response, the challenge presented to sustainable design research focused on individual products and materials from the 2000s onward has shifted attention to more social, cultural, and systemic perspectives. From the second half of the 2000s, we saw the growth of new design programs exploring "emotionally durable design" as elaborated by industrial designer Jonathan Chapman (2005), which attempts to "enhance the emotional tie between the user and the product," and the somewhat different "design for sustainable behavior" approaches (Lilley and Wilson, 2017), which seeks to nudge consumers toward making better design decisions. Ceschin and Gaziulusoy argue that both approaches have advantages in that they acknowledge that green material innovations and extending product lifespans in themselves will not deal with further cultural problems "of boredom, frivolity or psychological obsolescence." The holy grail of both these approaches is to design products that "continuously provide the user meaning and value over time" while ensuring that "product replacement can be delayed or avoided" (Ceschin and Gaziulusoy, 2019: 27).

This is an influential agenda, and emotionally durable design has developed a sophisticated ongoing research program that expands from research on heirloom products or products that improve with age to the role that personalization, customization, and codesign can play in stimulating attachment to outcomes. As Ceschin and Gaziulusoy observe, how people value objects can vary according to culture, users, the nature of the goods (personal or utilitarian or not), and general subjectivity. "The same product features that enhance product attachment in a particular user might not work for another user." Many commercial enterprises do not necessarily have the incentive to increase product attachment. The product-level focus of emotionally durable design can delimit its impact. There are financial disincentives for private firms to implement such strategies fully. These are clearly significant challenges. Such research agendas remain important and increasingly connected to sustainable design discussions focused on repair, maintenance, and reuse (e.g., Berger and Irwin, 2022) that have brought artists and designers into multidisciplinary conversations with historians, anthropologists, sociologists, activists, and incipient social movements and struggles. However, as Joanna Boehnert has noted, to the extent that such approaches remain primarily focused "on a critique of consumption rather than a critique of production (and in particular the production of consumption), they remain limited in their remit and potential impacts" (Boehnert, 2018: 21).

Perhaps the mid-2000s contributions to eco-design that have obtained the most extensive popular traction in popular press and business circles are William McDonough and Michael Braungart's *Cradle to Cradle* (2002) and the biomimicry approaches to design popularized by Janine Benyus (Benyus, 1997). Both of these approaches start from the proposition that paying close attention to natural systems can offer inspiration for eco-design strategies that move beyond the ambition to minimize harm and toward modes of design that actively create positive, regenerative, or restorative outcomes. Biomimicry is, of course, a practice as old as design itself (as we discussed in Chapter 2), but over recent times has stimulated real innovations in product design—from antibacterial surfaces inspired by shark skin, adhesives inspired by the feet of geckos, bullet train designs inspired by kingfisher beaks, and Velcro inspired by burdock burrs, to take a few famous examples. The extent to which such innovations necessarily contribute to socially and sustainable outcomes is rather another matter: biomimicry, after all, can inspire the design of regenerative materials as well as sidewinder missiles.

Drawing inspiration from industrial ecology, LCA, and biomimicry, as well as developments in sustainable materials and green chemistry, McDonough and Braungart's *Cradle to Cradle* offered a broader approach to eco-design. Focusing attention on reducing waste and toxicity and on the need to "design for disassembly," they argued that if products were to be feasibly recycled at their end of life, attention needed to be paid not merely to recycling but rather, "upcycling," and to shifting from disposable use of products (a "cradle to grave" linear economy) toward a model of continuous recycling and recovery. *Cradle to Cradle* suggests that a paradigm shift in design from a focus on eco-efficiency to "eco-effectiveness" could open up all manner of possibilities for design contributing to regenerative abundance—for example, through the redesign of material and energy flows in production processes so that wastes from one process can either become constructive inputs to others, or else be rendered regenerative and additive to the environment. All this being said, *Cradle to Cradle* was primarily aimed at rethinking the design of products, from carpet tiles to office chairs and so on. If its basic ambitions resonate with more contemporary and scaled up attempts to envisage circular economies and circular modes of design for buildings and cities, McDonough and Braungart's further contributions to these discussions have been somewhat stymied by their own attempts to develop proprietary and exclusive certification schemes, limited voluntary uptake by companies pursuing such innovation in the absence of meaningful compliance-based regulations, and mismatches between hyperbole and reality in some of the execution of McDonough's architectural projects.

So, what is the way forward for design for sustainability, given that the experience of the 1990s and early 2000s would suggest that voluntary, design-centric, consumer-facing, and purely market-based approaches to expanding eco-design have produced at best patchy results? In the wake of the Kyoto Protocol in 1997, as more attention got directed at environmental and climate impacts, we can see a rise in regulations and compliance mechanisms and a focus on national-level impacts, not just consumer choices. The European Commission's Eco-design Framework Directives (EC, 2005, 2009) established mandatory minimum energy efficiency and performance standards and required the use of LCA in energy-using and energy-related products made in the EU, with a view to harmonizing sustainable design standards across the European Union. Such directives recognized the importance of regulatory frameworks for moving design for sustainability forward, though initially they have also had challenges in terms of uneven enforcement and compliance by member states.

More recent work on the circular economy, which is generally defined as "an industrial system that is restorative by intention and design" (Ellen MacArthur Foundation, 2013[4]), has expanded the focus beyond the design of individual products, instead tackling the larger systems of production, including supply chain impacts. Since the early 2000s, there has been a wealth of scholarship on circularity broadly and the circular economy concept in particular, going beyond consumption-level impacts to look at material flows at the scale of entire sectors and urban systems. The publication in 2013 of the Ellen MacArthur Foundation's *Towards the Circular Economy* report series arguably helped the circular economy break through to a business-oriented audience and become a policy priority across a number of European institutions. The report laid out three principles of a circular economy: it (i) “aims to ‘design out’ waste”; (ii) “introduces a strict differentiation between “consumables,” which “in the circular economy are largely made of biological ingredients … that can be safely returned to the biosphere,” and “durables” like engines or computers, which “are designed from the start for reuse,” and (iii) is powered by renewable energy.

More recently, mandatory compliance frameworks requiring the tracking and tracing of materials from raw material through end of life targets or caps on per capita resource consumption, and extended producer responsibility requirements that makes manufacturers responsible for full life cycle costs have advanced across the EU, placing greater compliance requirements on producers and manufacturers, rather than consumers, and attempting to reign in rebound effects.

The circular economy, then, is the paradigm that currently represents the clearest path to a design-friendly, innovation-friendly, and legislative meta-framework to force systemic and multi-scalar change, toward a more integrated closed loop system that can open a fundamental rethink of systems of extraction, production, consumption, and waste disposal. The Chinese Communist Party enacted a *Law for the Promotion of the Circular Economy* in 2008, attempting to integrate circularity criteria across the entirety of its industrial policy, while the European Commission has passed increasingly stringent circular economy policies, such as the Circular Economy Action Plan into its European Green Deal of 2020, going beyond its initial directives of the previous decade. Within the EU, durability and repair provisions have recently entered into force as part of its Right to Repair Directive along with its ESPR—the 2024 Ecodesign for Sustainable Products Regulation. As legislative initiatives increasingly shift from directive to regulation, with mandatory compliance frameworks that contain specific targets and enforcement mechanisms, the variability of national interpretations that limited earlier efforts becomes less of an issue. The advantage of the circular economy is that (in theory) it can incorporate the best insights of life cycle assessment, design for disassembly, emotionally durable design, and regenerative and restorative design approaches with product systems service design approaches while finding ways to avoid the numerous limitations that any of these design approaches can encounter when pursued in isolation.

These discussions, which started with energy-using appliances and electronics, but then spread into textiles, furniture, packaging, and construction materials, have also had a profound effect on the design and construction of the built environment in certain architectural circles. How has the overall move toward design for sustainability played out in architecture and its allied fields? As we’ll see, it spurs a good deal of innovation within architecture, but also produces unexpected reactions and currents of resistance.

Green Buildings, Architectural Education, and the Resistance of the Architectural Avant Garde

> [A]n essential part of the disciplinary superstructure of architecture repels environmentally related change in the realm of design practice. It is difficult to embrace a problem that radically puts your practice in question …. As one comes

> to the realization that many of the cherished principles driving architectural conception may be proved aberrant by environmental concerns, it is naturally tempting to keep these at bay as long as possible. (Pedro Gadanho, 2022: 51–2)

In the last chapter, we saw a number of promising threads of environmentally attuned architectural design cropping up in countercultural currents throughout the 1960s. How did environmental design, and later design for sustainability, play out in the world of architecture, starting in the 1970s post-countercultural moment and the subsequent turn toward markets? While the lineage of earlier alternative technology experiments formed something of a foundation, the energy crisis of 1973 and the subsequent raft of regulations and innovations were the catalyst that set off larger-scale transformations in architecture and the built environment. As we shall see, 1973 created deep reverberations, enshrining "energy efficiency" as the dominant frame for responses to environmental concerns that would continue to shape the architecture field for decades.

Many projects in the later 1970s, 1980s, and 1990s responded to the 1973 oil shock by incorporating innovations in energy efficiency and renewable energy. Solar PV may have remained extremely expensive for much of this period, ensuring that it was financially impractical to use on buildings unless massively subsidized by grants or high-income clients, but a number of projects nonetheless incorporated it in an experimental fashion. Architects such as Michael Reynolds incorporated solar and energy efficiency into his Earthship houses in New Mexico in the 1970s and 1980s, which were built using local and recycled materials such as tires, cans, and bottles to reduce the embodied energy of construction. The *Rocky Mountain Institute*'s first headquarters in Snowmass, Colorado, which the Lovinses constructed between 1982 and 1984, became an influential demonstration project in energy efficiency, featuring extra-thick walls, super-insulated windows, solar panels, and a tropical greenhouse. US, Canadian, German, Austrian, and Scandinavian architects made great strides to incorporate solar thermal systems, passive cooling systems, heat recovery ventilation, and green roofs into buildings, all with the goal of reducing energy demand. The emergence of the Passive House movement in Germany in the late 1980s and early 1990s built on those innovations in energy efficiency and energy conservation.

While some of the innovations in energy efficiency and passive buildings drew on long-standing alternative technology conversations, the external pressure of codes and regulation arguably moved the needle further. Starting in 1975, energy standards began to be published in the United States by engineering groups like the American Society of Heating, Refrigerating, and Air-Conditioning Engineers (ASHRAE), were subsequently adopted by most states as part of building code. In 1978 the US Congress mandated that states employ energy efficiency standards for new buildings. These minimum energy standards changed the practice of architecture, with engineers, consultants, and specialized firms becoming a key part of design teams, alongside architects.

It is interesting, though, how many of these developments were treated by architectural avant-gardes—in the United States in particular—as a curio, a fad, or a professional threat to the underpinnings of the discipline, with certain elite centers of architecture and critical architectural theorists/historians emerging as vocal and persistent resisters to making any accommodations to environmental concerns.

Hostility to the proposition that architecture should center environmental issues, ecology, or climate has been extensively discussed *within* architecture over the past three decades. At the more superficial end of the spectrum, as Lance Hosey notes, citing AIA Gold Medalist Cesar Pelli, architects value "the looks" of a building, and as Pelli said, "sustainability doesn't necessarily photograph" (Hosey, 2012). Perhaps more common is the fear that environmental and climate issues impose a set of unreasonable technocratic limitations on the artistic agency of architects. The pioneering ecological architect and key instigator of the term "ecological design," Sim Van der Ryn, reflecting on the limited take-up of his ideas by the profession, has suggested that architecture has continued to be undone by mythologies of architecture as art and "the Howard Roark Syndrome," the Ayn Randian vision of "...the lone creative genius that is determined to foist his great ideas on an unknowing public" (Van der Ryn, 2013: xiv). For Van der Ryn, postmodernism and deconstructivism in architecture, in particular, ushered in "... an empty colorful stream of architectural fantasies" (Van der Ryn, 2013: xiv), which are all about "design is for the 1%; the people, corporations and government that can afford it." The proposition though that "what is needed most is design for the other 99%" (Van der Ryn, 2013: xiv) disappeared from the discussion.

At the most basic material level, it could be noted that for many architecture firms operating in a highly competitive industry with small margins of error, designing higher performance buildings adds costs and can result in expensive "add-ons." It clashes with the need for architects to produce buildings that are cheap, on time, and photogenic. The architectural historian Tahl Kaminer (2016) adds a set of nuanced and useful refinements to this argument, noting the growing difference in conditions of production between architecture, design, and planning that takes place in the increasingly neoliberalized United States, UK, and other "Anglo-Saxon countries" compared to conditions of production in the Netherlands, Austria, Denmark, and Spain, which were also impacted by neoliberalism but at least maintained some thread of past social democratic commitments to planning, the provision of public housing, and ecological modernization from the 1990s onward. Notably, Kaminer suggests that in the increasingly neoliberal Anglo-Saxon world, "Architecture with a capital A occupies the coveted status of a cultural practice yet is (socially) marginalized, 'deployed' in the design of museums, libraries, luxury hotels and the like." This is an architecture that is increasingly focused on producing "rarefied objects and luxury artifacts" but is hardly involved at all in the design and construction of housing (whether private or public). In contrast, architects in many continental European countries remain involved in designing mass housing, "which necessarily engages architects with societal issues of habitation and daily life" (Kaminer, 2016: 17) and, we might add, questions of resource use, environment, and climate.

We can also see the many demands that are starting to unfold within the environmental design realm, which places many design fields in tension with the dominant neoliberal political economy that they increasingly find themselves needing to serve in the 1980s and 1990s. Felicity Scott has argued in *Architecture or Techno-Utopia* that with the collapse of modernism and the implosion of the New Deal, the Great Society, and the New Left, architecture in the United States post-Reagan increasingly gets incorporated either into the corporate structure or into art (virtuosity for the sake of virtuosity that is never intended to be built, commonly referred to as paper architecture). As she notes, at one end of the architectural spectrum you see "the unwitting integration" of architecture "into the dominant technological and social forces" and at the other you see the architectural academy engaged in a "withdrawal into autonomous disciplinary domains of form and/or semantic investigations" (Scott, 2007: 248–9). In the quest to defend architecture as its own art form, rather than a service profession responsive to outside forces like politics or economics, or a practical craft responsive to demands of engineering, the pursuit of autonomy prizes investigations into the internal, timeless logics of the discipline, such as form, pattern, and spatial explorations. The turn to semantics/rhetoric/ and semiotics in the 1990s and the 2000s—ironically often informed by Manfredo Tafuri's pessimist Marxism and Adorno-influenced critical theory—causes architecture to further systematically draw in on itself. Even with a looming environmental crisis, such a project, as Scott notes, is seen as "too boring for the radicals, too much of an imposition for the elitists." If what was left from the social and economic crisis of the 1970s was a good deal of paper architecture and paper radicalism for many more practitioners (who had made their peace in various ways with neoliberalism), then the inclusion of sustainability, ecology, environment, or environmental justice stood as an impediment to the road back to respectability and proximity to power.

We might also consider how the inward turn within the architecture academy was occurring just as energy code was proliferating and making inroads within the architectural practice. This "external" transformation also exacerbated a growing tension between architects and engineers, two distinct fields that had grown apart over the previous half-century and now found themselves in competition over whether energy innovation or artistic originality would now constitute good architectural design. New energy efficiency requirements resulted in the introduction of courses on "environmental systems" into architecture education, such as classes on passive heating and cooling, solar orientation, daylighting, and embodied energy, along with the engineers qualified to teach such classes. While architecture schools with greater proximity to engineering programs accommodated a greater technical engagement with such expertise, in many architectural programs such classes were treated as peripheral technical coursework, taught by adjuncts from outside the core faculty, and rarely integrated into design studios. Pursuing these conversations in a technicist fashion has often siloed "the environment" into what Kiel Moe refers to as "an equipment-centric, managerial epistemology for environmental topics in architecture" (Moe, 2018).

Finally, it could be wondered whether one of the key tension points between the architectural avant-garde and environmental design emerges from the observation that it is particularly threatening to traditional architectural ideologues—whether of the Left or the Right—precisely because it can decenter the building, the architect, and the architectural critic at the privileged center of the design process. For all those forces seeking to open up new conversations about energy use and environmental impact, adaptive reuse and landscape, labor and race, or systems of power and broader urban form—or even simply complying with the minimum legal limits of increasingly stringent building codes—can be read from an architecturally purist position as potentially dangerous and destabilizing to the so-called autonomy of architecture (see Schumacher, 2024).

Stewart Brand's *How Buildings Learn* (Brand, 1995) offers an interesting elaboration and contestation of these arguments. The building-centric ontology that increasingly grounds architecture in the 1990s, Brand argues, is primarily concerned with permanence. However, the problem is that buildings evolve and develop character and new purposes long after the architect has departed the scene, and urban form changes over time. Architecture is stuck with the idea of permanence, but:

> It is an illusion. New usages persistently retire or reshape buildings. The old church is torn down, lovely as it is, because the parishioners have gone, and no other use can be found for it. The old factory, the plainest of buildings, keeps being revived: first for a collection of light industries, then for artists' studios, then for offices (with boutiques and a restaurant on the ground floor), and something else is bound to follow. From the first drawings to the final demolition, buildings are shaped and reshaped by changing cultural currents, changing real-estate value, and changing usage.

Brand suggests that the epistemology of the discipline has been far too shaped by the proposition that architects deliver the last word on a building, often buying into Bauhaus- or Frank Lloyd Wright-type mythologies that they can deliver "total design." But architects quickly move on after construction and rarely, if ever, return to do post-occupancy evaluations of whether buildings meet clients' expectations. The proposition that clients or the public may customize structures to fit their own needs or tastes has no status in avant-garde architectural discourse, even though "adaptive reuse is the destiny of most buildings" (Brand, 1995).

Brand ultimately takes his critique of architecture in *How Buildings Learn* in a somewhat conservative direction. Little is said about Mierle Laderman Ukeles's observation that a great deal of general maintenance is gendered, classed, and racialized work (Ukeles, 1969), and Brand ends up replacing the architect with the real estate developer as the real author of buildings. However, the basic argument that he makes is not dissimilar to the core propositions of architecture professor Jeremy Till, who suggests that architecture depends on many forces beyond architecture for its execution and functionality, and these are all in flux. Markets, capital, taste, fashion, function, social purpose, site, servicing, the broader landscape, environment, and political climate can all change. However, till suggests much of architectural education and practice is trapped in visions of permanency or myths of control. As Till has it, this offers an incomplete account of what the division of labor in building and construction work that architecture depends on. This failure with architecture academia to acknowledge the many forms of labor and expertise that go into the construction and maintenance of buildings generates an inability to think about change and politics in sophisticated ways and an inability to think about the possibility of dynamic buildings or dynamic urban forms (Till, 2009).

Not everyone in the academy resisted the widening scope of concern that both energy efficiency and the wider incorporation of environmental topics brought in. Some of the more innovative architectural programs were pushed to open up engagements with broader interdisciplinary design research in areas such as low-carbon materials, hyperefficient buildings, design for disassembly, building for a circular economy, and so on. And while environmental course requirements for architecture schools in the 1990s focused on loosely defined concepts of "environmental stewardship," in subsequent decades they have evolved to cover "ecological systems," "building performance and sustainable design," and rather specific requirements around technical systems, building integration, energy modeling, building materials, and climate resilience.

The proliferation of such engagements has also given rise to calls that history/theory courses reach beyond a building-centered art history to engage with environmental history, political ecology, and the political economy and sociology of design–society relations in ways that could offer "different understandings of how architecture and the environment have been co-produced" (see Förster et al., 2022). In particular, we can see a cluster of scholarship that explores designed projects not merely as isolated physical objects, but as constructions embedded within a system of material flows and labor relations, contingent on the landscapes from whence their building materials are derived. For example, in *Reciprocal Landscapes: Stories of Material Movements* (2021), the landscape architect Jane Mah Hutton reframes some well-known designed landscapes as stories of extraction of materials from discrete raw materials, made possible by specific acts of labor. Exploring, for example, the granite that was used in the construction of Central Park, or the tropical hardwood in the High Line, she traces "the movement of materials as they pass from production landscapes (plantations, quarries, factories) to designed constructions (buildings, landscapes, infrastructure) through demolition and disposal or re-use." In *Empire, State & Building* (2017), the architect Kiel Moe frames the iconic Empire State Building in New York as a set of energy and material flows, tracing territorial patterns and labor impacts. Drawing on ecologist Howard Odum's concept of the solar "emergy" embodied in manufactured products, and broader currents of political ecology from world systems theory to Brenner and Schmid's planetary urbanization thesis (Brenner and Schmid, 2014), Moe suggests that while the building is treated as an object by architects, its laborers and contractors see it as a set of material flows, which are drawn from a discrete territory by state and economic forces, and embody specific quantifiable thermodynamic characteristics.

Not dissimilar to Stewart Brand's and Jeremy Till's critique of architecture, Moe suggests that modern architecture has fetishized *the appearance* of the building facade, but more broadly fetishized the building as a static Cartesian object. The result of this has been a discipline that is not only addicted to "the look" and "the new" but has systematically failed to explore how buildings come into the world in energetic and material terms, how labor and materials flow through the building from many other sites and spaces or end up as waste elsewhere. Almost like neoclassical economists who start their inquiries ignoring how economies are embedded in broader ecological relationships and processes, treating all side effects of this thinking as "externalities," building-centric modern architecture has proceeded in the same fashion (see Moe, 2017; 2018). These kinds of systemic explorations resonate with a rather more historical and politicized set of approaches to thinking about the relations between processes of urbanization and their hinterlands that we can see in the interventions of Neil Brenner, Christian Schmid, Nikos Katsikis, and Daniel Ibañez's works on planetary urbanization. Much of this work has provided a theoretical bridge between architecture, design, and urban political ecology/critical geography (Brenner and Schmid, 2014; Katsikis and Brenner, 2021; Ibañez et al., 2019). In expanding the limit of concern within architecture to incorporate not just the finished building but its embodied energy, Moe was making energy and building performance palatable for the architectural avant-garde academy, building on prior generations of architectural theory that similarly explored buildings as mechanical systems, machines for inhabitation made possible through technology, such as Rayner Banham's *The Architecture of the Well-Tempered Environment* (1969). Gradually, as environmental concerns escalated in the early 2000s, systems thinking overcame resistance from within the academy, taking its place within architectural theory.

From LEED and the Rise of Green Certification, to Building Performance and the Carbon Lens

While energy code pushed the building sector to reduce its environmental impact over the course of the 1980s and 1990s by raising the "floor" of building performance through minimum code compliance, a handful of organizations began introducing voluntary certification schemes to create more ambitious targets that projects might reach for. The 1990s saw the launch of BREEAM (Building Research Establishment Environmental Assessment Methodology) in 1990 followed by the EPA's Energy Star system in 1992, and LEED (Leadership in Energy and Environmental Design) in 1998. SITES (the Sustainable Sites Certification System for landscape architecture and planning) and the Living Building Challenge (LBC) followed another decade later. All these certification systems provide performance benchmarks for design practice with varying levels of ambition and success.

Passive House (or Passivhaus) is another influential certification system, where rather than offering a general set of sustainability criteria and features like LEED or BREAM, the focus is tightly trained on a small set of performance variables that result in extremely energy-efficient buildings. Passive House standards were developed in Germany in the 1990s based on the North American experimental passive solar buildings of the post-'73 oil shock era, such as those profiled in Ed Mazria's *The Passive Solar Energy Book* (Mazria, 1979), and as such measured their environmental impact primarily in terms of energy savings. Passive House buildings are evaluated on their energy consumption and airtightness, and typically use super-insulation and triple-glazed windows to create a very tight seal and advanced heat recovery systems to reduce the thermal losses from ventilation. Passive House buildings also tend to be all electric, cutting out gas heating to maintain good indoor air quality. While such features can result in significantly higher costs compared with standard construction, a recent resurgence of innovation in the Passive House architectural community has resulted in Passive House affordable housing projects that have the additional benefit of reduced utility bills—such as the Sendero Verde affordable housing project in

New York's East Harlem neighborhood, by Handel Architects, which recently became the largest Passive House–certified building in the world with over 700 affordable units and boasting 40–70 percent reductions in energy use.

At the same time, a shift was underway within architecture, spurred by a growing recognition of the urgency of addressing climate change, which entailed going beyond increasing the energy efficiency of buildings' operations (their operational energy) to also tackle the impacts of the manufacturing of their materials, their transportation to the construction site, and their assembly during construction. As we've just seen with Hutton and Moe, this embodied-impacts conversation in architecture, as with eco-design for products, demanded a shift in focus away from the designed object itself, toward the system of flows and processes that it sets into motion across its entire life cycle.

Whereas initially architects began to talk about the *embodied energy* of building materials, looking for ways of avoiding energy-intensive materials the same way they avoided energy waste in buildings, a subsequent paradigm shift has decentered energy in favor of a focus on carbon. This stems from a rather recent recognition that since the underlying driver of the climate crisis is *carbon dioxide*, rather than tracking energy (measured in watts or megajoules), designers should be tracking *carbon* (measured in tons of CO_2 emissions). This shift reflected a growing recognition that the measure of "energy" in and of itself doesn't line up neatly with carbon emissions, since some energy grids are more carbon intensive than others, and some are cleaner from a carbon emissions perspective. Some states and municipalities have older and dirtier power plants, and some have deployed more clean renewable power. As an example, while the smelting process uses the same amount of energy, aluminum produced in northwest China in smelters powered primarily by coal-fired electricity is going to produce more carbon emissions than aluminum produced in Iceland in smelters powered by clean hydroelectricity. More energy use on a cleaner grid doesn't necessary mean more carbon emissions, meaning that as the energy grid transitions to cleaner energy generation, architecture must look beyond energy efficiency and track the "thin thread of carbon" (Carlisle and Pevzner, 2021).

This realization has crystallized into a growing focus on embodied *carbon* in architecture and construction, rather than just embodied energy (Carlisle, 2017). The shift in focus aligned with the urgency of the climate challenge: whereas operational emissions accrue over the years that a building is operating post-construction, the embodied carbon is released all at once, in large pulses before and during construction, from the mining, smelting, and heating of materials during production, and the land disturbance and heavy equipment on site, creating a carbon debt that even the most energy-efficient buildings can take decades to pay back. The focus, then, shifts to the carbon impact and other environmental impacts of building materials, a development that the various green certification schemes have incorporated to varying degrees.

While some of the buildings that achieved the highest rank in the LEED classification system—those with "Platinum" status—did genuinely push the envelope on energy efficiency for their time, the LEED certification system process itself has been endlessly critiqued over the past three decades for its "tick the box" qualities, which can lead to a gaming of the system, for its focus on pre-occupancy plans rather than post-occupancy outcomes, and for its resolutely building-centric focus. LEED has demonstrated some capacity to address its critics and make incremental improvements. In 2007, LEED for Neighborhood Design (LEED-ND) incorporated more holistic urban design metrics that rewarded compact and walkable districts, and LEED V5 in 2022 targeted the reduction of embodied carbon emissions from building materials and construction processes, which marked a shift from energy efficiency and "local materials" to carbon accounting and more directly tracking climate impacts.

Other certification systems have tried to be more holistic, beyond energy efficiency, such as the Living Building Challenge (LBC), which offers a more comprehensive and challenging approach that pushes practice to think beyond reducing a building's negative environmental impacts to consider the possibilities of regenerative architecture, from buildings that *produce* their own energy, to those that handle their own water and waste, to those that avoid toxic building materials (Red List Chemicals). The list of LBC-certified projects worldwide ranks in the mere hundreds, compared with over 195,000 certified with LEED and over 600,000 certified with BREEAM, but this is by design—the goal of the International Living Future Institute (ILFI) that established LBC was to make achieving LBC certification extremely challenging, thereby pushing the bounds of practice; buildings are certified only after they demonstrate their performance after construction, rather than awarding certification based on mere plans. Their recent introduction of Zero Carbon Certification is designed to focus even more squarely on climate impacts, rewarding only buildings that demonstrate 20 percent reductions in both operational and embodied carbon, which in practice means all-electric buildings powered by renewable energy. The US Green Building Council has also launched LEED Zero Carbon Certification for operational energy, carbon emissions, water, and waste, while the European Union and China have leaped ahead to develop Zero Carbon Codes.

Supporters have argued that voluntary certification systems have at least legitimized a version of sustainability in private practice and industry, providing some means for ratcheting building performance expectations from a very low bar. Arguably, the green certification boom did incentivize attempts to upskill knowledge in the architecture professions, provide methodologies and benchmarks that helped to inform sustainable building design, construction, and operation and create some space within architecture, planning, and design for the emergence of organizations that have pushed for the profession to decarbonize more rapidly: *Architecture 2030* and the 1.5 Degrees C Communique (which has attempted to mobilize the profession to focus on the urgent necessity to decarbonize practice rapidly). We have seen growing ongoing engagement between architecture, the building sector, and the UN COPs as a collective result of this work (Gadhano, 2022).

Critics of the voluntary approach note that LEED, BREEAM, and the Living Building Challenge are effectively for-profit enterprises that charge fees for their green building certification programs. Green building conferences and trade shows (such as Greenbuild, organized by the same organization that created LEED) routinely focus on showcasing "green products," driving a form of "green consumerism" for architecture. None of these voluntary building-level certification systems allow architects or the construction industry to think in the aggregate about energy, material, and social structural drivers of unsustainable production and climate emissions that range across the built environment, nor do they have anything to say about the labor conditions of the "design-production networks" that architecture is produced through (Baxi et al., 2018). As voluntary, market-driven certification systems that operate as a means by which private actors can showcase their "sustainable best practice" without government mandates or regulations getting in the way, they are interventions premised on the hope that good examples will ultimately crowd out bad actors. At the same time, not dissimilarly from Amory Lovins's embrace of market drivers, these voluntary certifications have demonstrated that there is some market demand and a willingness to pay for greener building design and construction.

Voluntary initiatives have also created space for more aggressive voices to emerge, with organizations like *Architecture2030* that have pushed the profession to decarbonize more rapidly. Its 2030 Challenge, issued in 2006, asked the architecture community to set a 70 percent reduction target for new buildings' energy consumption and carbon emissions, ramping up to a target of full carbon neutrality by 2030; this was taken up by the AIA in its AIA 2030 Commitment. Ahead of COP 26 in Glasgow in 2020, a network of over 400 architecture firms and organizations formed *Architects Declare*, issuing a declaration of a "Climate, Justice, and Biodiversity Emergency" to advocate for attention to the linkages between the profession's climate impacts, labor practices, social justice, and environmental harms. Around the same time, sixty of the largest architecture, engineering, and construction firms and most professional organizations signed the *1.5°C COP26 Communique* demanding more aggressive national-level targets. These early commitment programs by architects have spurred other allied disciplines, like MEP2040 by mechanical engineers, SE2050 by structural engineers, and the ASLA Climate Action Plan by the American Society of Landscape Architects.

For this newer generation of building designers, the primary concern is not marginally better energy efficiency, but climate justice. As such, the key unit of performance is not the megajoule or kilowatt, but the tons of avoided carbon emissions. This conversation engages much more broadly with how buildings, landscapes, infrastructure, and products are made, where they come from, who benefits and is harmed across their full life cycle, and the systems of power, privilege, and inequalities embedded in global supply chains. It is a critique of building performance that is much broader than the efficiency of a fan coil unit, and its goals can't be met with a switch to greener products alone. The goal is not to use incrementally less fossil gas and petroleum products in projects, but to phase out the buildings' dependance on fossil fuels entirely. To accomplish this aspiration requires deeper engagement with both building performance modeling and data on building materials that can reach upstream to bring into view raw material extraction and production, and downstream into use, demolition, and disposal. The assumption here is that designers can't really understand or reduce the impacts of their projects unless they can understand them across the full life cycle and that they can't minimize what they can't measure.

Tracking embodied carbon is clearly a difficult task to do, and to assist architects, engineers, landscape architects, and contractors, we have seen the emergence of a whole ecosystem of software tools, employing "whole building LCA," which tracks both operational and embodied carbon, or using Environmental Product Declarations from the producers of various building materials. As carbon accounting and LCA become more mainstream, and data on the climate impacts of the buildings and infrastructure becomes more clear and widely understood, the practice has enabled a recent wave of building sector policy, incentive programs, and regulations globally, such as limits on carbon intensity for the concrete used in public procurement in states like California and Washington.

What follows from this recognition of the construction industry's carbon intensity? Some architects have focused their attention on minimizing their use of high-carbon materials such as concrete and steel in favor of designing lighter structures that need less structure. Some have worked on sourcing lower-carbon concrete from companies working to produce concrete mixes with low-carbon cements, cement replacements, or CO_2 injection. Other designers have turned their attention to wood and bio-based building materials, chief among these being "mass timber" construction (Ibañez et al., 2019). Bio-based building materials hold out the promise of being carbon negative, *absorbing* carbon rather than emitting it, since they're derived from plants, which naturally do the work of carbon sequestration.

Architect Alan Organschi helped articulate the polemical argument that if cities were built with enough wood construction, entire cities could even become a carbon sink (Churkina et al., 2020), and subsequently published a textbook and teaching materials for designers, *Carbon: A Field Manual for Building Designers* (Kuittinen et al., 2022). Mass timber high-rises have gone up in cities like Portland, Oregon, Vancouver, British Columbia, and Milwaukee, as well as in Vienna, Melbourne, Bordeaux, and multiple cities in Norway, Sweden, and the Netherlands. Wood construction is certainly lower carbon than standard building methods like steel and concrete, but as the enthusiasm for mass timber construction has grown, alarms are ringing about ensuring that the wood used for mass timber projects is harvested from sustainable sources, with a fear that an unregulated boom in mass timber could fuel the illegal wood trade and drive deforestation (Searchinger et al., 2023[5]).

Some design offices have emerged as leaders in advocating for sustainable sourcing and foregrounding the labor side of mass timber construction, as well as considering what happens in the forests where this wood is grown. For example, ZGF Architects, led the design for the PDX Airport's Main Terminal Expansion in Portland, Oregon, with the dramatic use of mass timber. They ensured that timber was carefully sourced from local small forest landowners, Tribal forests, and community forests and combined this with innovative off-site manufacturing techniques and modular construction processes.

But private industry can only lead so far; on the regulatory side, local fire and building codes often complicate mass timber construction, and on the national and international level, transparency and material traceability in the wood industry is still being argued over. Stronger regulatory controls that would bolster the sustainability arguments for mass timber are needed before this pathway can win over actors across the environmentalist coalition, who mistrust the timber industry because of previous environmental battles over clear-cutting and endangered species. Production of high-quality mass timber is regionally variable and still minuscule relative to the size of the overall construction industry. And the clearing of regulatory and technical hurdles is often happening ad hoc, municipality by municipality, by individual design offices struggling with individual projects.

Other groups have taken a broader approach to advocating for plant-based building materials, beyond mass timber. Loose-knit advocacy organizations, such as The Bio-Based Materials Collective (which is composed of smaller actors from across the building material manufacturing supply chain, including foresters, farmers, and manufacturers alongside architects, designers, and builders), and small construction companies like New Frameworks in Vermont are seeking to support this emerging low-carbon building materials sector, overcome regulatory hurdles, and increase the share of building components that are made from fast-growing crops and agricultural byproducts, like insulation made with hemp or compressed panels made from straw, all of which could draw down and sequester carbon and potentially replace fossil-based or higher-carbon materials.

Of course, the focus on embodied carbon does not let operational emissions off the hook. Architectural historian Daniel Barber has suggested that reductions in both will be needed. Alongside a focus on low-carbon building materials or developing new performance standard codes, he suggests that an agenda for low-carbon architecture also needs to more directly grapple with the global inequalities of thermal comfort, in which the automatic whir of air conditioners or radiators in the Global North exacerbates acute climate harms in the Global South, where thermal control is not universally available. For Barber, an architecture "after comfort" (Barber, 2019) would not only have to start finding alternatives to the carbon-intensive worlds of building made with glass, steel, and concrete but also be open to reconsidering what constitutes

Figure 7 Portland Airport's Main Terminal by ZGF Architects. Photo: Nicholas Pevzner.

a normatively comfortable building environment, even playing with discomfort to make it "designed, managed, and made desirable." The provocation to think "After Comfort" raises important questions about what the architecture avant-garde might learn from vernacular architectural practices that could facilitate new methods of cooling and shading, guides us toward exploring suppressed or sidelined traditions of passive cooling in modernist architecture, and challenges architecture to think beyond reliance on the modernist crutch of air conditioning and "toward engaging with the limits of comfort." The challenge here from a just transition perspective would be to bring these critically important architectural discussions of thermal design into direct engagement with the classed, raced, and gendered politics of discomfort produced by substandard housing, the deadly consequences of energy poverty, and the need of architecture and planning to directly address the vast uneven vulnerabilities to heat exposure that a changing climate will portend for all manner of working people in the field, the factory, the mine, the call center, and throughout the city, in both hemispheres.

Garden City High-Rises

If the technically sophisticated movement to cut both operational and embodied carbon in architecture represents a tangible, if still somewhat niche, subset of action within the architecture profession, what the general public tends to see are the flashy green buildings that tout their sustainability contributions. Following the inroads made by design for sustainability and in response to green neoliberal calls for sustainable development, the late 1990s and early 2000s saw a flurry of experiments in commercial green buildings, many choosing to forego the certification route. The first wave of these projects, such as Foster + Partners' Commerzbank Tower in Frankfurt (1997) and Swiss Re headquarters in London (2004), excited some architectural critics for their innovations in curtain wall systems, natural lighting, and ventilation. So did the San Francisco Federal Building by Morphosis (2007), the first US office tower to go without air conditioning on most of its floors. Sauerbruch Hutton's elegant design of the KfW Westarkade building in Frankfurt (2010) has been celebrated for its insulated double-skin facade that enables daylight penetration and natural ventilation without sacrificing energy performance, alongside geothermal-powered heating and energy-saving features (see Hosey, 2016).

More recently, we have seen a number of the largest architectural firms in the world—from Skidmore, Owings & Merrill (SOM) to Gensler and Foster + Partners to more midsize practices such as Bjarke Ingels Group (BIG)—make bold claims about their offices' climate goals. SOM recently proposed the "Urban Sequoia™"—a prototype building concept that is designed (at least on paper) to remove three times its total carbon contributions over a hundred-year lifetime, by engaging in carbon drawdown through direct air capture technology. Perkins&Will has pledged to eliminate embodied carbon from every commercial interior space they design by 2030, in line with the net-zero timeline set by Architecture 2030 and the AIA 2030 Commitment. BIG, in characteristically grandiose fashion, has proposed a (presumably polemical) global-scale energy masterplan, aptly named "Masterplanet," to "solve" climate change with a planet-sized supergrid (BIG 2020[6]). Of course, the devil here lies in the details.

For critics who view the idea of "sustainable high rise" as a contradiction in terms, there is little to be seen here in this endless wave of PR pronouncements beyond greenwash. To be sure, there is plenty of exaggeration in the world of architecture in the service of marketing. Mehaffy and Salingaros (2017), for example, have noted that post-occupancy evaluations of some of the most celebrated sustainable buildings of the 1990s showed a gap between promised energy savings claims and real performance, whether because of problems with fancy curtain wall assemblies or energy usage choices by tenants deep in the interior of the buildings far from the high-tech windows. All things remaining equal, an improvement in energy efficiency in a green building may do little if users are more profligate in their energy use. The additional embodied carbon in tall buildings requiring much more structure (carbon-intensive steel and concrete) easily swamps marginal carbon savings from energy use. Rebound effects and the Jevons Paradox are as real for green buildings as broader energy systems.

One of the challenges of judging many "sustainability" claims in architecture has to do with how poorly this term is defined within the architecture discipline. MIT's Mark Jarzombek has complained about the open-endedness of the term, saying "On the one hand, we have design-build projects, self-sufficiency houses, solar houses, eco-villages, and even so-called health houses. On the other hand, we have new high-tech materials, sophisticated software programs, and the emergence of various types of international, 'green' consultancy companies. The word sustainability can encompass everything in this spectrum" (Jarzombek, 2006: 7). This vaguely defined agenda for architecture is far from the precise measurement of energy savings and weathertightness prized by technical building performance agendas such as Passive House, whose success is judged by whether or not it works. It is also quite far from the specific definition of "sustainability" popularized by the 1987 Brundtland Report, which defined sustainable development as that which "meets the needs of the present without compromising the ability of future generations to meet their own needs" (World Commission on Environment and Development, 1987)—which if taken at face value would present a rather specific charge when considering both resource consumption and climate impacts, and which would suggest a much more radical transformation of the building sector to do no harm, full stop, rather than merely be less harmful.

Sustainability claims also raise the question of where one ought to draw the limit of responsibility for sustainable architecture. The headquarters of the Abu Dhabi National Oil Company (ADNOC), designed by HOK, may have received praise for its energy-saving technical innovations. However, there is no configuration of smart energy-efficient building technologies on the planet that will mitigate the current greenhouse gas emissions of ADNOC, let alone the mass of emissions guaranteed by ADNOC's commitment to opening up new oil fields for the foreseeable future. Similarly, expanding the argument from individual buildings to the building sector broadly, even if the average building is much more energy efficient than twenty years ago, carbon emissions from the building sector are still rising. As Barber has continually emphasized, "Some recent buildings emit less greenhouse gas than they might have otherwise, but the global carbon footprint of the industry continues to expand" (Barber, 2024).

If most climate concerned architects would agree that the high modernist energy-guzzling buildings made of glass and steel have had their day, pragmatist nevertheless argue high-density buildings still have an important place in the design of global urban futures—and from a social, community, environmental, and climate perspective, there are better and worse ways to design and construct tall buildings. Philip Oldfield (2019, 2022), for example, has argued that while building higher undoubtedly has energy and climate costs and supertall buildings make little sense, in specific contexts where land is scarce and housing needs are pressing, both mid- and high-rise structures can offer "greater density, putting more people in close proximity to low-carbon public transit, and the civic infrastructure of the city" (Oldfield, 2022). Moreover, from retrofitting existing buildings to making different material choices, reconsidering glazed facades and maximizing permeability, using atriums, and passive cooling and shading systems, there are better and worse ways of building high (Oldfield, 2019, 2022). The virtues of urban densification, transit-oriented development, and walkable and livable cities have largely become a new orthodoxy for many progressive planners, whether these sentiments are framed through advocacy for new urbanism, smart growth, green urbanism, ecological urbanism, or something more radical. UN ventures such as Agenda 21 or the SDGs, with their focus on local authorities as key drivers of sustainable planning and the rise of place-based environmental justice struggles, additionally offered the field of urban planning a new source of legitimization in turbulent times.

Finally, there's the social and urban dimension of what green buildings do, not only for the climate and environment but also for people and communities. The two residential towers of the *Bosco Verticale* project in Milan, designed by Stefano Boeri Architetti, captures many of these tensions of the Green High-Rise. The striking tree and plant-covered facades of the two residential towers have become a totemic example of the new biophilic green architecture since the towers were built in 2014. With its over ninety species of trees, shrubs, and other plants growing across its facade, Bosco Verticale echoes earlier experiments in green buildings, like Ken Yeang's bioclimatic architecture with vertical gardens and passive climate control systems, seen in Yeang's projects like his Menara Mesiniaga in Malaysia (1992) and his National Library of Singapore (2005).

With its image endlessly reproduced in architectural magazines, *Bosco Verticale* has spawned dozens of imitators and, as Owen Hatherley (2025) has noted, further inspired a great deal of solar punk urban imaginaries over the past decade. Not only do the structures look remarkable, indeed evoking a "vertical forest," but the living facades of the two buildings have been conceived to provide ecosystem services: provide natural insulation, create more spaces for urban biodiversity, and contribute to summer cooling and, as claimed, CO_2 absorption more effectively than typical heat-trapping high-rise buildings with glass facades. At the same time, the steel-reinforced planters and extra-thick concrete balcony slabs that are needed to hold the additional weight of the soil and planting increase both the embodied carbon of the building and its construction costs compared to the "typical" alternative. High-rise gardening, pest

Figure 8(a, b) Trudo Vertical Forest Eindhoven front and side view. A social housing project that accommodates 125 social housing units on 19 floors with a green façade that contains 135 trees and 5.200 smaller shrubs and plants and other vegetation. Architect. Studio Boeri. Copyright Studio Boeri used with permission.

control, and fire safety provisions generate maintenance challenges to the upkeep of these structures. Most of these "vertical forest" projects at present have been built for high-income residents. Critics have observed that while the development did involve some degree of participatory process, and residents reportedly appreciate the beauty of the buildings, *Bosco Verticale* has also facilitated further gentrification of the locality and seen the management of the public park surrounding the development handed over to private developers (Di Paola, 2021). Nevertheless, Stefano Boeri has seen a huge spike in demand for these vertical forest high-rises, with similar tree-covered projects built or projected in Treviso, Lausanne, Eindhoven, Paris, Utrecht, Nanjing, Cairo, Tirana, Huanggang, and Dubai.

Landscape Architecture, Urban Planning, and the Struggle for the Just Green City

If architecture has often dominated the green design conversation, others have argued that building-centric approaches alone are not going to do the job; more integrated approaches that can grapple with climate challenges at a variety of scales and sites are required. Landscape architecture, with its historical ties to architecture and land use planning, ecology and horticulture, hydrology, soil science, civil engineering, and forestry, has arguably possessed some of the broadest range of resources within the architecture disciplines to engage with the challenge posed to design by the climate and biodiversity crises. Years before the first Earth Day, the creation of the EPA, or the major wave of environmental legislation that occurred in the 1970s, a group of landscape architects assembled by the Landscape Architecture Foundation in 1966 issued a declaration of concern about declining environmental quality in the United States. It asserted the need for landscape architecture as a field to address the ecological crisis. Out of this moment came Ian McHarg's foundational text, *Design with Nature* (1969), which inspired a generation of architects to look more closely at natural processes taking place below their construction sites, and which in turn provided disciplinary legitimization for the later discussions around urban ecology in design (Spirn, 1984), new urbanism (Katz, 1994), and landscape urbanism (Corner, 1999).

The Landscape Urbanism movement within landscape architecture in the early 2000s (Waldheim, 2006) saw landscape architects taking on urban design and planning more explicitly, and the selection of top landscape architecture firms to lead the redevelopment of former industrial and brownfield sites (in New York's Fresh Kills landfill and Brooklyn Bridge Park, or Toronto's Lower Don Lands and Downsview Park, for example) thrust landscape architecture into the design spotlight. The Museum of Modern Art's high-profile *Rising Currents: Projects for New York's Waterfront* exhibition in 2010, which asked architecture and landscape architecture teams to imagine New York City adapting to sea level rise, further raised the profile of a crop of emerging landscape architects as contributors to a critical discourse that could speak to ecological integration, urban adaptation, and climate change. More recent climate disasters have seen landscape offices turn to designing for heat waves and flash floods, with shade trees and public open space becoming prized resources for designing more just and pleasant neighborhoods. Many firms now take on environmental topics such as ecological restoration and climate adaptation, even if most projects do not yet account for the carbon emissions that are produced from their construction. We have seen the rebranding of wetlands, urban forests, and parks as "green infrastructure" that can provide "ecosystem services" and the growing engagement of landscape architecture with projects focused on urban climate resilience and decarbonization agendas. Whether the contemporary project for landscape architecture has been framed in terms of ecology, sustainability, or resilience, the field has demonstrated a capacity to engage with contemporary socio-ecological issues and offered more tools for imagining and building dense, livable, and ecologically functional urban places ready to weather climate impacts, beyond the biophilic appeal of singular green buildings.

The extent to which these kinds of projects can systematically reorient landscape architectural practice, which, as Billy Fleming has observed in the United States, is still very much shaped by a fairly conservative client-dependent small private practice model, is an open question (Fleming, 2019; Weller and Fleming, 2016). Certainly, landscape architecture has not escaped the impact of neoliberalism, as debates around Landscape Urbanism's alignment with real estate development make clear and as the impact that high-profile projects like the High Line in Manhattan have had on driving gentrification demonstrates.

In many ways, the disciplinary opening that Landscape Urbanism was leveraging to enter the multidisciplinary urban design conversation in the early 2000s was the result of the precipitous decline in the power and influence of modernist urban planning, which, at least in the Anglo-sphere, lost face thanks to overzealous urban renewal programs that tore apart communities (Fullilove, 2004) and wiped out the spaces for urban life most beloved by critics of urban renewal like Jane Jacobs. Through the urban turmoil and social crises of the 1960s and 1970s and the rise of neoliberalism onward, sustainable urban planning and ecological urbanism did open up opportunities for urban planners to reposition themselves, rebrand their skill sets, and reclaim their public legitimacy. Sim Van Der Ryn and Peter Calthorpe's *Sustainable Communities* (1986) or, later, the projects and writings of Jan Gehl (see Gehl, 1987, 2010) became canonical attempts to reclaim a post–Jane Jacobs vision of human-scale planning. Both texts were just radical enough (in a formalist sense) to draw interested younger professional designers and planners into a discourse that could gain some traction in more progressive and liberal-leaning communities, but apolitical enough to avoid scaring the powers that be.

Participatory and advocacy planning championed by radical planners as an alternative to high modernist planning may have left a mixed legacy, which we will explore further in Chapter 5. Nevertheless, the virtues of urban densification, transit-oriented development, and walkable and livable cities have largely become a new orthodoxy for many progressive planners, whether these sentiments are framed through advocacy for new urbanism, smart growth, green urbanism, ecological urbanism, or something more radical. UN ventures such as Agenda 21 or the SDGs, with their focus on local authorities as key drivers of sustainable planning and the rise of place-based environmental justice struggles, additionally offered the field of urban planning a new source of legitimization in turbulent times.

Again, several decades of this activity have clearly generated a mixed bag of results. To be sure, urban livability agendas coupled with other urban policy agendas in select cases have helped urban revitalization efforts in certain spaces and places, partially reversing the decline of inner-city cores, which was often seen as inevitable during the 1970s. Defenders of such approaches to urban revitalization point to the successes of model cities such as Copenhagen, Melbourne, Paris, or Geneva. At the same time, critics observe urban revitalization efforts converging with a throughgoing embrace of neoliberal urban policies have also facilitated the growth of a good deal of "cappuccino urbanism" (Krivý and Ma, 2018) and "glitzy green urbanisms" (Anguelovski and

Connolly, 2021) in already affluent urban centers and neighborhoods across the globe. To the extent that such interventions have shrunk the focus of planning to the pedestrianization, redesign of the street, bike lanes, and the need for creating lively "people centric" downtown spaces while outsourcing broader urban planning issues such as the need for affordable urban housing, good quality public schools, well paid jobs, and policies that can address the cost of urban living to the vicissitudes of the free market has resulted in suboptimal results. Notably, we have ended up with increasingly segregated "livable green cities" that are good for high-income members of the creative class but increasingly unlivable for the rest. As research by Isabelle Anguelovski and James T. Connolly, Hillary Angelo, Ken Gould, Tammy Lewis, and their colleagues has demonstrated (see Anguelovski and Connolly, 2021), to the extent that green urbanism has become entangled in neoliberal urban schemes narrowly focused on "urban livability," it has also contributed to "green gentrification" and house price inflation. Of course, such developments are not just given in the nature of things, nor a topic purely for design to tackle. They are the product of policy choices and policy outcomes. As Oscilowicz, Hajtmarova, and Anguelovski detail, modes of socially conscious green urbanism planning that guarantee housing and land rights *before green design projects proceed* can produce better outcomes. As they outline, a programmatic policy agenda based on public institution–led planning practices that directly engage working-class communities and creates opportunities for inclusive zoning and rent control, urban policies that force developers to submit and fund anti-displacement agreements and community benefits agreements (which mandate minimum amounts of green space and minimum numbers of affordable housing units), robust advocacy for the ongoing preservation, improved maintenance, and upgrading of social and affordable housing can make an enormous difference (Anguelovski and Connolly, 2021).

Adaptive Reuse, Repair, Refusal

While the mixed legacy of numerous prominent green buildings suggests caution, it is safe to say that new waves of regulations on both the operational carbon and the embodied carbon front are now pushing up the baseline environmental and climate performance of the architecture field, much like they did decades before with ASHRAE energy codes for energy efficiency. And yet architectural and engineering engagement in low-carbon construction might still be regarded as a niche practice within the profession. More radical voices now suggest that the climate crisis requires much bolder breaks from typical practice than mainstream neoliberal green design can or will deliver. Looking at the scale of our current crisis, and the embodied emissions in new construction as incompatible with international decarbonization targets, some radical voices in architecture have argued that if we take the carbon numbers seriously, what is required is something more akin to a complete moratorium on new construction. Architect and urban designer Charlotte Malterre-Barthes (2025) has recently elaborated such a call, in part inspired by Bruno Latour's observations that the momentary interruption during the global COVID lockdown generated a space where everything stopped and everything could be questioned. Malterre-Barthes has proposed a moratorium on new buildings in the United States and Europe, China, and South Asia as well as the MENA Region (the Middle East and North Africa) (see Malterre-Barthes, 2025: 32). Such a moratorium, she argues, would highlight the destructive climate and labor treadmill of construction that the building industry is committed to and will not stop without intervention. Malterre-Barthes argues that a pause, moreover, would allow public discussion about what we are building and why we are building. It would draw attention to the vast amount of unused space that currently exists in the built environment of many high-income countries and the ecological and social turmoil embedded in the extraction of building materials, and would allow reflection on the staggering waste of resources that exist within the existing built environment. It would elevate adaptive reuse and renovations, and "building with what exists, inhabiting it differently, and caring for it." A pause would force discussion on the kinds of carbon, resource, and labor standards that need to be in place to commence new building.

Malterre-Barthes's book-length elaboration of the moratorium call provides a meticulous, detailed, and careful critique of the ecological, material, labor, and climate failings of the architectural profession and the building construction industry more generally, and lays out a case for the urgency of the need to act. However, the extent to which the solution proposed helps or hinders climate politics is another matter. For example, this avowedly anti-capitalist, feminist, and decolonial proposal does not explain which workers or publics would support such a ban, or what the millions of people across the globe involved in the building, design, construction, and building products industries are going to do to sustain themselves and their families while such a moratorium is in place. Nothing is said about how a building moratorium facilitates alliance building between design workers, environmentalists, and labor unions, or how a moratorium avoids simply reinstalling the "jobs versus environment" binary at the center of climate-design politics. Similarly, the concept of a global moratorium does not distinguish between modalities of extraction that are wasteful versus modalities of extraction necessary for sustaining human life and between forms of new construction that are involved in building hospitals, schools, public housing, health care centers, or climate-resilient urban infrastructure versus the new construction of suburban sprawl, speculative real estate, or luxury condos for the 1 percent. How the terms of a global moratorium between developed and developing nations would be negotiated or policed (nationally or globally) without recourse to new imperial relations is not made clear (Will China and India just stop all new construction because architectural radicals in the high-income world suggest so?). As a provocation, talk of a moratorium on new construction certainly grabs attention. However, as practical politics, it does need to be asked whether the end result of these kinds of degrowth climate maximalist demands and think pieces are more likely to fuel climate backlash, pushing building workers further into the arms of a Populist Right than creating the basis for climate–labor alliances that speak to the interests and concerns of working people.

If calls for a full-on building moratorium are unlikely to deliver any time soon, the more modest proposition that repair, maintenance, and adaptive reuse are going to be—and need to be—ascendant design strategies shaping how we adapt to life on a warmer planet perhaps captures the reality of where we stand and where "good design" may be heading, at least in in the affluent world.

Adaptive reuse and retrofit for many decades may well have been positioned in the hierarchy of architecture and design academia as somehow a "lesser" practice to the "real inventiveness" of constructing new buildings, but times are changing. The AIA estimates that adaptive reuse and retrofit now account for some 50 percent of architectural buildings in the United States. The interventions of Ellen Dunham-Jones and June Williamson (2009, 2021) are notable here for exploring how obsolete suburban sites—such as derelict office parks, ghost malls, and parking lots—could be retrofitted to address urgent challenges, including environmental justice, public health, social equity, and climate resilience. More recent advocates of adaptive reuse in the journal *intAR* have demonstrated over the past decade vast opportunities that exist in all manner of urban contexts, brownfields, rust belt, and already-developed urbanized land as well as suburban land that could be creatively and imaginatively repurposed and retrofitted for housing and social services and "re-densified" to generate walkable neighborhoods that could ease the pressure on greenfield sites. The convergence of adaptive reuse conversations with growing interest in circular building practices, embodied carbon discussions, and design for disassembly has stimulated further innovative discussions about the importance of developing material banks and material product passports that can optimize the reuse of materials.

Of course, purely design-focused discussions of adaptive reuse that have nothing to say about the politics of class, race, and gender are deeply limited. Retrofitting older industrial facilities into more luxury apartments for speculative investors may save some materials and embodied carbon compared to new construction, but is not going to open up new pathways to just transitions. Much like "urban livability agendas," without being combined with a range of further mechanisms—such as community land trusts, cooperative housing, wealth taxes, or robust state support for public and affordable housing to constrain financial speculation in land—adaptive reuse can become as much a tool for urban gentrification, speculative practices, and displacement of communities as new buildings. Yet there are interesting developments afoot seeking to address these issues. *House Europe*! is a European wide campaign aimed at securing new EU laws that "prioritize renovation and adaptation over demolition and reconstruction" and do so in ways that bolster urban policies defined by "equity, resilience, and community-building" that are designed to assert the public good over private speculation.[7]

Adaptive reuse and retrofit received perhaps its most public recognition within the elite architectural design conversation to date when the 2021 Pritzker Prize, the top honor in architecture, went to Anne Lacaton and Jean-Philippe Vassal, who had made a career of careful retrofitting of older modernist housing blocks with an eye to residents' desires and needs. Their retrofitting of a 1960s tower block, the Tour Bois le Prêtre in Paris, has been widely celebrated as a notable example of a successful building retrofit of public housing. Not only did the renovation expand the size of the units, removing the old facades and creating new glazed "winter gardens" and balconies around the building's exterior, but tenants were consulted about these changes both before and after the renovation took place and were allowed to remain in place along with all their belongings while construction work went on. Demolition "is a waste of

Figure 9 Before and after photo of the transformation of 530 social housing units that are part of the 'Cité du Grand Parc' in Bordeaux. Architects. Lacaton & Vassal + Frédéric Druot + Christophe Hutin. Photograph Philippe Ruault. Copyright Philippe Ruault. Reproduced with permission.

many things—a waste of energy, a waste of material, and a waste of history," Lacaton and Vassal are quoted saying in their Pritzker Prize announcement. By working in ways that minimize demolition while attending to the craft of architecture, they point the way forward for sensitive, careful adaptive reuse work as an aspiration for the field. While they don't make the embodied carbon argument explicitly, such projects are a near-perfect distillation of many of the most promising trends in green architecture, with attention to energy efficiency, avoided emissions, minimization of extraction and waste, and compassion for people.

Beyond Greenwash

In this chapter, we have attempted to map the evolution of (largely) Euro-American design-environment discussions from the energy crisis of the early 1970s to the circular economy discussions of the 2020s. As we have seen, this is, in part, a story of the DIY politics of significant currents of the counterculture environmental scene of the 1960s and 1970s either morphing into form-based, material, or process-based design experimentations, heading off to Silicon Valley, or embracing attempts to outline a progressive green neoliberalism through design. It is a story of the United States achieving dominance in solar and wind manufacturing but then giving it away. It is also a story of the increasing disaggregation of architecture, design, and planning understood as fields engaged in a common project. As the Keynesian welfare state project retreats, the institutional spaces for these fields to encounter each other and work on common public projects fractures, a concept of "design for the public good" fades, and all the design and planning fields are increasingly siloed, fragmented, and reshaped by neoliberalism. In this context, a recalcitrant architectural avant-garde in the United States, having lost status and public confidence (due to the exhaustion and disasters of urban renewal), attempts to preserve the "purity" or "autonomy" of architecture from the grubby threats of environmental innovation, labor, engineering, or other design and planning disciplines. "Critical Theory" is pronounced as the solution to this crisis. But ironically, much of the theory produced in this moment has nothing to say about the housing crisis, the climate crisis, urban inequality, or how design might serve the needs of working people. Rather, we end up with more and more baroque forms of theory providing an intellectual gloss to "starchitecture" and the rise of the Real Estate state (Stein, 2019).

We have also seen in this chapter many currents of design, architecture, and planning attempting to push back against these currents (see Textbox 1 for a summary). None of these currents is able to entirely escape the pull of green neoliberalism, but we can locate flickers of possibility in discussions around adaptive reuse, retrofit and repair, embodied carbon, and in experiments with restorative and regenerative building practices. We can start to see what the contours of a more ecologically and socially rational material culture might look like in design for sustainability discussions, in attempts to think about design for disassembly, design for circularity and emotionally durable design. We can start to see what the contours of a more just urban form might look like in the attempt to ally landscape architecture with environmental justice, green infrastructure, biodiversity conservation, in planning for just green urbanisms and sustainable public housing. The forces of inertia and resistance that continue to propel business-as-usual thinking and practice within design, such as the force of commercial interests, the real estate industry, and capital, remain enormously powerful. Hence the extent to which a particular innovative set of green design interventions dissolve into greenwash is an ongoing problem. But so are proposals which sound good as provocative think pieces to pitch at architecture biennales but that are largely detached from the practical political question of how we can construct popular design agendas that address the aspirations, desires, agencies, and capacities of working people in general to live better, richer lives and contribute to climate solutions. The question then looms of how a progressive ecological and labor-friendly architecture, design, and planning could institutionally reground itself? We turn to consider different answers given to this question in the next three chapters.

Text Box 1: Nine Approaches to Climate and Environmentally Focused Architecture, Design and Planning

	Approach	Description	Key Advocates/ Examples
1	Hyper-Efficient Low-Energy Passivhaus Design	Focuses on minimizing energy consumption through airtight envelopes, advanced insulation, and high-performance windows. Emphasizes measured energy performance over points-based certification.	Passive House Institute
2	Net Zero Energy Buildings (NZEB)	Buildings designed to produce as much energy as they consume annually, typically via on-site renewables (solar, geothermal). Combines extreme efficiency with renewable energy generation.	U.S. Department of Energy, International Energy Agency, EU
3	Embodied Carbon-Focused Design and Architecture	Addresses the carbon intensity of building materials and whole buildings across their life cycle. Prioritizes low-carbon materials to minimize total carbon impacts beyond operational energy use.	Carbon Leadership Forum, C.Scale, Tally, Carbon Conscience
4	Bioeconomy and Hyper-Local Materials	Prioritizes local, renewable, and bio-based materials (bamboo, straw, timber). Promotes buildings as carbon sinks.	Alan Organschi, New Bauhaus Earth, Built by Nature
5	High-Density Ecological Urbanism	Seeks to decarbonize urban systems through high-density development, transit-oriented design, district energy, upzoning, and urban infill, creating efficient and resilient neighborhoods.	Alex Steffen, Peter Calthorpe
6	Urban Rewiring and Retrofitting	Focuses on upgrading existing buildings by electrifying systems (heat pumps, rooftop solar, batteries, EVs), improving insulation, and integrating smart technologies for demand flexibility and grid interaction.	Saul Griffith, Rewiring America

7	Designs for Adaptive Reuse, Modularity, and Prefabrication	Encourages retrofitting over demolition, modular construction, design for disassembly, and prefabrication to extend building life and reduce waste.	Retro First (Architects' Journal)
8	Design for Resilience	Emphasizes adaptation to climate risks (flooding, sea level rise, wildfire, blackouts) through strategies like living shorelines, levees, floating structures, and home hardening.	SCAPE, Kristina Hill, Arcadis, One Architecture
9	Regenerative and Circular Architecture and Design	One can identify here a range of approaches to circularity and regenerative design from approaches that seek to restore ecosystems and close material loops with green roofs, living walls, rainwater harvesting, urban agriculture, and spaces for environmental education and biodiversity to more comprehensive attempts to bring the full carbon and material impacts of design into view across industrial and urban systems.	William McDonough, Sim Van Der Ryn, Pamela Mang, Bill Reed

4

Labor-Centered and Worker-Centered Design for Just Transitions?

"Are you an environmentalist or do you work for a living?" is the name of a seminal intervention in environmental history made by Richard White (1995). The essay title was drawn from a bumper sticker widely circulated in the Pacific Northwest of the United States in the late 1980s and early 1990s among loggers and extractive industry workers. The sticker marked the rise of a particular kind of divisive "hippy versus hard-hat" sloganeering that captured the beginning of a new wave of corporate-funded antienvironmental backlash politics. The essay went on to press an uncomfortable point, though. While unfair and unwelcome, Richard White argued, the slogan did touch on some deep tensions in the cultural politics of certain kinds of romantic environmentalist discourse circulating in the United States circa 1990. Specifically, he suggested that the *proposition* that "Nature seems safest when shielded from human labor" prevented many environmentalist currents from fully investigating the relations between work and labor, class and race, history and ecology. Certain kinds of environmental discourse had a tendency to downgrade or even erase the *knowledge* and *pleasure* gained from productive work through the transformation of nature. Notably, not only did this frame force an unhelpful dualist "humans versus nature" binary on environmental politics, but it reinforced the dominance of upper-class *white environmental narratives* at the expense of interventions that explore the environmental histories of the multiracial working class and indigenous people. This, in turn, delimited the possibilities of such groups seeing environmental struggles *as their own struggles*.

Since Richard White wrote his essay, there has been a remarkable wave of scholarship in historical and political ecology, environmental history, environmental justice, indigenous studies, and world ecology that has sought to recover the *labor histories* of environmental transformations from romantic and Malthusian environmentalisms. We have come to see all kinds of landscapes, from the national parks to the Amazon rain forest, not as an empty space of "Nature" that has been kept in some static state of balance for millennia but rather in some part, dynamic *historical and social ecologies*. Such scholarship has argued that these landscapes are best understood as historical ecologies that have taken on different forms at different times, and have been disturbed and transformed by a range of social and ecological climate forces (for summaries see Mann, 2005; Kimmerer, 2013; White et al., 2017; Escobar, 2018). Yet acknowledgment of the complex ways in which diverse kinds of labor—waged and unwaged, enslaved and indentured, peasant, industrial, and indigenous—have made and remade ecosystems in different ways across space and time, with sometimes remarkable and sometimes deeply problematic outcomes, has often been missing from the core of environmental studies. Indeed, all too often US environmental studies been drawn to singular stories of (generic) humans as "environmental degraders" but had little to say about the *reconstructive agency of labor within the web of life* (Moore, 2015; White et al., 2017).

Much the same could be said about the relationship between design, ecology, and labor. Modern design involves the organization and coordination of many kinds of design labor. These forms of coordinated labor are, in turn, involved in extracting, transforming, reworking, repositioning, and deconstructing all manner of social ecologies to produce the design world we live in. And yet, as Peggy Deamer (2015, 2020), Sérgio Ferro (2024), Namita Dharia (2022), and Michelle Arevalos Franco (2024), among others, have underlined, design, architecture, and planning education has, for the most part, avoided reflecting on the labor ecologies that the very act of designing rests upon. As Ferro and colleagues in the emerging field of production studies (Thomas et al., 2015; Ferro, 2024) have argued, the building site has been completely absent from architectural education, as has the field, the factory, the mine, and the waste dump. How then should we approach the labor question in relation to environmental-design concerns?

In this chapter, we proceed through various circuitous routes attempting to excavate some largely forgotten labor-environmental histories of design. We begin by recovering a range of moments from reflections on "the factory as it might be" to movements of architects, planners, and trade union associations embarking on projects to build high-quality housing and urban form for workers in the 1920s and 1930s in Amsterdam, Berlin, and Vienna to traditions of worker-centered design in the 1970s and 1980s. We suggest here that the proposition that the workplace, the home, our material economies, and our urban and rural relations could be redesigned in more labor- *and* environmentally friendly ways have been at the forefront of some startling schemes and dreams weaved by various designers, social theorists, and activists across the past hundred and fifty years to think the world differently.

Second, we argue that if we shift our horizons beyond telling stories of credentialed designers and the objects they fashion to focus more attention on the vast amount of gendered, racialized, classed, and other modes of *invisible design labor* that has been involved in the design and maintenance, care and repair of our material culture (see Barca, 2012; Dharia, 2022; Akama and Yee, 2019), we can open up many further vantage points for thinking about the labors of design.

Finally, we review some contemporary promising clusters of design work that are attempting to bring environmental and labor concerns into more proximate alignment with design and struggles for just transitions today, notably through: (i) worker-centered human computer interaction design and platform cooperatives; (ii) calls for labor and community-friendly designs for circular economies.

Redesigning the Workplace: William Morris's "A Factory as It Might Be"

Let us begin here, though, in a fairly conventional and Eurocentric fashion with the iconic figure of William Morris (1834–1896), whose life and politics were continuously struggling to reconcile the relation between design and craft, labor and nature, meaningful work and useless toil. In the essay "A Factory as It Might Be" (1884), Morris proposes designs for a workplace for "the days when we shall work for livelihood *and pleasure* and not for profit" with the view that there is no inherent reason why manufacturing processes or facilities should alienate human labor from each other or from the natural world. Rather, Morris presents us with a design proposal for a future workplace that is "built with pleasure by the builders and designers" and designed with a range of further facilities such as a "dining hall, library, school, places for study of various kinds, and other such structures" to further expand the cultural lives of labor. The work occurring in the factory will be "useful and therefore honorable and honored" because it will not turn out trash. The "manufacture of useless goods, whether harmful luxuries for the rich or disgraceful makeshifts for the poor," would end. This workplace of the future would strike a balance between automation, "machines of the most ingenious and best approved kinds will be used when necessary, but will be used simply to save human labor" and craft. Work would be "neither burdensome in itself nor of long duration for each worker; but furthermore, the organization of such a factory, that is to say of a group of people working in harmonious cooperation towards a useful end, would of itself afford opportunities for increasing the pleasure of life." He suggests that industrial facilities should be situated in the broader contexts of living landscapes and beautiful gardens that are cooperatively maintained by workers on a voluntary basis. This reconfigured workplace would "make no sordid litter, befoul no water, nor poison the air with smoke."

"A Factory as it Might Be" provides us with a utopian proposition that never entirely resolves the tension running across Morris's thinking between craft and efficiency, between his celebration of the freedom that comes from non-alienated work through craft/handmade production and the freedom from toil and the possibilities of plenty that could emerge from design allied to advanced manufacturing. His assertion that "a house, a knife, a cup, a steam engine must be either a work of art or a denial of art" is a stunningly demanding vision of what a liberatory eco-aesthetic might look like. How to put this into practice and avoid a design economy that merely produces "nice things for rich people" is another matter. How we might identify ways of socially organizing design, engineering, and manufacturing to combine the gains of technological innovation *and* craft production without fostering alienation, deskilling, and ecological pollution is a question that troubles all manner of design schools from this point on: from the Bauhaus to the Garden City movement, the Russian Constructivists to contemporary advocates of peer production and the commons.

Nevertheless, and for all its limits of time and place, Morris's essay can hold up a mirror to current times. "A Factory as it Might Be" comes close to anticipating what we would now recognize as the basic project of cradle-to-cradle design or circular design that we engaged with in the last chapter. Yet, unlike such approaches, Morris asks that we actually center the workplace and its power dynamics as an object of reflection. From this, he suggests we think about modes of manufacturing that are devoted to an emphasis on the quality and not simply the quantity of production, better rather than simply more, in ways that are largely missing from contemporary environmental engineering. He asks us to make far-reaching political and aesthetic judgments about the kinds of material culture that could contribute to a good human life.

"A Factory as it Might Be" provides broader guidelines for the design of the humane workplace that bears some resemblance to the kinds of practice that are still offered in certain elite spaces of higher education or certain privileged tech-worker spaces like the Googleplex with their generous food and leisure facilities, time off for personal projects, and so on. Yet, Morris also asks that we see the right to the humane workplace as a right for all, not simply, as we have come to see it today, as a benefit offered to certain special kinds of elite workers. His proposition that the design of workplaces of the future should aspire to bring the garden into the workplace has resonances with the kinds of ecological architecture interventions that have been proposed and built by progressive artist and landscape architects like Michael Singer to introduce living systems into the architecture and design of the sustainable workplace.[1] Morris's call to bring nature into the workplace moves this discussion further forward by forcing us to also ask questions about the political organization of work, the political geographies of factories and their extended estates that they are intimately and reciprocally connected to. This lead us to the observation that calls for the ecological redesign of the workplace at home are going to be of little value if they are built on the backs of labor exploitation and environmental degradation at the plantation, colony, mine, factory, and sweatshop. Finally, running through Morris's essay "A Factory as it Might Be" is the proposition that we should take the technical and organizational knowledge that workers can gain from working on nature seriously (Wark, 2015).

Designing Workers' Palaces: Amsterdam, Vienna, Berlin

Morris's call that designers should place the alienation of labor front and center in their thinking and do all they can to design in ways that affirm the dignity of labor and capacity for workers' self-governance may have had little resonance for a mainstream design industry increasingly focused on commercial production, but it did serve as an important point of inspiration for a number of early- and mid-century architects and designers in Northern Europe. For example, the Amsterdam School of Architecture emerged out of a group of Dutch architects, led by Michel de Klerk, Johan van der Mey, Guillaume la Croix, and Piet Kramer, clustered around the publication *Wendingen*, took up Morris's challenge.

Wendingen was an avant-garde magazine that saw architecture, design, music, dance, and the decorative arts as a single field. Inspired by such a proposition but seeking to add more explicit political content to this project, de Klerk and his colleagues proposed a radical architecture and design that would aspire to build "workers palaces" premised on de Klerk's assertion, "Nothing is beautiful enough for the laborer who has had to live without beauty for so long."

The Dutch Housing Act of 1901—and subsequent housing legislation in the Netherlands provided the opening here, responding to a chronic housing crisis. The act allowed municipalities, cooperatives, and housing associations (Catholic, Protestant, Jewish, and socialist) to apply for funding from the state to build public housing. This opportunity opened up a significant wave of experiments in constructing municipal housing, and the Amsterdam school used this moment to embark on building a range of housing projects, villas, public buildings, street furnishing, and facilities around the city—in collaboration with trade unions, socialist housing associations, and the social democratic dominated municipal government.

Influenced by socialist and anarchist politics, German expressionism, art nouveau, the garden city movement, and the arts and crafts movement, as well as modernist currents and the decorative arts and architecture of Indonesia, the Amsterdam School aspired to construct low-cost but high-quality and finely designed housing (both in terms of interiors and exteriors) for working people. Their projects varied in size and shape from larger housing complexes such as *De Dageraad* and *Het Schip* to smaller villas and some commercial buildings, as well as making numerous further contributions to the decorative arts, producing clocks, lampshades, and furniture "for the workers." While the exteriors of their buildings were mostly red brick and tile, giving rise to their work being described as "brick modernism," the construction materials were often further sculpted into expressive organic forms and curves inspired by Gaudí. Unemployed local craftsmen, carpenters, sculptors, and iron workers were put to work to add further decorative elements—such as stained-glass windows, wrought iron gates, and decorative doors. Figurative sculptures were built into many building facades, and Amsterdam school buildings contained many additional decorative elements, including towers, spires, courtyards, and gables. As well as attending to exteriors and interiors, public housing complexes such as *Het Schip* had a post office and telephone booth built into the housing complex, a library/small meeting hall, a central courtyard and space for some private gardens. Attention was also paid to crafting fairly unique and distinct street furnishings surrounding the housing complexes.

Between 1910 and 1923, over 16,000 housing units were built in Amsterdam by either Housing Associations, cooperatives, or the municipal government. As public housing, urban improvement, and better facilities for workers became central commitments in the political programs of socialist and social democratic parties in inter-war "Red" Vienna and Berlin, such projects expanded further. As Sophie Hochhäusl has noted, municipal social democrats, led by the German architect and chief planner of Berlin under Weimar Germany, Martin Wagner, built over 20,000 housing units in the interwar period, which Wagner hoped could provide the basis for the development of "co-operative cities." During the 1920s, Wagner sought to develop a housing model where "where individual settlement cooperatives managed their facilities under the umbrella of the larger organization of tenants and construction unions" (Hochhäusl, 2021: 77–8). The Berlin-Britz housing estate, better known as the 'Horseshoe Estate' due to its crescent shape, was designed by the utopian socialist architect Bruno Taut and housed 3,000 members of gemeinnützige Heimstätten-,

Spar- und Bau-Aktiengesellschaft (GEHAG), a building society set up in 1924 by Germany's trade unions. Informed by garden cities' ideas as well as modernist architectural ambitions, Taut constructed a series of simple, but multicolored brick-built modernist apartments with shared communal green areas. Taut's estate remains highly functional, well-regarded, with beautifully tended gardens and communal spaces to this day—even if the gentrified cost of much of the housing has now escaped the wages of workers.

Evaluating the legacy of the connecting threads between late nineteenth-century utopian socialisms and fin de siècle/interwar social democratic projects is complicated. The Italian Marxist architect and architectural critic Manfredo Tafuri is probably the most well-known critic of these ventures. In *The Sphere and the Labyrinth* (1987), he argues that Wagner's attempt to build cooperative cities was a failure in Berlin, and it impacted similar attempts in Vienna. Notably, for Tafuri, the attempt to build socialist islands in a sea of capitalism was, much like Howard's Garden City project, not only doomed to failure but ended up providing housing projects that became "tools to control workers." It is certainly true that we can observe certain paternalistic currents moving through the public housing projects of Amsterdam, Berlin, and elsewhere. The slogan "high-quality housing for workers," even among labor-friendly designers, could shade toward moralist and environmental determinist aspirations to build housing to "elevate workers" and "teach them to be good citizens." As Esther Kreikamp has noted, some Dutch garden village experiments—such as the Dutch garden village Asterdorp—were specifically constructed to house "problem families" where it was hoped that such new surroundings would help them to become good citizens (Kreikamp, 2022). Rent collectors at *Het Schip* also checked on how workers were maintaining the neatness of their homes.

Ironically, some of the more socialist public housing that followed more lavish building designs in Berlin pushed rents beyond the affordability of many manual workers. However, more recent interventions by Sophie Hochhäusl have sought to push back in part against Tafuri's pessimism.

Hochhäusl has argued, contra Tafuri, the attempt to establish cooperative settlements at least in Vienna "were indeed based on social movements with diverging class and professional interests, which their organizers sought to unify" and this did in many respects achieve "the attainment of a working-class cultural hegemony" which was "anticipated and at least partially realized." If the construction techniques followed by Berlin largely followed Taylorism, Hochhäusl notes "in both Vienna and Berlin—but not in Frankfurt—settlement administrators championed cooperative labor processes and thought about restructuring the economy by reshaping the construction industries and tenants' organizations, and with it the architectural profession" (Hochhäusl, 2021: 85).

In contrast to the sometimes paternalist elements of much design inspired by utopian urbanism, the economist Otto Neurath in Vienna, in dialogue with the architect Josef Frank, sought to develop a labor-friendly system of building and construction that aimed to employ both potential tenants and construction workers in the on-site construction of building materials—such as a brick-making operation. At one of these new housing sites in *Siedlung Hoffingergasse* Hochhäusl details how much of the communal infrastructure around the housing development was created by such a cooperative system. The housing site then was not simply focused on experts building housing for workers, but it involved workers and inhabitants who were paid to build the social and cultural infrastructure of the site, which "included a kindergarten and playgrounds, two ponds for the breeding of ducks and fishes, and a clinic with an adjacent sun, light, and water bath" (Hochhäusl, 2021: 87). As Hochhäusl notes, the focus was not on passive and deskilled workers having beautiful settlements made for them but rather developing a system where "settlement inhabitants were able to collaborate in imagining and building long-lasting communities and landscapes of housing based on self-management of their labor" (Hochhäusl, 2021: 88). It should be noted that such projects often involved humanely addressing issues of disability and incorporating the contributions of the many disabled veterans into these projects. In the winter of 1921 and 1922, the ÖVSK further provided educational enrichment experiences for workers, which involved

> hands-on instruction on gardening techniques and construction alongside courses in architecture, economy, social organization, and the law for cooperative leaders, organizers, and inhabitants. In general, these courses offered material-theoretical instruction aimed at communal knowledge building and dissemination. Themes of instruction were based on the settlement movement's socioeconomic character and potential and encouraged inhabitants to become more fully immersed in the movement. In this manner, Neurath hoped, the ÖVSK could indeed become a fully self-managed movement granting poor citizens opportunities to become its leaders and experts. (Hochhäusl, 2021: 87)

The aim then of this particular approach to labor and design was not simply to place "the technological means of production in the hands of an urban citizenry," rather, through rethinking the very nature of work and labor, it sought to affect the very way work was conceived (Hochhäusl, 2021: 87).

Worker-Centered Design for Industrial Democracy and the Rights to the City

As we can see, then, labor-centered design can give rise to some very different political understanding of how such projects can and should be constructed. We can identify a tradition of thinking within labor-friendly design and architecture that has been concerned with delivering plans, designs, and building "for the workers." This is not the only tradition that can be excavated here, and we can also identify attempts to think about a labor-centered design that seeks to work *with labor* to design new futures, acknowledging the skills and contributions that lay and credentialed labor can make to the design process.

It is perhaps too easy to draw a technocracy/populist line in such approaches, which neatly divides such approaches into "good guys" and "bad guys." Both approaches have their strengths and drawbacks. Nevertheless, if a great deal of postwar public housing was largely informed by the former approach, which, despite its accomplishments, could shade toward paternalism, a focus on efficiency, and an embrace of Taylorism (scientific management), which sometimes resulted in the instrumentalization of workers and inhabitants, we can also identify more experimental streams of thinking within architecture, planning, and design that sought to mobilize the design insights and skills of working people in different ways.

The proposition that design might aid struggles for industrial democracy is one of the original motivating forces for triggering the emergence of participatory design. The postwar participatory design movement has its origins in Scandinavian dialogues and practices that occurred within the trade union movement during the 1970s. Participatory design—in its initial radical formation—was specifically focused on attempts to implement labor-friendly modes of design and innovation at scale. The aspiration was that modes of design could be developed and propagated that would not only be capable of resisting Taylorism and deskilling but also actively expand industrial democracy in the workplace.

The movement converged at the high point of Scandinavian Social Democracy as a wave of "democracy at work" legislations were passed in Norway and Sweden. This legislation allowed for: (i) worker representation on company boards; (ii) the right by trade unions to stop production dangerous to worker health, and (iii) the Swedish Joint Regulation Act of 1977, which (in theory) outlined for workers and trade unions the right "to co-determination in production issues such as design and use of new technology" (Ehn, 1990: 256). The 1970s and 1980s in Norway and Sweden thus opened up a fairly unique moment for publicly funded designers to work with organized labor to think not only about "freedom from management prerogatives" (Ehn, 1990: 254) but also modes of work-oriented designs that could aid worker-friendly redesigns of industry. As Pelle Ehn documents in his remarkable intervention *Work-Oriented Design of Computer Artifacts* (1990), two projects here are significant.

DEMOS (Democratic Planning and Control in Working Life—On Computers, Industrial Democracy, and Trade Unions) was established in 1975. For four years, DEMOS drew together a set of interdisciplinary design teams with workers and trade union representatives at four different Swedish enterprises. Influenced by, but significantly adjusting, Enid Mumford and the Tavistock Institute's "socio-technical" approach to humanizing work, Harry Braverman's labor process theory (1974) which focused on alienation and deskilling and Paulo Freire's *Pedagogy of the Oppressed* (1970), a central aim of this research project explored the ways in which unions and workers could influence the design and use of technology in the workplace. Rather than construct the research agenda in terms of humanizing work via "gifts from management," the focus of the Scandinavian approach was to explore "the importance of the employees themselves having the right to determine the content of humanization by real and meaningful participation" (Ehn and Goranzon cited in Ehn, 1990: 269). A follow-up, the UTOPIA Project, explored whether the design process could be democratized through workers and designers developing modes of collaboration via "designing tools and environments for skilled work and high-quality products and services" (Ehn, 1990: 279).

Both projects had to address multiple complexities, management resistance, and disappointments. Nevertheless, they produced openings (Ehn, 1990: 279). They demonstrated that under certain circumstances, workers through their trade unions could be brought into the design of a product through the establishment of "mutual learning processes" between workers, social scientists, and engineers (Ehn, 1990: 332). DEMOS and UTOPIA also led to further discussions in Sweden and Norway about the need for a national technology policy for the democratization of industry and the need for designers and workers to receive broader forms of education and training to make industrial democracy a reality. The projects helped consolidate the proposition that employers had to negotiate with local unions before making "major changes in production."

DEMOS and UTOPIA, in some sense, can be seen as produced by the unique political conditions of the time. The call for worker-oriented design in Scandinavia emerged from a broader political culture underpinned by very high levels of unionization in Sweden and Norway, a hegemonic social democratic party, and corporatist trade union relations that allowed for centralized negotiations with the government and employers (Ehn, 1990: 254). DEMOS and UTOPIA are, of course, examples of *labor*—rather than *ecology*-centered design projects. The Lucas Plan is interesting here to turn to for considering a project to develop industry democracy that emerges at a very similar time to the Scandinavian movement, but extends such concerns more explicitly into the labor-ecology space.

Lucas Aerospace was one of the largest aerospace component suppliers in the UK, and in the late 1970s, it proposed a restructuring program that would sack 4,000 workers across its 17 factories. Rather than simply accept this fate, trade union shop stewards opened up a comprehensive discussion with workers between 1975 and 1976 concerning how the company might transition from making defense industry weapons to socially and environmentally useful production. As Mike Cooley—a design engineer at Lucas Aerospace details—through a Combine Committee that linked all the union representatives of the factories and via shop steward committees, union members were asked to contribute ideas to how an industrial conversion could take place, and within a month, as Cooley notes, "We had 150 ideas of products with the existing tools and skills we had" (Cooley, 1980: 118).

The proposals were subsequently refined and produced six volumes of technical and engineering drawings and proposals, which became known as *The Lucas Workers' Plan for Socially Useful Production*. As Cooley notes, the Lucas Plan rejected the "tendency of the alternative technology movement for products to be provided that are little more than play things for the middle classes in their architecture-built houses." In contrast, we can find a range of class-conscious but also ecologically conscious designs and proposals that seek to honor the skills and intelligence of the workforce and address the diverse needs of working people and communities.

For example, Lucas workers collaborated with design students and faculty at the North East London Polytechnic to develop designs for low-energy houses that would facilitate more collaborative modes of building. Communication design students wrote an assembly-instruction manual that "would allow people, working side by side with skilled building workers, to go through a learning process and produce at the same time … ecologically desirable housing" (Cooley, 1980: 121). Drawing from the background of workers in aerodynamics, they proposed a range of wind generators. Connecting with the Open University, the Lucas workers also designed and built

Figure 10 Peter Kennard "Alternatives are Possible." Poster promoting arms conversion in support of the Lucas Plan. Poster used with permission.

Figure 11 The Road-Rail Bus. One of the projects proposed by the Lucas Aerospace Workers Combine. Illustration Clifford Harper. Reproduced with permission.

heat pumps to be installed in public housing and proposed combined heat and power systems for home heating. For environmentally friendly mobility, they proposed hybrid power packs made up of petrol and electrical engines to replace the internal combustion engine, gaseous hydrogen fuel cells, and road/rail vehicles. In addition to this, the plan advocated for Lucas to massively expand production of kidney dialysis machines and further proposed numerous designs for disability such as hobcarts for children suffering spina bifida.

The Lucas Plan advocated for, and anticipated in embryonic form, many of the technologies that we would view as central to decarbonization over the next four decades. More generally, it is the collaborative labor process that the plan drew from and developed further that is particularly interesting. Hilary Wainwright and Dave Elliott note that at least up until the mid-1970s, one of the features of the aerospace industry was that its production processes were marked by a good deal of small batch production which also often involved further customization of products to meet the specific needs of clients. The dynamic and iterative nature of this production process often departed from a standard mass production model. Rather than following a command-and-control model, Lucas already had a process in place where "close contact between designers and shop floor workers was normal" with "shop floor workers trying out and suggesting modifications to designs which they would then discuss with the technical workers from the design office" (Wainwright and Elliott, 1981: 35). A history of regular exchanges between technical office workers and blue collar shop floor workers aided further collaborative engagement during the process of consultation. It also helped stimulate discussions amongst workers concerning how management efficiency drives might be challenged by new design methodologies and labor processes that "could give full vent to the creativity of the hands and minds of the workers, and that could be carried out through non-hierarchical forms of industrial organization" (Cooley, 1980: 133). As Cooley notes, the substance of such discussion was not to deny the importance of technical and coordination skills in the production process but ask questions about how such skills could become part of the labor process and how technical and shop workers could be encouraged to have a great role developing these skills rather than have management use such tools to police workers.

Advocates of the Lucas plan were fully cognizant of the limits of a narrow workerist design agenda and sought to bring labor concerns into dialogue with broader communities, local trade unions, political leaders, and more. As Cooley notes:

> The Lucas workers were deeply conscious that if the debate was limited to industrial workers … it would represent a new form of elitism. We therefore made strenuous efforts to involve wide sections of the community at large in discussions around these issues in order to interact with them and learn from them. (Cooley, 1980: 131)

A lack of support from management, trade union leaders, and the rise of the Thatcher government ultimately undercut the Lucas Plan. No matter how good the ideas were, the very proposition that organized workers should challenge and even seek to determine production priorities was viewed as totally unacceptable by management. Similarly, the national trade union bureaucracy in the UK saw bottom-up demands for worker control and self-organization running through the spirit of the Lucas Plan as potentially undermining their leadership role. Nevertheless, news of the process and the plan circulated in trade unions, friendly media, and technology circles around the world. A number of trade unions did seek to encourage demands for workers to be involved in developing similar plans in their sectors, and it did help generate some of the most extended discussions of the possibilities of industrial democracy since the 1930s.

Invisible Design Labor: Gender, Race, and Socialist Feminist Design Histories

Historically, the professional fields of design, architecture, and planning emerged out of deep social and historical divisions between mental and manual labor, power-laden struggles to distinguish "lower" modes of craft making from "high" modes of designing activities, and multiple modes of erasure that hide the labors of design. In the United States such conceptual divisions are impossible to disentangle from the ways in which education access has been deeply racialized and gendered through the attempt to delimit African American education opportunities post-Emancipation to trade schools, the enforced assimilation of native Americans into boarding schools focused on the eradication of their own culture and traditions and the development of basic skills, agriculture, and "moral instruction," and the effective delimiting of access of the working class, Jews, and women in many elite centers of design education well into the twentieth century. As virtually every field of design, planning, and architecture went through a process of professionalization and compartmentalization in the late nineteenth and early twentieth centuries, it was clearly influenced by such structural forces and the elaboration of specific fields of inquiry and domains of expertise, which allow for functional specialization and the accumulation of expertise. For much of the twentieth century in the United States, non-white, female, and working-class design workers were often consigned to quite specific back-office spaces or administrative roles within design firms with "lesser" forms of blue-collar design labor, often experiencing super-exploitation.

For example, Sonja Dümpelmann (2022) has argued that the professional field of landscape architecture in the United States largely came into being through a racialized demarcation and consolidation of the distinction between "landscape gardening" and "landscape architecture." Landscape gardening for much of the first half of the twentieth century in the United States was the field taught in historically black colleges and technical schools, whereas landscape architecture emerged within the contained world of the white elite university, dominated by white men. Such histories carry over today in the United States landscape architecture and design industries—as we have seen earlier—where the organization of labor remains deeply marked by these racial, class, and gendered histories (Franco, 2022, 2024).

If one aspect of a credible worker or labor-centered design has to acknowledge the erasure of a certain kind of design labor from design history, a second move needs to recover the visions and potential of this invisible design labor. Dolores Hayden's *The Grand Domestic Revolution-A History of Feminist Designs for American Homes, Neighborhoods and Cities* (1982) is of value here for the ways that she documents how between the 1860s and the 1930s, numerous materialist feminists in the United States raised critically important environmental and labor questions by exploring the design-politics of cooking and cleaning, food preparation, and childcare and through asking broader questions about how the social, political, and financial subordination of women to men could be overcome. Central to the rise of these material feminists was not simply the call for economic remuneration but many proposals for "…a complete transformation of the spatial design and material culture of American homes, neighborhoods and cities" (Hayden, 1982: 1). Hayden's material feminists adopted the earlier communitarian socialist focus on spatial analysis "to accompany economic analysis" (, 1982: 8), but as she notes, they moved beyond the village as site of utopia to embark on probing and imaginative investigations of the future of urban form more generally. The "overarching" theme of much of this materialist feminist advocacy was focused on overcoming the "split between domestic life and private life created by industrial capitalism as it affected women." Hayden argues two key aspects of industrial capitalism received critical scrutiny here: (1) "the physical separation of household space from public space"; and (2) "the economic separation of the domestic economy from the political economy" (Hayden, 1982: 1).

Feminist activists across this period argued for wages for housewives, abolition of the private kitchen in favor of communal kitchens, and community restaurants. They championed socialized childcare and proposed designs for communal workplaces. As urban and municipal infrastructures of water, fuel, and electricity began to develop across US cities, many material feminists further argued that women should own these services to give them financial independence from men, and use them as their base of economic power (Hayden, 1982: 12).

Many feminist campaigners had a nuanced view of automation. They were deeply concerned with the eradication of toil and the need to address socio-technical changes that would address the loneliness and isolation of the domestic sphere. As Hayden notes: "Material feminists believed the solitary housewife doing her ironing or mixing dough could never compete with the group of workers employed in well-equipped laundries or hotel kitchens … Neither could the isolated home compete with the technological and architectural advantages offered by larger housing complexes …" (Hayden, 1982: 16). A very different view of the configuration of more social and ecological domestic technologies could also have beneficial impacts to increase residential densities, generating social and ecological gains, but also further facilitating power relations. As she notes:

> Devices such as elevators, improved gas stoves, gas refrigerators, electric suction vacuum cleaners, mechanical dishwashers, and stream washing machines which were designed for use in large enterprises such as hotels, restaurants and commercial laundries, could also be used in large apartments. Because this technology was first developed at the scale suitable for fifty to five hundred people, any group interested in mechanizing domestic work simply had first to socialize it and plan for collective domestic consumption by organizing households into large groups inhabiting apartment hotels, apartment houses, model tenements, adjoining row houses, model suburbs or new towns. What was unique about the material feminists was not their interest in these technological and architectural questions, which also attracted inventors, architects, planners, speculators, and efficiency experts, but their insistence that these economic and spatial changes should take place under women's control. (Hayden, 1982: 17)

In this respect, Hayden argues that the particular forms that suburbanization took on in the United States in the mid-twentieth century can be viewed now as not simply reinforcing racial and gender divides but providing the carbon- and energy-intensive infrastructures that we have to deal with today.

The design-politics proposed by material feminists in the late nineteenth and early twentieth century, as mapped by Hayden, had their limitations. As Hayden fully acknowledges, the racial politics of many late-nineteenth-century, predominantly white material feminists were not always enlightened. She notes that turn-of-the-century material feminist traditions had an insufficient analysis of power, class relations, or race relations. Nevertheless, the most developed discussions did outline how changes in social relations, design forms, and ownership structures could facilitate social emancipation.

Invisible Labor, Colonial Ecology: Sérgio Ferro

The writings of Brazilian architect, painter, and historian Sérgio Ferro have made some decisive interventions in labor/design conversations. Yet he has only attracted the attention of an Anglophone audience very recently, as his work has been translated into English from French and Portuguese. We have already observed early in this chapter that the foundational observation of Ferro scholarship is that architecture emerges from its disentangling from the construction site. What is required to rectify this are accounts that place the building site and building workers at the center of the inquiry. His first-hand account of the building of Brasília offers an important contribution to a labor-centered architecture and design.

Ferro took part in the construction of Brasília as a young architect in its "early stages" (Ferro, 2024:273). The project, Ferro notes, reflected a desire to "forge a new Brazil" and was embodied by the high modernist urban plans of Lucio Costa and the architecture of Oscar Niemeyer. At one point, it seemed to offer the opening for a new populist-egalitarian modernization project for the Global South. And yet, as Ferro argues drawing from his own experiences gained from visiting the construction site, and his discussions that occurred with many construction laborers that he shared prison cells with after they all got caught up in the military coup of 1964, the discrepancy between the look of Brasília, the rhetoric of modernization that surrounded it, and the super-exploitation of labor that was entailed in building the new capital was staggering. As he notes:

> when I was arrested, I shared my cell with workers who had participated in the construction. They told of a suffering we could have barely imagined then: numerous suicides, workers throwing themselves under trucks, attacks of dysentery on an almost daily basis, being fenced in and unable to leave. (Ferro, 2024: 273)

Ferro argues that Lucio Costa's plan for Brasília was "valid as an initial proposal," and Niemeyer offered many buildings of staggering beauty. However, his plan for the new capital was proposed and then implemented without any serious democratic discussion (Ferro, 2024: 276). The plans aspired to building a kind of utopia, but Ferro argues from first-hand experience that working conditions for the large migrant worker population, "the candangos," were marked by "unbridled exploitation" (Ferro, 2024: 274). Ferro is careful to acknowledge that one cannot set all these issues at the door of design or at the door of Niemeyer. Many forces came together in the construction of Brasília which gave rise to an authoritarian and brutal design and building culture: business interests committed to an extractive economy, political elites committed to the violence of modernization "at any cost," bureaucratic corruption, and a very powerful military presence ready to assert itself at any moment all played a role (Ferro, 2024: 283). However, Ferro does argue that the design-politics of the site and architectural decisions and choices did not help matters. Notably, Niemeyer's architecture was often visually stunning, but its prioritization of form and virtuosity over feasibility carried a human cost. Ferro outlines that some of Niemeyer's most iconic buildings, such as the palace of the National Congress (Congresso Nacional), were aesthetically remarkable, but the bowl-shaped chambers that frame the building were so difficult to construct that some workers ended up being crushed under the steel structures that supported them. Similarly, Ferro has observed that the stunning Cathedral of Brasilia (*Catedral Metropolitana Nossa Senhora Aparecida*) with its crown-like structure tapering up to the sky generated a high-risk construction site that required workers to dangerously swing between the concrete ribs during the construction process with little training, poor equipment, and generally no regard for safety. The results were magnificent buildings defined by building practices that were "unnecessarily cruel," constructed with materials such as poured concrete that exacerbated health impacts to workers and produced harmful side effects. More importantly, there were other paths that could have been taken.

An alternative starting point for Brasilia would have ensured much more public consultation in the planning process. As Ferro argues:

Figure 12 Marcel Gautherot. Workers constructing Brasilia.

> Lucio's design for Brasilia, though perhaps valid as an initial proposal or a subject for debate, should have gone to the streets, been subject to criticism, force to digest corrections in the spirit of democracy. And after that, his design should have been allowed to drift towards its own fulfilment, to shift according to its own dynamics, to head its own critical transmutations. (Ferro, 2024: 278)

More generally, he suggests better choices could have been made with materials that offered ways to build that minimize risk, that allow for the worker's skill and joy in building to be better expressed. There were elementary better choices in materials that could have been made that would have respected producers and attended to the health and safety of workers. A labor-friendly mode of design could have emerged from this that was more open to a dialogue between the designer and the workers, which honored the skills of the workforce and facilitated more room for initiative on the building site. But the moment was missed.

Pro-Labor Computing, Worker Centered Human–Computer Interaction Design, and Platform Cooperatives

So, what relevance do these past attempts to develop labor- and worker-centered designs have today?

One of the most explicit attempts over recent times to build on Scandinavia's traditions of worker-centered design can be found in the Human-Computer Interaction HCI research community, which are committed to exploring and elaborating "pro-labor computing" as well as "worker-centered HCI" design.

Worker-centered HCI design has drawn together scholars emerging out of the overlapping worlds of computer science, design research, software studies and design, anthropology, tactical media, and science and technology studies, many of whom have sought to work with labor organizations and worker advocacy groups to conduct their research. Much of this research seeks to understand the roles that design and digital and information technologies have come to play in the contemporary workplace and labor markets as labor relations and technologies have been remade by four decades of market liberalization, privatization, and outsourcing, marked by an erosion of labor protection and benefits, increased precarity, and informalization of work (Irani and Silberman, 2016; Fox et al., 2020; Spektor et al., 2023; Fox, 2024).

Specific concerns of worker-centered design range broadly from exploring the ways in which particular designs for the workplace as well as deployments of AI and other ITC systems have expanded the algorithmic control and surveillance of workers by management to the ways in which we have seen the spread of design interventions, and particular deployments of automation that deskill workers and undercut their own agency, insights, and knowledge. The field displays further concerns with the expansion of uncompensated labor and wage theft. It documents the ways in which workplace design can undercut workers' health and safety. It has also highlighted the often gendered, racialized invisible labor that props up all manner of digital services.

Worker-centered designers have also sought to develop digital tools and make interventions that might reorient technologies and platforms in the light of worker feedback or that open up spaces for workers to communicate with each other about shared concerns without management interference, that facilitate organizing workers or connecting workers to labor lawyers and labor advocates. For example, Irani and Silberman's "Turkopticon" developed a web application and browser add-on that they built to help workers evaluate and sometimes refuse work that is offered by Amazon's Mechanical Turk's crowdsourced platform (Irani and Silberman, 2016). Other examples of more recent worker-oriented design would include apps that allow workers to track their time, so that they have documentation that can help them resist employer wage theft. Worker-centered designers have worked with housekeepers to understand the ways in which algorithmic management systems can lead to the deskilling of workers, speed up their work, or undercut their own labor-saving practices, and the ways in which worker-informed adaptations of modified systems could improve their functionality.

Much of this work in worker-centered HCI has a great deal of complementary overlap with the important scholarship and interventions of Trebor Scholz and the Platform Cooperative Consortium to encourage an institutionalization of the development of worker-owned platform co-ops as an alternative to the neoliberal sharing economy (see Scholz, 2016). The work of climate-design activist currents working in the areas of design visualization could potentially build more direct links between climate and labor struggles (Boehnert, 2018). As we have seen in the last chapter, green performance indicators and data visualizations are playing an increasingly important role both within environment and climate concerned design, architecture, and planning as tools to track the energy, environmental, and carbon impacts of buildings from life cycle assessment (LCA) visualizations (which can represent energy use breakdowns or material flows to carbon footprints) to Building Information Modeling (BIMs), which can incorporate energy performance data into its visualization. The *Who Builds Your Architecture* group has also provided powerful visual aids to connect architectural practice to labor justice, mapping the ways in which the whole architectural industry is embedded in material and labor relations, from migrant workers to architectural offices (Baxi et al., 2018). If much data visualization work to date has tended to focus on energy, carbon, or labor, one question that looms is whether further developments of these platforms could integrate these labor-ecology questions, bringing into focus the social, ecological, and labor impacts of energy, material flows, and carbon emissions. Deamer et al. (2015) have suggested that one of the impacts of computer-aided design and building management systems is that in theory they have brought into view "the design ingenuity" of fabricators, environmentalist engineers, and contractors and made architects aware that "these players don't just execute design, they initiate it as well. The vertical structure with the architect visionary at the top and contractors and subs below is rapidly disappearing, as is the idea that they do 'work' and we do design." This observation could just heighten competition between different design workers, but it could also open up discussion about the collective work of design and the ways in which this might be more effectively organized.

Worker-centered designer Sarah Fox has argued that a further critical terrain of future engagement for worker-centered design surrounds the question of how we might make interventions in AI or automation technologies that "complement human labor rather than displaces or degrades it" (Fox, 2024). Clearly, complexities emerge in attempts to align labor-centered HCI with struggles for climate and environmental justice. At one level, advanced computation, ICTs have been and will remain central to the development of high-performance computer models that simulate the Earth's climate system, and remote sensing technologies and advanced imaging technologies play a key role in monitoring the Earth's surface and ocean temperatures (Bratton, 2015, 2019). ICT and AIs could play all manner of important roles in facilitating smart grid technologies, optimizing energy efficiency and coordination of power supplies, providing ways to track greenhouse gas emissions, and providing early warning energy or climate forecasts. At the more prosaic level, the use of robotics for demolition of buildings using machine learning, for example, could, in theory, eradicate a lot of dangerous drudgery and improve the quality of sorting and maintaining of materials (Fox, 2024). Strategies to decarbonize the household and maximize the efficacy of decentralized renewables have increasingly looked to the opportunities that smart devices and the Internet of Things could play in helping to network local energy-producing and energy-saving devices through the construction of virtual power plants.

At the same time, there are ways of designing AI and ICT systems that augment workers' skills and those that deskill. We have also come to understand that the hidden rebound effects of ICT and AI, in particular, are very significant, with estimates suggesting that they could account for anything between 2 percent and 6 percent of greenhouse gas emissions (Freitag et al., 2021). If we consider here that training a single natural language processing model "may emit 626,155 pounds of carbon dioxide, equivalent to what five cars produce during their useful life" (Sharma et al., 2023), that e-waste is now the fastest growing waste in the world (Little 2022, 2023), or that the turn to AI is likely to lead to an enormous increase in energy consumption and greenhouse emissions of already-existing energy data centers, the challenges of energy-hungry data centers are enormous.

Broader challenges posted by the consumer electronics industry to environmental-labor issues would include as Sharma, Kumar, and Nardi note, "the constant need for novelty, the accompanying throw-away culture around consumer electronics, and the glorification of disruption for its own sake." Both Google and Microsoft look like their pledges to be carbon zero by 2030 are likely to be undercut by their growing energy needs through their AI research. How to address these vexing tensions is still a matter for considerable debate. A broad path could focus on strategies that increase the energy efficiency of products, expand smart technologies to create broader efficiencies in domestic, work, and production spaces, and seek to encourage digital and ICT industries to directly address their own expanding energy needs by building on site through renewables, invest in virtual power plants or through deploying on-site small nuclear reactors. Environmental justice opens up a second set of approaches, suggesting the need for comprehensive regulation and ultimately the eradication of e-waste through the deployment of circular economy practices. A third intervention that has opened up of late is focused on opening up consideration of "digital descaling" or digital degrowth of the ICT and computer industries. Such discussions could include everything from the energy, carbon, and health virtues of taking digital sabbaths and promoting a culture of "digital sufficiency" to "salvage computing" focused on reuse, repair, and upcycling of ITCs and/or developing open source or commons-based or locally controlled approaches to computing (Girard et al., 2024).

Labor- and Community-Friendly Designs for Circular Economies

As we saw in Chapter 3, the circular economy has over the past two decades become one of the most dominant mainstream sustainable design frames through which the insights of the many subfields of ecological design, industrial ecology, and some variations of ecological economics have been reorganized. Promising the possibility of a "regenerative" or "restorative economy" through the comprehensive embrace and integration of the promise is that it will offer a "win-win" solution to climate and environmental problems that can integrate the full range of design possibilities closed-loop production and consumption systems, regenerative design for disassembly, repair and retrofitting, extended producer responsibility, eco-industrial parks and more. The assumption here is that this would allow for green growth—that is, a decoupling of carbon emissions from GDP growth and decarbonization.

Business-friendly and largely neoliberal circular economy agendas have been relentlessly promoted by The Ellen MacArthur Foundation over the past decade, and its key claims have generated a mounting array of skeptics and critics. For example, Gregson et al. (2015) have argued that use of the Circular Economy concept in both practitioner and academic literatures "tends to be approbatory, uncritical, descriptive and deeply normative." We have seen others argue that the kinds of visions of the circular economy promoted by the Ellen MacArthur Foundation are marked by hyperbolic claims but also a high degree of strategic avoidance of certain key issues: notably the political economy of planned obsolescence, the issue of rebound effects, issues of product quality, and difficulties of recovering high-volume, low-quality waste streams (Ceschin and Gaziulusoy, 2016; 2019; Corvellec and Stål, 2019). Concerns have been raised that the vagueness of many circular mandates can often operate as a kind of greenwash, allowing companies to develop somewhat decorative adjustments while avoiding/forestalling more comprehensive regulation (Gregson et al., 2015; Hobson, 2015; Lazarevic and Valve, 2017). Perhaps most damning, Valenzuela and Böhm (2017) have declared the zero-waste-circular-economy "a fetishizing narrative within a capitalist order, which offers a win/win technocratic ecomodernism promising green growth forever while keeping all else intact".

If we acknowledge the validity of many of these criticisms, what is the positive agenda for moving forward? If at some level, some kind of circular economy is going to be necessary to produce a just transition, where are the reconstructive horizons for doing things differently? Recent interrogations of the circular economy by Martin Calisto Friant and his colleagues (from Friant et al., 2020; Friant and Vermeulen, 2024) have suggested that a focus on the very different political, economic, social, and design possibilities that could exist in a "circular society" opens up the discussion in more productive ways.

Friant and his colleagues provide us with a four-fold speculation of future possibilities. They suggest we can imagine two future elaborations of the circular economy that would very definitely embed unjust transitions: notably, a technocentric circular economy—essentially green neoliberalism and a regressive, survivalist, Malthusian/Nationalist "Fortress Circular Economy." They also suggest two rather different possible circular futures that might exist for embedding the principles of the circular economy in more labor and community-oriented projects: notably a green social democratic "Reformist Circular Society" and a "Transformational" and postcapitalist circular society. Calisto Friant's intervention then seeks to remind us that the future of the circular economy is not given, and while it presently leans toward a technocratic or fortress vision, other possibilities remain worth fighting for.

Circularity discourse typology Friant and Vermeulen, 2024. Table reproduced with permission of author.

		Approach to social, economic, environmental, and political considerations	
		Holistic	**Segmented**
		Reformist Circular Society	**Technocentric Circular Economy**
Technological innovation and ecological collapse	**Optimist**	- **Assumptions**: socio-technical innovations can enable eco-economic decoupling to prevent ecological collapse, so a reformed form of capitalism is compatible with sustainability. - **Goal**: human prosperity and well-being within the biophysical boundaries of the earth. - **Means**: technological breakthroughs and social policies that benefit humanity and natural ecosystems.	- **Assumptions**: technological innovation can enable eco-economic decoupling to prevent ecological collapse, so capitalism is compatible with sustainability. - **Goal**: economic prosperity and development without negative environmental externalities. - **Means**: economic innovations, new business models, and unprecedented breakthroughs in CE technologies-

	Approach to social, economic, environmental, and political considerations	
	- **Example concepts**: natural capitalism, cradle-to-cradle, performance economy, the natural step, the blue economy, regenerative design, sound material-cycle society, doughnut economics. - **Proponents**: various international organizations, academics, large foundations, and some governments.	- **Example concepts**: industrial ecology, reverse logistics, biomimicry, industrial symbiosis, extended producer responsibility, cleaner production, bioeconomy. - **Proponents**: some academics, many corporations, various national and city governments, and international organizations.
	Transformational Circular Society	**Fortress Circular Economy**
Sceptical	- **Assumptions**: socio-technical innovations cannot bring absolute eco-economic decoupling to prevent ecological collapse, so capitalism is incompatible with sustainability. - **Goals**: a world of conviviality and frugal abundance for all while fairly distributing the biophysical resources of the earth. **Means:** complete reconfiguration of the current socio-political system and a shift away from productivist and anthropocentric worldviews.	- **Assumptions**: socio-technical innovation cannot bring absolute eco-economic decoupling to prevent ecological collapse, but there is no alternative to capitalism. - **Goal**: maintain geostrategic resource security in global conditions where widespread resource scarcity and human overpopulation cannot provide for all. - **Means**: innovative technologies and business models combined with rationalized resource use and migration and population controls.

	Approach to social, economic, environmental, and political considerations	
	- **Example concepts:** conviviality, steady-state economics, permacircular economy, degrowth, eco-**anarchism, Buddhist economics**, buen vivir, ubuntu. - **Proponents:** many academics, social movements, bottom-up circular initiatives, and indigenous peoples.	- **Example concepts:** the tragedy of the commons, the population bomb, overshoot, disaster capitalism, capitalist catastrophism. - **Proponents:** survivalists, a few academics, some geostrategic think tanks, and state policies.

What might be the opportunities then to develop a just circular economy—whether reformist or transformational? This is a conversation that is still at its early stages, but there are promising collaborations emerging through the EU-sponsored *Just2CE* research network,[2] which has begun to explore just circular economy frameworks. *IndustriAll,* a trade unions federation representing manual and nonmanual workers in the metal, chemical, energy, mining, textile, clothing, and footwear sectors has also critiqued the technocentric circular economy in its recent policy documents while arguing for a circular economy strategy incorporating just transition principles, social dialogue, and industrial policy.[3] What specific opening might emerge in this space for design and architecture orientated toward just transitions?

Pedro Gadanho (2022) highlights a number of promising examples of attempts to address labor and circular design issues in the design field. The Brussels-based Rotor Design, founded in 2005 by Maarten Gielen, Tristan Boniver, and Lionel Devlieger, is a cooperative design practice that is involved in design, architecture, interior retrofit, and design research. As a design practice it offers interior design projects "crafted around reclaimed elements." It also runs a worker owned co-op—Rotor DC—which salvages building components.[4] Rotterdam's Superuse Studios is an international architecture collective experimenting with circular and sustainable design. Further examples of urban planning and architectural studios engaging with the circular economy frame would include the Frontiers of Circular Architecture festival, organized by the European Bauhaus in 2022, and the Aarhus School of Architecture's Bank of Materials[5] project.

Olga Ioannou (2023) has further suggested that circular design could pose disruptive aesthetic, organizational, and political-economic challenges to the design and construction industries, which could be generative for advancing designs for just transitions. For example, she has observed that circularity presents interesting aesthetic and philosophical challenges to traditional form-centered design aesthetics. This is so because in a circular design context, "Materials are not simply what architects employ to consolidate form; instead, form is perceived as a temporary configuration of materials and products, all of which will outlive the building." What follows from this, then, is that "In the Circular Economy paradigm, the survival of the building's various parts is far more important than the building form itself. Aesthetics now comes closer to ethics; what is beautiful needs to last" (Ioannou, 2023:157). But, in addition, and as she notes, socially and organizationally, design for disassembly only makes sense "if it is adopted by the entire chain," moving from suppliers and workers to consumers and supported by regulation. The small scale of many construction companies and the lack of coordination also impact the effectiveness of circularity. As she notes:

> The fate of materials and building products now depends on the synergies of all the stakeholders involved, with the designer only one amongst many who are responsible for the buildings service life. In the CE framework, repair subdivides accountability. The various figures involved include the designer, manufacturer, owner, supplier, and sub-contractors, as well as the ongoing maintenance team that will service the building. (Ioannou, 2023: 157)

As such, how a successful and just system of circularity could be brought into being requires a much more expansive account of the forms of planning and coordination, coalitions, and partners that would have to be brought on board to achieve a transformative agenda. We might say then the attempt to think about how we might move a circular economy/circular society forward draws us back to the question of how different forms of design-labor can be coordinated, a question that MacKenzie Wark (2013, 2015) has persistently argued is one of the critical questions that any kind of red ecology for the Anthropocene has to address.

Unionizing Climate Labor and Design Labor

Let us conclude here with the matter of unionization. If a great deal of energy in the labor-environment design space in recent times has been given to thinking about designs that might highlight empowerment of workers, a growing body of voices have argued that the single most impactful way in which the structural conditions of environmental–labor–design relations could be improved would be for environmental and climate campaigners to support the unionization of workers—energy workers, design workers, and indeed all workers in the climate labor space. For example, if we take the green jobs and green technology sphere, a growing body of research and writing done by Dustin Mulvaney, Sarah Knuth, Anthony Levenda, Jesse Goldstein, Matt Huber, Lee Harriss, Seaver Wang, and others has argued that the development of the renewables industry in the United States has been marked by all manner of troubling labor and human rights issues. Research by Sarah Knuth, Anthony Levenda, Jesse Goldstein, and their colleagues have demonstrated that key players in fossil, financial, and green capital have become increasingly intertwined across the US energy economy this past decade through buyouts and mergers, and the bonanza of rent seeking that has followed from this as "clean tech" has become an increasing site of venture capital funding and state largesse (see Knuth, 2018; Knuth et al., 2022; Goldstein, 2018). Matt Huber has argued that the configuration of the "actual existing decentralized renewables industry" in the United States is marked by a mix of "deregulated markets, tax shelters for corporations, a rentier development model, and an anti-union industry dependent upon a transient and insecure workforce" (Huber, 2022). Lee Harris has largely substantiated such concerns detailing how the lives of many solar works in the United States takes the form of unpredictable gig work and "at will" contracts with a mobile solar workforce, rolling from job site to job site, often receiving little to no training to do dangerous installation work with modest if little benefits or insurance. Todd Vachon (2021) has further argued that solar and wind jobs in the United States not only pay considerably less than unionized coal or nuclear jobs, but that once solar and wind has been built it has relatively little need for maintenance compared to nuclear or fossil fuel power stations that offer high-paying permanent employment. If we consider here that in the United States, the Bureau of Labor Statistics informs us that solar panel installers and wind turbine technicians are two of the top five fastest growing occupations in the United States (Bureau of Labor Statistics, 2022), the expansion of a casualized and low-paid green job workforce not only undercuts the very premise of a just transition but provides ample opportunities for backlash politics. That the terms and conditions of solar and wind jobs often compare poorly with better paid and more secure union jobs in the oil, gas, and nuclear sectors is a huge ongoing problem that is going to delimit the capacity to enact just transitions unless it is squarely addressed.

On the green jobs front *the Labor Network for Sustainability* and *the Green Workers Alliance* have all called for clean energy jobs receiving federal tax credits to pay the prevailing wage and to fight for policies to ensure that all low-carbon-energy jobs are "high road" union jobs with national project labor agreements, marked by community workforce agreements, targeted hires, and proper health and safety training. Both have further argued for just transition agreements which at a minimum offer full opportunity for apprenticeships, job retraining, bridge benefits, and community support for fossil fuel workers, and fossil fuel–dependent communities caught in the crosshairs of energy transition are absolutely vital for obtaining union and broader community support for public power. The fight for full unionization of solar, wind, and nuclear workers as well as workers in the building trades is clearly essential.

On the design front, the interventions of Peggy Deamer, Michael Toste, Danielle Aubert, and members of the Architecture Lobby have sought to draw attention to ways in which labor in the design and architecture industries—particularly in the highly neoliberalized design economies of the United States and the UK are also a key space of anti-union and anti-worker practices also marked by the growth of casualization, outsourcing, wage theft, and unpaid overtime and technologically directed deskilling. Much of this work has sought to remind designers and architects that they *actually have a labor history*. As Peggy Deamer has noted, during the height of the New Deal in the US architects unionized through FAECT: The Federation of Architects, Engineers, Chemists and Technicians. Danielle Aubert has detailed how graphic designers in Detroit in the Detroit Printing Co-Op joined the IWW (Industrial Workers of the World), while J. Dakota Brown has documented the history of designers who joined the International Typographical Union. Many architects in the United States and the

UK may see their field as having little connection with unionization, yet in other places, for example, Sweden, they have long been unionized, and even with Scandinavia being impacted by neoliberal labor market reform, 80–85 percent of the profession today still remains unionized.

The project to unionize design workers undoubtedly confronts very significant challenges. Peggy Deamer (2020) has argued that historical tendencies of design workers to obscure the labor of design through adopting professional identities as "creatives," artists, or members of the professional middle class have not helped matters. Much of the industry is organized around small firms or independent practitioners, with freelance and increasingly isolated virtual labor increasingly common. Discussions of the need to unionize design labor have almost been entirely missing from ecological design discussions for much of its history. Nevertheless, there are good reasons it is essential in its own terms and essential for linking the cause of labor to the broader struggle for the just transition.

5

Critical Design Studies, Redirective Practices, and the Labors of Co-Creation

If many mainstream currents of architecture, design, and planning ignored, slow walked, or resisted centering the climate or ecological crisis over the past two decades, perhaps one of the most notable attempts to move past these legacies with demands that design urgently place the socio-ecological crisis at its core has emerged from the transdisciplinary space of critical design studies.

Critical design studies has evolved through various programs and projects over the past two decades—from the call that a climate-focused design for a "destabilized world in crisis" has to now foreground multi-scalar "redirective practices" that will open up livable futures (Fry, 2009) to manifestos championing "design when everybody designs" (Manzini, 2015). It has given rise to bodies of work that argue design for a just post-carbon future must fundamentally break from the "one-world world" of neoliberal globalization and rather reorientate design toward building "design for a pluriverse," which would welcome "a world where many worlds fit" (Escobar, 2018), to the claim that ecological design has to abandon its obsessions with the new and rather help cultivate prefigurative social practice and projects that could ultimately contribute to system wide socio-ecological transitions (Irwin, Kossoff and Tonkinwise., 2015).

This chapter seeks to map and appraise this field of critical design studies. We suggest here that a good deal of work that has emerged in this field has often been genuinely inventive (in conceptual and propositional terms) and makes important contributions to the search for designs for just transitions. Notably, it is a field that has sought to destabilize taken for granted disciplinary silos and bounded design ontologies. It has sought to draw design into trans-disciplinary and post-disciplinary conversations that have often been productive and generative. It is also a literature which raises the perennial question about how proximate, professional or credentialed designers should relate to "design by society" (Woodhouse and Patton, 2004) and in doing so, it champions a range of innovative approaches to codesign. However, we also identify some institutional, labor, and scalar gaps particularly concerning how some currents of critical design studies think about localism, statecraft, planning, and public institutions, the labors of cocreation, and the labors of design. We conclude with some thoughts on the potential benefits of drawing the insights of transition and labor-centered design together.

The Rise of Critical Design Studies

The roots of critical design studies can be located in a number of places, spaces, and interdisciplinary design conversations. We might track here a range of foundational points of inspiration and departure. Many critical design currents emerge draw inspiration from many of the ecological design discussions that we explored in Chapters 2 and 3—and specifically the left-libertarian and communitarian traditions of design that run from the *Whole Earth Catalogue* to Bookchin, the accenting of noncredentialed, vernacular, indigenous, Southern, and lay design knowledge accented by Bernard Rudofsky and Colin Ward, Victor Papanek, and Ivan Illich as well as the defense of the potential of distributed energy systems made by Amory Lovins. Most forms of critical design have also drawn inspiration from the participatory critique of high modernism urban design that evolved from the mid-1960s onward in urban studies through Jane Jacob's seminal polemic *The Death and Life of Great American Cities* (1961) and attempts to reground planning in the aftermath of the critique of "urban renewal" along the lines of advocacy planning, communicative planning, community design, and the non-plan, the Scandinavian and Latin American traditions of participatory design and Italian traditions of design for self-organization and autonomy (see Jones et al., 2005). Third, many currents of critical design studies have also been further influenced by theories of ontological design, which are in debt to conversations emerging out of research conducted at the interface of Heideggerian phenomenology, linguistics, and computer science led by American computer scientist Terry Winograd and the Chilean engineer and polymath, Fernando Flores who had played a significant role in launching Project Cybersyn in Allende's Chile (see Chapter 2). In particular, the recursive and relational view of human-interaction design outlined in Winograd and Flores's key text *Understanding Computers and Cognition: A New Foundation for Design* (1986), which argues "in designing tools we are designing ways of being" and hence the kind of design interventions we make not only alter society but human experience, has been widely influential. Yet such conversations have been further transformed over the past two decades by a number of additional developments.

We can identify four important further threads here: (i) the broader impact of digitization and computation culture on design practice, which not only gave rise to user experience design but has stimulated further discussions about the potential of interaction design, systems design, and product service design contributing to new modes of sociality and new perhaps dematerialized modes of need meeting; (ii) a more general orientation to the proposition that design concerned with exploring and aggregating prefigurative social/socio-ecological practices, design that aligns with and seeks to augment design-leaning social movements (tactical urbanism, everyday urbanism, hacktivism and maker movements, regenerative agro-food movements), or design that starts from the need to think about maintenance and repair, retrofitting, and adaptive reuse offer much more productive pathways to materializing just post-carbon than design informed by techno-solutionist ecomodernist design obsessed with the materialization of the new; (iii) the incorporation of design strategy and design thinking into organizational management, business school discourse, and the corporate and governance worlds often through the particular influence of IDEO's tool kit approach to design and the growing interest in social innovation and social entrepreneurship methodologies that follow from this; (iv) attempts to move design education and design practice beyond its Euro-American centricism (Rudofsky, 1964; Muzaffar, 2024) and foreground decolonial and indigenous interventions, social design knowledge "from below," and bottom-up design disruption represented by developments in the Global South as varied as *jugaad* and frugal innovation, "design for the 99%" and Lo-TEK agendas (Watson, 2019).

The range of design currents that occupy this general space is very broad as are the politics that follow from this (which sometimes pull against one another). Nevertheless, we can identify certain general theoretical commitments that are shared by those located at the more critical end of the design spectrum.

Much critical design theory (i) rejects the modernist master builder vision of the role of the designer and starts from the premise that designing is a plural activity that has always enrolled many more skilled participants than the very small groups of individuals who identify professionally as designers. A post-Eurocentric account of design history that takes the history of noncredentialed designers seriously is seen as necessarily affirming such a position. Hence, different kinds of critical design have been very drawn to ideas of cocreation, although how this is understood, staged, and implemented, and what this implies for the relations between "professional" and "lay" designers can be quite varied. (ii) Pushing back against technological determinism and whiggish history, critical designers further argue that there are always political choices in the modes of designing and worldmaking that we embark on and political consequences from choosing one design pathway over others. Different design choices can facilitate the distribution of power and agency in different ways. All forms of critical design thinking see *design as inherently political.* (iii) A great deal of critical design thinking of late has been influenced by postcolonial and decolonial theory. Hence, there is a growing sensibility that the history of design has to be systematically reappraised and reconsidered across different times, spaces, and cultures (Dilnot et al., 2015). The largely Eurocentric and self-aggrandizing modernist tales of the rise of the design industries and the Eurocentric histories of particular design professions and the objects/processes/services they have fashioned as such are seen as deeply limited (Abdulla et al., 2019). (iv) Virtually all critical design currents whatever their differences seek to reclaim the role of design as a *future making activity* and design advocacy as a speculative activity that should be constantly arguing for, promoting, and envisaging different futures. However, unlike a great deal of modernist architectural utopianism which has been drawn to tabula rasa imaginaries, where the past is erased and the plan imposed, often in an act of colonial imposition (Muzaffar, 2024), critical design studies has paid much more attention to the proposition that post-carbon futuring may well involve building the new to some degree, but this may also involve locating, augmenting, or stitching together existing agencies or drawing out the potential of certain *embedded practices.* Hence the attraction to processes of repair, maintenance, redirection, repurposing, and retrofitting. Much critical design theory is struck by the *ubiquity* of the designed world we now reside within, whether viewed through the discourse of climate change, ecological crisis, planetary boundaries, the Anthropocene, or the rise of planetary-wide computation and global digitalization. This often runs with the claim that design is ontological in the sense that it describes those aspects of the human condition involved in prefiguration, anticipation, and worldmaking. While many figures have contributed to this project, the writings of Tony Fry and Ezio Manzini have had a significant influence on the field.

Design as Redirective Practice

Perhaps one of the most persistent voices arguing for the redirection of design theory and practice over the past two decades has emerged from the writings of Tony Fry. Across a range of interventions, Fry has sought to recover design as a "field of political action" (Fry, 2010). Fry's prolific output is marked by three starting convictions: (i) a refusal to locate design in a single design discipline but assume the need for "working across visual communication, industrial design, architecture, and urban design" (Fry, 2009); (ii) the assertion that what we think of as designing needs to be expanded well beyond design as service profession for contemporary capitalism; and (iii) recognition that the climate and broader ecological crisis that intersects with the crisis of coloniality places the entire human project in peril. As a product of "our" human-centeredness, human beings have treated the planet "as an infinite resource at our disposal" (Fry, 2009), and this bodes disaster. A reconceptualized and expanded understanding of design as a socio-material, socio-ecological, and socio-technical form of *redirective practice* is critical for bringing into being survivable post-carbon futures.

Fry argues that our starting point for any kind of design politics today is recognition that we live in a made world, indeed a (mal)designed world. As he puts it, this is a world where "Nature alone cannot sustain us: we are too many, we have done too much ecological damage and we have become too dependent on the artificial worlds that we have designed, fabricated and occupied" (Fry, 2009: 3). An eco-political imaginary relevant for contemporary times *must render our designed world visible* rather than retreat into the dualist society/nature binary that has long informed romantic Western environmentalism. In this respect, Fry's account of a thoroughly designed world is generally proximate to Donna Haraway and Bruno Latour's view that we live in hybrid socio-ecological worlds marked by all manner of boundary breakdowns and border crossings. In contrast though to Latour's rather unbounded anthropomorphic view of agency, Fry foregrounds the act of design as the act of making and remaking; it "names our ability to prefigure what we create before the act of creation" (Fry, 2009: 4). This move allows Fry to think about design as naming a material but also as a political process of making and remaking—*redirective practice*—and he argues modes of redirective practice to build sustainable futures can be conceptualized as potentially working at many scales.

Redirective practice can occur in the personal or intimate sphere (how do I redesign and redirect my habitus toward lower carbon outcomes?), but it is a way of making visible all manner of other kinds of cultural, material, creative, and political inventions that could redirect praxis and open up other futures, whether we are considering gardening, the making of products, systems and services, or broader material interventions such as architecture, planning, and so on. Indeed, our capacity to embark on multiple modes of redirective practice through multiple modes of design is one of the fundamental characteristics that make us human (Fry, 2009). Design is nothing less than a "world shaping force" (Fry, 2009: 3). In a debt to Marx and Heidegger's account of humans as a self-designing species, it is underlined that design has played a role for millennia in shaping our species' being. The worldmaking potential of design, then, is revealed by understanding design as ontological; we design our world, and our world, in turn, acts back on us.

Design, then, is central to *the constitution of our species being*, but it is of course fundamentally *Janus-faced*. If designing has taken on many different forms across human history and been institutionalized in many different ways (Fry, 2013), in its most recent commercial, capitalist manifestation, design is suffused with instrumental rationality; it has become technocratic, narcissistic, profoundly de-politicized, and central to what Fry refers to as the current *de-futuring project* of modernity. For Fry, de-futuring refers to the propagation of systems of production, consumption and lifestyles that, through their environmentally destructive forms, are literally stealing or radically narrowing our future possibilities (Fry, 2020). Design has continually failed to acknowledge its complicity in building and sustaining a de-futuring material culture through its uncritical support for the hyper-consumer economy. There has been a fundamental failure across the professional worlds of design to acknowledge not only how much it is implicated in de-futuring but the extent to which design takes on a life of its own, since "designed things go on designing (be they designed to do so or not)" (Fry, 2009: 38). This inability to think in sophisticated ways about the many temporalities of the environmental crisis has also generated a failure to see that unsustainability or de-futuring is structurally embedded in our current capitalist political economy but also in the very designed "habitus" we have come to occupy, from our homes and transport systems to our materially embodied and designed forms of work and leisure and beyond. This will all have to change.

What is to be done then? Fry argues that design is also a social and cultural praxis that is enacted by many more people than the small field of professional designers. It is a practice concerned with material things, and it is critical to the resolution of socio-ecological questions. There are two tasks that need to be addressed that are of central importance: first, to slow the rate of de-futuring and, second, redirect ourselves toward more sustainable modes of planetary living (Fry, 2009: 6). The climate crisis then needs a recovery of design understood as systematic project of retrofitting and redirecting our personal habitus, our culture, our homes, our cities, and our broader socio-ecological systems to reclaim the future through a politics of redirective practices. This will need to move from considering the different ways in which we might live better and more sustainably through new modes of care of the self to new material interventions at much more comprehensive levels. We are talking about a politics that is concerned with the material, social, political, and ecological assembling of our material culture. It will all have to change.

Fry posits a range of ways in which such relationally interconnected modes of redirective practices could contribute to a serious shift to a futuring society. At a certain level, this requires incorporating and updating some of the visions of a closed loop quality and new modes of product service design—themes that have long run through industrial ecology and product systems/services design literatures and practice and more recent discussions of the circular economy. But Fry's vision is more encompassing than this. Design needs to involve not just making but also *unmaking*. Circularity is not enough. We will need to practice *eliminative design*. There are vast amounts of goods, products, and services that will have to be rendered sustainable, but there are many more that will have to be literally designed out of existence. (Think of the vast material and visual culture and the modes of material extraction, exploitation, and ecological degradation that have been built around constructing throwaway goods and services that contribute little to human or nonhuman flourishing.)

In the light of climate change and the likelihood that vast numbers of people will be unsettled and literally on the move (2015), Fry further argues that we need dynamic, adaptive, iterative, and proactive modes of design. So, for example, we need to think about how we might have to engage in comprehensive urban retrofitting—"metrofitting," in some cases, possibly abandon urban forms or rebuild them elsewhere (Fry, 2014). More generally, there is much more attention given to re-futuring as a hegemonic intellectual project: designers need to be made aware that "all that design brings into being remains in process" (Fry, 2009: 30). We need more generally to realize among publics but across all institutional actors, "a new design intelligence"; this is a mode of literacy acquired by every educated person that would "deliver the means to make crucial judgments about actions that would increase or decrease future potential" (Fry, 2009: 41).

Design When Everyone Designs?

If Tony Fry's project could be seen as primarily an intervention in metatheory working through Heidegger, Schmidt, Marx, anthropology, and decolonial theory to confront the civilizational and epochal challenge of the climate crisis through calling for in many respects a revolutionary messianic critical design-politics, the writings and interventions of Ezio Manzini have tended to evolve more pragmatically, bringing meso-level theoretical propositions and insights into conversation with specific case studies and design experiments in sustainable living and social innovation. Manzini's *Design, When Everybody Designs* (2015) and *The Politics of the Everyday* (2019) proceeds from a general reading of the contemporary moment as marked by a climate and sustainability crisis but also informed by Zygmunt Bauman's liquid modernity thesis and Lash, Beck, and Gidden's focus on detraditionalization, a crisis of modernity marked by social fluidity and detraditionalization. If commodified neoliberal landscapes, "major economic forces" such as international finance, atavistic nationalisms, racism, and the reversion to exclusive communities mark one largely disastrous response to the contemporary world, Manzini argues that design has to move on and let go of many of its old commitments and moorings. In these times of crisis, the modernist project of design professionals making more objects for passive publics must be transcended by design approaches that locate, work with, and amplify "collaborative communities" and "creative and enterprising people."

Manzini defines designing as "making a critical evaluation of the state of things, imagining how we would like them to be, and having the necessary relational systems and tools at hand to transform them—all this in terms of their practical functioning (problem solving) and their meaning (sense-making)" (Manzini, 2019: 58). The designer here can be distinguished from the engineer that Manzini suggests (and following Lévi-Strauss) is a field that emerges from a reductionist, mechanical standpoint and that claims a position of total control and the capacity to bend reality to their will. Rather, the sensibility of design is closer to Lévi-Strauss's bricoleur, that is, an approach that is inherently dialogical, acknowledges complexity, and is involved in reassembling preexisting objects, reworking, and reinterpreting their meaning. Designing then as opposed to engineering or architecture is an approach that accents listening, to things and to people, seeing the potentiality in pathways forward (Manzini, 2019: 51), making ethical choices about the design projects we embark on, and looking for collaborative possibilities (Manzini, 2019: 56). A designer for Manzini is *a project maker* and this understanding of designers encompasses credentialed and noncredentialed designers. The argument here overlaps with Fry in claiming that all human beings are endowed with "a design capacity" and have the potential to exercise this at different levels. However, Manzini argues that this design sensibility is particularly pronounced at the moment in diverse collaborative communities, which are located either in small-scale production, mutual aid groups, care communities, digital spaces, or the urban commons that are continually coming up with new solutions and *transformative* social innovations that open up the potential for reconnecting people with communities and places (Manzini 2019, 9). These "fluid communities" are communities of interest, communities of choice, place or purpose, they live online and offline and often are located in the hybrid spaces in-between but most critically these are communities that are—voluntary, light, open not the "closed identitarian communities" that are "harking back to an imaginary past…confined within their asphyxiating localism" (Manzini, 2019: 21) or narrow atomized individualism that the market offers. The aim then is to align such communities with projects which open up the potential to form "designing coalitions" (Manzini, 2019: 23). The project is to locate and aggregate these forces for transformative systemic changes at a local bottom-up level driven by collective collaboration to open pathways for a sustainable future.

What are the kinds of projects that interest Manzini? Manzini's edited collection *Sustainable Everyday: Scenarios of Urban Life* (2003) (coedited with François Jégou) explores a range of experiments in collaborative living. Such a focus is not on communes or kibbutz but on neighborhoods that choose to be involved in sharing services or mutual aid. It might be as loose as neighborhood restaurants or formalized as cohousing. Such enterprises might be linked to service design and collaborative platform design that facilitates the growth of farmers markets, experiments in slow food, neighborhood gardens, car sharing and carpooling, community-based fabrication labs or urban gardens, and new public spaces. The aim is to build design-enabled networks that could envisage and start to realize an alternative, sustainable everyday life. *Design, When Everybody Designs* (2015) similarly argues we can draw on traditions of participatory design to bring publics directly into the process of cocreating and redesigning systems, services, and new institutions, but also new cultures, experiences, and ways of living. It is through these cocreated cultures and institutions that we might build regenerative forms of social and ecological innovation that meet diverse human needs in more sustainable ways. *Design, When Everybody Designs* notes that from health care to care for the elderly, food cultivation and provisioning to social solidarity, all current institutional forms are under profound stress and will have to change rapidly. Yet, Manzini also maintains that social life is malleable, communities always have untapped resources and capacities to deploy, and the point of a design imaginary is to look for and see *pre-figurative socio-institutional forms* that could start to address the crisis in the here and now. Teaching school children to grow food and tend a local garden proximate to a school brings schools together with local gardening institutions, but expands the educational worldview, has the potential to revitalize urban space, teaches children about ecology and biodiversity, and creates cross-generational relationships. Designing an elder people's home that is proximate to a university and offers free accommodation to students for services could allow for mutually beneficial relationships that address the loneliness epidemic. In short, then Manzini suggests service design, community mapping strategies, and participatory design could be key tools to facilitate active institution- and culture-building, the building of new services and facilities which are cocreated, agile, and adaptive to new contexts. Professional designers can play a critically important role here, not as master builders but as conduits through which design tools can facilitate contexts that can unleash the context of citizen designers. Games, community mapping, prompts, video sketches, citizen storytelling, solution cards, scenario-building, and even prototyping can all be used to discover untapped community resources, build on existing community resources, develop collaborative communities, and open up citizen-based visions of possible futures. The broader aim of design for social innovation is experimental in its orientation and aimed at allowing multiple modes of social innovation projects to aggregate up, to create a new commons or infrastructures for "sustaining multiple connected initiatives" (Manzini, 2015: 5).

Transition Design and Designs for the Pluriverse

The kinds of socio-ecological imaginaries advocated by Fry and Manzini are only one moment of a broader field of socio-material and socio-ecological engagements. The program of Transition Design established at Carnegie Mellon (Irwin et al., 2015; Kossoff, 2015; Irwin, 2015) has a great deal of overlap with the general project of Fry and Manzini, further linking such currents to traditions of ecological, sustainable, and green design that we surveyed in Chapters 3 and 4 (from Bookchin influenced social ecology approaches to ecological systems theory), more recent currents of socio-technical transitions theory, and social practice research to update the case for ecological design. Transition design is an approach to design that is explicitly attempting to consider how design strategies, tools, conceptualizations, platforms, and services might identify, augment, amplify, and work with groups, citizens, businesses, or public institutions experimenting with sustainable and regenerative prefigurative practices, institutions, visions, and activities. Once again, like Manzini, the theory of change informing transition design is the proposition that some of the more radical ideas for how we might live richer, more convivial socio-ecological lives in post-carbon futures can already be glimpsed in the practices of various creative communities and social movements (such as transition towns, slow foods, rights to repair movements, local currencies and time banks, experiments in cohousing, and workers' and community cooperatives). And in this respect, transition design seeks to encourage dialogical approaches to design for transition that can gather together different kinds of communities and stakeholders to understand the diverse problems that socio-ecological change throws up and facilitate reflections on the existing resources and capacities that communities may have to start to address these issues. It also seeks to bring a grounded speculative futuring into such discussions by amplifying, augmenting, and sharpening possible prefigurative ways of living and working to ensure that when the crisis really hits home, new ways of meeting needs are available but also by opening up discussions about how in changed circumstances, other ways of living and other pleasures, and other ways of being in the world might be possible (Kossoff, 2015; Irwin, 2015). The key proposition here then is that a transformative design imaginary needs to attend to how everyday life can open up leverage points for change (Irwin et al., 2022).

Arturo Escobar's *Designs for the Pluriverse* (2018) can also be read as inspired by and seeking to elaborate the project of Fry, Manzini, and transition design, drawing further inspiration from the Zapatistas of Chiapas and the indigenous, Afrodescendant, and peasant communities of Colombia. Escobar's contribution to this literature has been to seek to further align the critical design project with the decolonial insights of his own work in post-development theory with indigenous political ecology and indigenous design, post-growth movements in the Global North and contemporary peasant, agro-ecology and land sovereignty movements in the Global South such as La Vía Campesina and other advocates of peasants rights and autonomy rural development. Perhaps Escobar's notable contribution to this discussion is his argument that designs for the pluriverse should embrace an ontology of *relationality* which is premised on acknowledgement of "the radical independence of all that exists" (Escobar, 2020 x). Notably, Escobar argues that the goal of design for the pluriverse is "the reconstitution, healing, caring for the web of interrelationships that make up the bodies, places and landscapes that we are and that we inhabit" (Escobar, 2020: xii). A relational ethics and politics "gives us a different sense of being at home in a world that is fully alive, and to design with this realization in mind" (Escobar, 2020: x).

The Virtues of the Critical Design Moment

So, what is the contribution that critical design studies make to both ecological design and designs for the just transition? We might map a number of contributions here. Firstly, it could be observed that critical design studies have clearly brought some productive and provocative advances to design in terms of the general conceptualization of space, time, and society–nature relations. Notably, and in marked contrast to many romantic ecological design imaginaries that emerge from the counterculture moment and the UN sustainable development era, which often assumed a sustainable society would be marked by a return to some kind of homeostatic vision of balance and harmony, Fry's writings in particular are notable for the assertion that the dream of return to a stable Earth (if it was ever coherent) has long slipped away. This is a compelling and productive starting point for thinking about design-futuring. The core position here, as Cameron Tonkinwise has elaborated, is that life on a hotter, more turbulent, and more dynamic planet will require design interventions that will have to be iterative and inventive, taking place again and again and again in climate-challenged worlds (Tonkinwise, 2003: 2015, 2024). The general proposition that design would do well to orientate itself toward a dynamic future "coming at us" (Fry, 2009) has much greater alignment with more recent developments in non-equilibrium ecology, climate science, and complexity theory than a great deal of equilibrium orientated ecological design of the past. The further proposition we are heading toward "an age of unsettlement" (Fry, 2014) and will need to design in the light of this is again a troubling and generative proposition. Current projections suggesting that we could be heading for 1.7 C to 2.7 C global warming under current policies by 2100 (Hausfather, 2025), with concurrent implications for sea-level rise, desertification, an intensification of extreme weather events such as wildfires and hurricanes, with the result that human and nonhuman natures "on the move," it would seem irreducible the case that many epistemological assumptions and the static formalist imaginaries that defined high modernism design, architectural and planning in the twentieth century will serve us less well in the future.

Here, critical design studies opens significant connections to climate design imaginaries that not only accent the importance of repair, maintenance, and adaptive reuse (see Berger and Irvin, 2023) but also climate design imaginaries that take mobility, movement, and flows seriously. For example, Fry's proposition that the design for urban form in the twenty-first century will have to be both much more resilient and much more adaptive to transformed socio-ecological conditions than anything imagined by twentieth-century ecological urbanist or architectural discussions is provocative and generative. If a decade ago, the proposition that we might need to embrace modes of urban "metrofiting" involving massive adaptation to new climate conditions, entertaining moving or even abandoning vulnerable cities may have seemed like it was engaging in survivalist fantasies, this proposition seems much less fanciful in the light of current and future climate impacts we have seen occur by Hurricane Katrina's impact on New Orleans, wildfire impacts on Los Angeles, or catastrophic flooding that has occurred from Ashville to Valencia to take three desperate examples. Fry's emphasis here that design for climate-changed futures will require designing structures, services, and ultimately urban forms that are resilient to a broader range of temperature extremes and that can anticipate the need for "managed retreat" from flooding or wildfire areas all makes good sense. Manzini and the transition design proposition to focus design on projects that can augment social connections, mutual aid networks and design competencies would also seem a vitally important task for design for transition to focus on in a world where mainstream design has contributed so much to digitally enhanced social isolation and loneliness, addictive apps, and mental stress. We have come to understand that not only resilience planning but thick social networks that are enabled by all manner of social infrastructure can be critical in a crisis. The more general proposition of transition designers that speculative design-futuring at the level of everyday life is going to be vital for thinking about how many kinds of social practices, adaptive skill sets, features of our material culture, and ways in which we interact will need to change in a warming world is important.

Second, it could also be argued that critical design studies present a range of productive ontological-political challenges to the high modernist design canon that stabilized itself around particular design disciplinary silos based on the self-contained building, the plan and the product that defined a good deal of design thinking in the Euro-American Academy (and beyond) in the mid-twentieth century. Rather, if we follow the claim that these materialized forms stabilized by design in service to the consumer economy of the twentieth century are better understood as complex hybrid assemblages situated within broader dynamically designed social, ecological, and technological systems, processes, and services and composed of material flows, carbon, human and nonhuman labor, and ecological relations and these hybrid assemblages need either radical redirection or elimination, the project of design is now called to respond to a very different project. Notably, and in ways that resonate with the critique of architecture posed by Stewart Brand and Jeremy Till or Jane Hutton and Kiel Moe we explored in Chapter 3 or the challenge to design posed by worker-centered designers and the focus on the invisible labors of design we explored in Chapter 4, we can see a range of forces that could potentially ally that seek to bring labor and ecology back to the center of the project of design. Moreover, this poses new questions for design such as how do we bring into view these dynamic relations? What would cultures of design and architecture look like that took design for repair, maintenance, and reuse, *as first principles of any design action or building practice rather than "after the fact" additions*? How might new systems and services relate to dynamic but stabilized forms? As Manzini's work points toward, what would it mean to engage in building construction that attends to the potential of overlapping and complementary systems and services, multifunctionality, the changing needs of users, ecological relations, landscape, changing climates, and the need for design for disassembly? If design is to lose its fixation on bringing about the new but seriously embrace the challenge posed by *eliminative design,* beyond the obvious attempts to find design and social interventions that could help eliminate single-use plastics or products made with nonbiodegradable ozone destroying or cancer-inducing "forever chemicals," could an eliminative design agenda help start a broader conversation over the kinds of design that we don't want—such as military drones and army robots, carceral and surveillance technologies, AI that deskills and dehumanizes rather than augments human capacities? Could we envisage a necessary transformation of how the fields of design, architecture, and planning are convened to address these challenges? Escobar's broader call here to foreground relationality clearly further brings nonhuman actors into the design process in ways that resonate with the growing interest in bio-inclusive, more-than-human, and multispecies design for transition (see Veselova and Gaziulusoy, 2022).

Third, in accenting matters of design innovation and cocreation, it could be argued that critical design studies offer some refreshing lines of pushback against architectural narcissism, Prometheanism, and design megalomania. Cocreation reinforces the observation that design work always draws from and appropriates many past and present knowledges and labor and always requires a coalition politics for its substantiation. Cocreation as a call for many different knowledges to be engaged in the present and future of worldmaking also touches on what Manzini refers to as the desire that people have to be authors and designers of their own lives. The argument then that design has to think beyond its inherent Eurocentrism can be read less as a simple moralism and more as an empirical observation that for design to merely engage with a very small slice of making across Euro-American history is to cut itself off from a vast repertoire of worldmaking that has been conducted across human history (see Graeber and Wengrow, 2021) and may well be needed for us to think about ways forward in radically altered contexts.

These are real gains, and a focus on comprehensive redirective practice, eliminative design, design as cocreation, and prefiguring can all play vital roles in thinking about how we might design for the just transition. However, perhaps the critical question that needs to be asked is if the proposition of ontological design greatly expands the potential remit of designing, and are the tools, interventions, and rule-of-thumb conceptualizations that critical designers bring to this task up for the job?

The Question of Labor in Critical Design Studies

Perhaps one of the first sets of questions that emerges from an engagement with critical design studies is how the *labor of design* is understood. As we have seen, the field of critical and transition design are very much focused on communitarian design, social practice, and the meta-systems thinking or relational ontologies that ecological design is often attracted to. However, and with some honorable exceptions (Boehnert, 2018), it has less to say about the specific historical and social systems that the design industry is located within, the power and labor relations that shape how these systems are structured, and the ways in which how this impacts *the work of design*.

For example, if we take the status of the *redirective practitioner* or *transition designer*, this agent of change is often framed as a figure that could play a large role in enacting or facilitating systems of change through their design work in the studio, the museum, the design firm, the community design intervention, the corporation, and beyond. This generally affirmative relationship between critical and transition design studies and the power of designers is reflective of the fact that a good deal of critical and transition design studies has been produced by design educators or design theorists who are in full time (mostly academic) employment and often situated in semi-functional if fragmenting welfare states. But critical design discussions have had much less to say about design workers *as workers* or the political economy of the design industry that most designers find themselves working within. Architects and designers who aspire to become something like transition designers, in actuality, find themselves thrown into design industry contexts that are deeply defined by the interests of *clients* and *capital* where their own labor conditions are often precarious and suboptimal. Here discussions about the classed, gendered, and raced labor of design that we have discussed earlier raise important questions. Such interventions have underlined the extent to which many designers find themselves learning in studio contexts that may well have low-carbon futures as the topic of discussion but are often further shaped by racial, class, gender, and sexual exclusions and a broader *atelier culture* where brutal and egotistical master/servant relations are often embedded in the student/professor dynamic. The exploitative relations in the design school are then often carried over to all-too-common-nonunionized employment relations in the formal design and architecture industries as energy-efficient buildings are designed via unpaid internships, proposals for circular economies are put together by design firms that are underpinned by cultures of poor pay and long working hours, digital designs for low-carbon-sharing economies are coded by deskilled "CAD monkeys," and all kinds of other forms of outsources design platforms such as CrowdFlower, Clickworker, Toluna, and Amazon's Mechanical Turk, which have massively expanded the capacity for design work to take the form of exploitative piece work and for designers to be situated as the vanguard of the precariat (see Deamer, 2020; Boenhert, 2018). Even for design workers who have relatively secure work in the affluent world, it is worth noting that their capacity to become redirective practitioners or transition designers in the contexts they inhabit without union protections or different kinds of work-management relations such as cooperative structures is circumscribed. For example, between 2016 and 2024 we saw significant waves of political mobilizations occur by tech workers, designers, and engineers in Amazon, Google, and Facebook around issues such as workplace harassment, ethical concerns about technology (e.g., military contracts, surveillance), and labor rights as well as the environmental and carbon impacts of their employees' practices. Such activities involved walkouts, public letters, and some attempts to unionize. Such requests that these firms embark on "redirective practices" were brutally suppressed by the tech industry, which has subsequently sidelined, dismissed, and, in the case of Amazon, shut down workplaces to avoid union drives.

It is important to underline here that exploitative and abusive working conditions are not simply a feature of the mainstream design industry. From Frank Lloyd Wright to Paolo Soleri and Bruce Mau, there is a long history of radical design and architectural projects being formulated around the cult of the charismatic (usually) male design guru who then moves such "sustainable projects" forward through a reliance on unpaid work and exploitative labor and gender practices (see Elzenbaumer and Giuliani, 2014; Julier, 2017). Similar problematic labor practices have been outlined in many social eco-design, transition, and community design interventions. Much of this work can be sustained by all kinds of forms of "sweat equity" or volunteering (Fleming, 2019). Such projects are often closed to working-class designers or designers with eldercare and childcare commitments.

If reflection on the laboring conditions of *design workers* is rarely foregrounded in critical and transition design studies, it could also be observed that these fields could also pay more attention to the laboring conditions of "publics." Manzini's proposition that we need to think about constructing design conditions where "everyone designs" provides a potentially compelling view of the public as "citizen designers" or "proximate designers," which could be valuable in many circumstances, but the publics that are addressed in Manzini's work are invariably the time-rich and commitment-light. Tony Fry's writings offer reflections on the conditions of possible design agency of a transhistorical human subject, but how this transhistorical subject relates to particular historical subjects who have to work to live in capitalist societies and whose agencies are shaped by the labor process, employment, and the demands of social reproduction is largely unaddressed. Similar observations could be made of transition design which draws heavily from the sociology of social practice and is largely focused on the domain of everyday life and "transition communities" but has had little to say about the workplace or processes of production that clearly shape the conditions of design agency in everyday life for most working people. Recognizing these issues does not invalidate Manzini's defense of the potential of codesign intervention or Fry's focus on the importance of redirective practice, but it does suggest that unless specific attention is given to the labors of design and design production networks, and careful attention is paid to the structural class, race, gender, and, in some cases, caste barriers that can enable and foreclose codesign and social practice, we end up with a design politics which speaks to the "usual subjects" (i.e., the time-rich, commitment-light greenish upper middle classes of the affluent world) while doing little to dislodge the hegemony of the post-material middle classes over environmental concerns or expand this coalition to engage with the broader concerns of working people.

Here, focusing on the labors of design presents challenges to critical and transition design as is. But it also could open up productive engagements with traditions of labor-centered design that we explored in the last chapter. Notably, it opens up the question of how critical and transition design currents might contribute to the reimagining and transition of the design industry itself? How might the organization of design workers enable unionization, or help establish worker cooperatives or facilitate codetermination polices transitions with design workplaces? If we foreground the labors of cocreation, what might be the institutional and public support systems that might diversify the range of people that can be involved in citizen design and prevent codesign from either being an extractive process or a burden for working people who are already involved in a great deal of visible and invisible labor, eldercare, home care, and the like?

The Question of Scale in Critical Design Studies

What should be the scalar imaginary of a radical and emancipatory design politics? This question has, of course, never received any kind of settled agreement in design. High modernism answered this question through a focus on advocating design interventions at the largest scales available, often embracing ideologies of "total design." As we have seen in Chapters 2 and 2, many currents of sustainable design can be seen as a direct attempt to reject this whole legacy of high modernism. But we have also seen that there have been many ecologically leading thinkers from Ebenezer Howard to Lewis Mumford that have argued it is the region that offers a scale of action that can mediate between megalomaniacal total design and localism. The rise of critical and transition design studies as a general sensibility within design is generally marked by a complex array of impulses here. At one level, the rhetoric of this movement sees the scale of the change faced by humanity as perhaps more threatening than at any other point in human history. As such, the role that design is ascribed to fulfil is world historic (Fry, 2009; 2013) and minimally aspiring to "system level change" (Tonkinwise, 2019). Yet, at the same time, and as we have seen, a good deal of critical design is not only very skeptical of legacies of modernist "design megalomania" and "total design" but also skeptical of a kind of instrumental rationalism that is seen as somewhat embedded in an engineering worldview. As such, in terms of the scales of the actual project that are highlighted and pursued by critical design currents, it is the installation, the design charrette, the community-focused workshop, the neighborhood garden project, and other modes of bottom-up social innovation and community impact that predominate. The claim that often follows from this is the notion that a Manzini-like focus on "small, open, local and connected" projects might ultimately proliferate and give rise to broader social movements (e.g., for transition towns, safe streets, or slow food) which can ultimately start to shift social practices.

Not all critical projects aspire to addressing issues at broader scales and nor should they necessarily. Critical and transition design, as think pieces, prompts, provocations, and non-repeatable experiments, performances, and interventions, should all have their role in facilitating a general zeitgeist of experimental low-carbon praxis. For example, the kinds of cosmopolitics or democratic experimentalism advocated by Manzini are already manifest in the huge energies that have been poured into citizen science, tactical urbanism, social innovation from below, DIY urbanism, self-build housing, creative communities, and coproduced maker spaces. Such interventions can provide sources of resistance, invention, institutional renewal, and revitalization, but a counter to this focus would be to observe that there are perhaps limits to how much these kinds of projects can move the needle when we consider how much decarbonization requires large-scale infrastructural change.

Tonkinwise (2019) here offers a sophisticated elaboration of how transition design thinks about and could illuminate this question of scale, size, and impact. As he notes, if we aggregate design impact along a scalar view of size, we can construct a trajectory that moves from letter form, image, white goods, car, interior, building, organization, neighborhood, and city, with the implicit assumption that greater impacts follow from larger scale. Following this logic, service design would be the most ephemeral mode of design in terms of climate impact, with urban or regional planning offering the most impact. However, if we think about designing in terms of volume of impact, we can see that an i-phone has impacted millions of users on an ongoing basis (and it might be argued that it has contributed to all manner of recursive social relations of surveillance, distraction, tracking, isolation, conflict) in ways that are far more consequential than the modest amount of people who are impacted by a single building (green or otherwise). Moreover, if we think about the potential impact that transition design might aspire toward, we can start to see a more layered account of the spaces for consequential transition design interventions. Notably, transition design would not necessarily give up on object making per se. But if we take it as given that part of the climate/environment dilemma for design is to move from a design culture based on disposability and conspicuous consumption to encouraging the growth of design for physical and emotionally durability, for repair, maintenance, and disassembly but also for design thinking that can foreground eliminative design and design intervention that also think beyond the object to facilitate interfaces, new systems, and services or broader opportunities for "problem reframing," we can potentially open up many more leverage points for change. Transition design here then stands less as a complete transcendence of "design as is" but rather operates as a current that asks different questions about how needs could be met or objectives achieved without immediate recourse to the object/building as a solution. For example, how to deal with a problem like bottled water could invite many responses. We need to proliferate reusable bottles is one way to go. Another would be to think about the return of large and small water fountains in all manner of spaces and places, from buildings to public squares that provide access to all. We could think about this water and relate it to recycling systems and food growing systems, or link it to broader attempts to develop public spaces, and so on.

This account of transition design as involved in a series of reframing interventions that seeks to work across many sites and scales is generative and useful, and it potentially opens pathways for critical design interventions to be more disruptive and impactful. It is an account of post-carbon innovation that is clearly influenced by Fry's account of redirective practices and the multilevel perspective that sees innovations emerging between the interplay between niche innovations, socio-technical regimes, and the broader socio-technical landscape (Geels 2002). But it raises further questions. To be sure, certain kinds of low-carbon lifestyles, social practices, and even technologies (such as solar panel adoption) can proliferate quickly through social movements and social contagion, through policy networks and via social entrepreneurs. At the same time, private markets and incumbents at the regime level also spend a good deal of time and money resisting or co-opting such developments. More generally, the broader energy, transportation, waste removal infrastructures and production networks that currently lock in many high-carbon practices and need to be transitioned are far more inert than consumption practices. Transitioning such infrastructures invariably requires building complicated alignments and coalitions between state actors, publics, labor, market actors, and multiple design fields from architecture and planning to urban design and engineering to accomplish the kinds of structural transformation that are necessary to encourage just transitions. And here it is not always clear that critical design or transition design has an account of the leverage mechanisms it could deploy, or the political coalitions it might look to, to make these changes possible.

The Institutional Gap in Critical Design Studies

The preceding discussion of labor and scalar issues in critical design essentially draws us to the question of how a critical design politics can relate to and be further enabled by public institutions, public agencies, critical public policy, and those aspects of the capitalist state that are subject to some degree of democratic control and direction while also avoiding co-optation.

The relationship between the state and design has long raised some tricky issues for design. As we have seen in Chapter 2, Le Corbusier happily flirted with fascists, syndicalists, and Stalinists if they showed any interest in helping him execute his projects. The relationship between the postwar welfare state and design, architecture, and planning generated a complicated set of fraught legacies that we are still working through today. The question of design's relationship to the state emerges again—as we have seen in Chapter 3—in the early 1970s as a critical bone of contention in the Global South as Papanek and Illich's defensive design from below is challenged by Maldonado and Bonsiepe's view that design and planning had to be part of a broader developmental state strategy to be impactful.

Critical and transition design currents over recent decades have generally tended to focus on social movements or community action as drivers of transformative praxis. As we have seen this has shifted focus to design for everyday life, but it has rather left questions about how design practice and imaginaries might contribute to structural changes in the broader economy (whether understood in informal or formal ways) to one side. Design for circular systems of consumption is seen as a space for design interventions, as are modes of design intervention that help script eco-conscious consumer behavior, but design interventions that might open up power relations in the domain of production—as suggested by the Lucas Plan—have faded from view in a good deal of critical design studies. Tactical urbanism is viewed as offering acupunctural spaces for urban interventions, yet larger moves to deal with green gentrification are much less evident. Tony Fry's call for "metro-fitting" is an exception here as a provocative call that urban design theory and conceptualization should move beyond its fixation on bringing new urban form into being and rather think hard about how urban form could be adapted, reused, moved, or abandoned (Fry, 2014). Yet, even in Fry's writing, it is unclear who or what the institutional, political-economic, and social forces are that are going to drive, support, and amplify metrofitting or redirective practices.

Part of the added complexity to this issue though is that when design has engaged with the (neoliberal) state over recent decades, the results have not always been to the good either. To take one example, following the global ascendance of neoliberalism, we saw the systematic contraction of planning agencies and architectural and design offices that existed within the local, regional, and federal levels for much of the mid-twentieth century and facilitated all manner of public service design activities from urban and rural master planning to social housing, municipal planning to civic architecture, community development to national infrastructure planning, and so on. Across much of the Global North and Global South, these design capacities of the public sector were outsourced, downsized, closed, or privatized under the influence of neoliberalism and the New Public Management view of public service provision. What followed from this was not just a loss of public design capacity, but a design consultancy boom fueled by the neoliberal state seeking to contract out or rebrand services to for profit or non-profit organizations (NGOs). This generated a significant shift in the design industry as we saw the rise of service design premised on aggressive marketing of business school "design thinking" offering universally appropriate toolkits that could provide "smart solutions" for wicked problems. Such approaches proved attractive to a whole range of governments from the 1990s onward, involved in withdrawing social support and downsizing public agencies.

The orientation of a great deal of service design and related fields of design for social innovation as Tony Blair's former advisor Geoff Mulgan has observed (2014) may have set out to improve user experience but also to improve productivity of welfare arrangements. But the attempts of Conservative and Progressive neoliberal/Third Way governments during austerity times to "empower publics" via encouraging them to participate in designing alternative welfare arrangements that would get citizens to "help themselves" frequently ended up with design-consultancy-participation ventures becoming a tool for austerity. Mulgan—one of the key champions of the possibilities of unlocking the insights of design for social innovation for government in the 1990s and 2000s—has more recently, and more soberly suggested that the design consultant meets public policy boom of the early 2000s did not produce the results that it claimed possible. Rather a frequent result of this moment produced designers as entrepreneurial neoliberal subjects "parachuting into problems they don't understand" to engage with "people and circumstances they know little about," and then they "disappear when the money goes" (Mulgan, 2014: x). More recently, economists Mariana Mazzucato and Rosie Collington (2023) have added substance to these observations suggesting that after nearly three decades of a consultancy boom in the public sector in the UK there is a good deal of evidence that suggests it has produced questionable results and often eroded public sector accountability and institutional memory. In the United States, Noah Smith (2025) has suggested that the growth of outsourcing public services to decentralized for profit and non-profit organizations over the last four decades—supported by Republicans and Democrats—has similarly eroded state capacity. The substance of these arguments suggests that a rebuilding of public sector capacity is going to be critical for implementing public service design for just decarbonizations.

Most critical and transition designers would of course be deeply critical of the instrumentalization and neo-liberalization of service design and design for social innovation that has taken place over the past decade. Nevertheless, it is a literature drawn to asserting the capacities of community-based movements and design transformations that might emerge from everyday life as the primary forces that will drive transition. Yet the often porous and sometimes problematic relationships that can develop between "community movements" and NGOs that profess to speak for communities but are not elected, often have quite opaque governance structures and often uneven efficacy is rarely scrutinized. More generally, what is often missing from this literature—which tends toward a general skepticism of all state and bureaucratic actors—is an account of the institutional and administrative support systems that could support, amplify, scale, and augment any kind of durable vision of design for redirective practice, transition design, or design when everyone designs.

The Ongoing Challenges of Designs for Cocreation and the Scalar Politics of the Just Transition

> the architectural and design disciplines could significantly enhance their capacity to make durable, progressive urban interventions by engaging more systematically

> with questions of institutional (re)design—that is, with the systems of collectively binding rules that govern the production, use, occupation and appropriation of space.
>
> Neil Brenner 2016, 135

In this chapter, we have seen that over the two decades and more, critical design studies has generated an exciting array of texts, manifestos, programs, and demands arguing that the whole field of design needs to confront the reality of the social ecological crisis, the crisis of democracy, and the legacies of colonialism and pivot accordingly. We have highlighted some of the gains that have followed from these moves. The great insight of much critical design studies is its ongoing insistence that both credentialed and non-credentialed designers offer a critical repository of skills and insights, schemes and possibilities for how we could live differently and that bringing these energies into alignment via codesign, redirective practice, or "design, when everyone designs" will be essential for building sustainable futures (Manzini). We have also seen that critical design currents have attempted to seriously grapple with the climate crisis (Fry, Tonkinwise) and the question of how to think with and beyond Eurocentric thinking (Escobar). However, we have also highlighted some of the challenges faced by the field. Notably, a great deal of critical design discussions emerge in the 2000s at a time when (as we have seen in Chapter 3) design, architecture, and planning have been institutionally disaggregated and the state is in retreat. Critical design studies then gives rise to a series of interventions that are both philosophically ambitious and make ambitious ontological claims for the world-changing force of design. But this account of design in practice is somewhat isolated from planning and architecture, heavily located in service design, and championing prefigurative projects that are often difficult to scale and difficult to maintain without the support of public institutions. We have seen that post-class and post-labor modes of critical design can have great difficulty thinking about the workplace, housing, and political economic or labor questions and are often premised on class-free accounts of design agencies. Much like the often-utopian transition politics of degrowth (see Alexander and Gleeson, 2019), such positions can often default into a post-materialist politics that treats the agencies and aspirations of middle-class environmentalists, bourgeois bohemians, and upper-class downshifters as a universal norm. Further issues have emerged from the ways in which the language of codesign over the past two decades has often ended up being co-opted and instrumentalized by both an "NGO industrial complex" that has become part of a "participation industry" employed to do "public consultation" as window dressing and progressive neoliberal third-way governments who have used critical design tools to implement neoliberal downsizing of the welfare state or undercut state capacity in the Global South.

Unsurprisingly, then, we have observed that critical design has its challenges. It's hard to do well particularly in the context of market fundamentalism and oligarchic rule. It is also important to acknowledge that practitioners, advocates, and activists involved in the very broad fields of critical design, codesign, and participatory design have hardly been oblivious to such issues. Indeed, many critiques of over-idealistic versions of critical design have come from *design practitioners* working in the field, and we can see growing attempts to address these limitations in these contexts.

For example, at the very beginning of the participatory and community planning movement, Sherry Arnstein in her seminal article on the "ladder of citizen participation" argued there was a very big difference between sham participation and real citizens' empowerment, with the many graduations in between (Arnstein, 1969). It is now widely recognized among many community designers and planners that calls for participation and codesign can be "ritualistic," "manipulative," and a "managerial exercise" (Cooke and Kothari, 2001; Arboleda, 2020; Muzaffar, 2024) and that the language of "community" can be asked to do a lot of work in a context when the communal reality might be marked by sharp ideological and political divisions.

The proposition that communities or solidarity is *made* not *given* has become a truism of much labor and community organizing which is in many spaces a politics that has become more intertwined and overlapping (see McAlevey, 2016). Erik Olin Wright has long argued that real empowered participation in the workplace *and* the community ultimately requires bringing power asymmetries out into the open and designing structural conditions and compensatory mechanisms such as providing citizens incomes that pay people for their time or a functioning welfare state that provides the kinds of childcare and eldercare and job security that would make participation meaningful and viable (Wright, 2010).

If we widen our lens to consider the very broad array of ongoing actors and forces practicing critical, participatory codesign today, a picture emerges of a dynamic field where discussions of labor, production, class, and work have returned to the field, and acknowledgment of the tensions between power, participation, and co-optation, accountability, and scales of impact are rich and ongoing. Our extractive digital and platform economies may often seem overwhelming but *design justice activists* continue to work with tech workers, labor, and community organizers to push back against "enshittification" and explore how design-enabled platform cooperatives and alternative ownership models might help us develop models for a very different vision of a future digital commons (Scholtz, 2016; Costanza-Chock, 2020). Manzini's recent interventions have acknowledged the "social disaster" of the neoliberal platform economy and the need to correct this by bringing the public sector into the platform economy (Manzini, 2019: 89). Service designers such as Lara Penin and Rashid Owoyele have turned their attention to the labor gap in their field by arguing for a *worker-centered service design* that encompasses demands for codesign of the workplace through workers' representation, unionization, and industrial democracy (Penin and Owoyele, 2026).

In terms of expanding the skills, expertise, and institutional gaps that can bedevil codesign projects, discussions in *production studies* in architecture over recent years have usefully highlighted the vital role that *intermediate institutions* such as technical advisory centers from the Community Design Build Movements in the UK to the Community Design Centers in the United States to Brazilian organizations such as USINA and Cooperation Jackson in Jackson Mississippi can play in providing an ecosystem of support to scaffold some of the most successful examples of community and labor-centered codesign (see Usina-CTAH/Arantes, Pedro, 2025; Akuno and Nangwaya, 2017). Such centers and community movements have often functioned at different points in the history of codesign to bring together credentialed designers and architects with diffuse designers and builders to provide the kinds of technical, legal, engineering, and institutional support for collective design upskilling. The Brazilian architectural collective USINA Centro de Trabalhos para o Ambiente Habitado ("Work Center of the Inhabited Environment") is interesting here for the ways in which it has developed a relatively successful model for offering technical, engineering, architectural, and financial support to workers engaged in the planning, building, and design of their communities since 1990s. Funded by public money but informed by a philosophy of self-management (autogestão), USINA has provided popular education on public financing, land reform, and architecture, offered planning advice, helped cultivate cooperatives, and contributed to the conception and execution of over 5,000 houses in Brazil for over three decades (see Usina-CTAH/Arantes, Pedro, 2025).

As work by critical design collectives and practices such as Muf Architecture and Assemble in London, the New European Bauhaus in Berlin, Atelier d'Architecture Autogérée in Paris, MASS Design in Boston, or Hector Design in New York further demonstrates, place-based community-focused design work in urban areas remains important. As long-time practitioners of urban and place-based codesign such as Damon Rich and Jae Shin of Hector Design outline, doing good participatory design in urban contexts in the United States is often a slow grind, which involves a great deal of careful coalition building with fractured communities, trust building with groups that often have other priorities, and dealing with chaotic public hearings or community meetings, as well as a great deal of further mundane intricate labor from balancing budgets, building permits, zoning codes engaging with questions of land and property ownership to negotiating with city governments, activists, local businesses, and local politicians to achieve a desired end (Shin and Rich, 2019). Much of this valuable and much-needed codesign today moves forward in the face of direct resistance from commercial forces, oligarchic elites, the "Real Estate-state" (Stein 2019), their lawyers, representatives, and heavies. Nevertheless, public-spirited urban codesign at its best as Damon Rich suggests (2022:276) is an attempt to work with tenants' movements, community groups, housing advocates, public officials, and sympathetic public agencies and institutions to think about how urban democracy could extend through the design of the physical world; it is an invitation for citizens to become masters and not servants of their locality. As the radical planner Sam Stein has argued community design and codesign, when approached in critical and thoughtful ways, can offer useful tools to help bolster and support the kinds of coalition politics that allow urban social movements to assert the demand that "the city must belong to those who build it, not those who buy it" (Stein, 2019:170). Here, we might read such codesign projects as functioning as Deweyian "democratic experimentalism," exercises in testing the possibilities of public agency as well as functioning as what the architectural critique and urban designer Colin Ward referred to as "seeds beneath the snow" (Ward in Wilbert and White, 2011). The "seeds" act as pilot projects that can be supported and multiplied at scale when a more promising moment arises.

It can also be argued that well-designed codesign projects can operate as an important means to engage communities that have been poorly treated by the technocratic state, if not actively subjected to subordination and elimination. Codesign here continues to offer tools for advocates of disability rights and migrant laborers, for enacting indigenous conservation and mutual aid networks, and for bringing into view the invisible labor, design expertise, and worldmaking capacities of those who have been marginalized, subordinated, dehumanized, and dismissed.

However, tensions remain in conceptualizing a scalar politics of design that tackles the climate crisis, ecological collapse, and just transitions. Notably, the tools, skills, and design knowledges that are optimal for facilitating place-based participatory approaches—such as codesigning parks, museums, community centers, neighborhood regeneration plans, cooperatives, and worker platforms—differ from the long-range planning, policy, engineering, and design skills needed to decommission fossil fuels and build decarbonized infrastructure for mitigation and adaptation on a warmer planet. Building out clean energy sources such as solar, wind, wave, battery storage, and nuclear power, constructing regional and interregional transmission that moves energy from where it is produced to where it needs to be, achieving the mass provision of affordable and secure low-carbon housing as well as adaptive reuse and retrofits of urban form, constructing high speed rail, electric vehicle charging networks, and mass decarbonized public transportation, and restructuring industrial processes so that they are circular and maximize decarbonization will require planning, design, and industrial policy that can work across multiple jurisdictions, manage a great deal of technical and engineering complexity at speed and scale, and bring in projects on time and on budget. This will require public planning institutions that can not only do regional and national planning effectively but can work through the trade-offs within and between political constituencies and economic sectors and mediate disputes within and between communities. What are some of the resources we can draw from the in history of design and planning to inform this work? These are some of the issues that we will explore in the next chapter.

6

The Green New Deal, Design, Planning, and the Public Imagination

Discussions of the Green New Deal and the Biden era of climate policy that followed it initiated probably the most extended public discussion in the United States around how the state, public institutions, and public policy might be used to systematically redesign political economy since the height of environmental consciousness was reached over five decades ago in the mid-1970s. In contrast to that earlier 1970s period of the environmental movement, which was primarily concerned with pollution and damage to specific ecosystems, this new moment was much more focused on global climate change and decarbonization. In particular, and for architecture, design, and planning, it was a moment that marked a significant break from the post-1960s community-libertarian "small is beautiful" traditions of ecological design that we outlined in Chapter 2 and the turn to market-driven visions of "natural capitalism" and green neoliberalism that became so influential in climate policy design and architecture from the 1990s onward (outlined in Chapter 3). Rather, in foregrounding climate change and arguing that this should be understood as an "existential crisis" which required society-wide the Second World War-style mobilization, it injected the question of speed and scale into environmental discussions. In drawing inspiration from the original New Deal of the 1930s, the Green New Deal questioned the role of markets and the private sector as the primary drivers of decarbonization, foregrounding instead the "entrepreneurial state" (Mazzucato, 2015, 2021) and the role of government and public institutions as key enablers of decarbonization. In design terms, it could be seen as a return to championing the virtues of large-scale design, building, and planning for the public interest and more generally advocating for the merits of public works, public architecture, social housing, and even regional planning, conversations that had been largely dormant in the United States for half a century and more. Finally, with its triple focus on decarbonization, jobs, and justice, it sought to popularize and advance a concrete program that might make common cause between climate activists, labor unions, indigenous struggles, advocates for social justice and racial justice, and champions of green manufacturing explicitly using the just transition as a guide for action.

In this chapter, we grapple with the various iterations of the Green New Deal and Biden-era climate policy. We dig backward to explore the design legacies of its inspirations in the original New Deal of 1933–9, and we outline how the ambitions of this era triggered a series of design-inflected Green New Deal conversations and debates around energy and public power, planning and public spaces, green social housing and the future of urban form. We consider how these discussions were very unevenly translated into the "government-led, private sector-enabled" climate and industrial strategy of the Biden Administration and how this moment was then brought to an abrupt halt in the United States with the reelection of Donald Trump to the White House in 2024.

We argue in this chapter that if we merely read this moment in a presentist fashion absorbed by the immediacy of US politics and its contemporary dramas, the discussion ends up as a foreclosing story. However, if we acknowledge that climate change presents a series of social, ecological, geopolitical, and design challenges that are going to be with us for much of the rest of this century and will play out in very different ways in different geographies, polities, and societies across the globe, thinking about what an agenda for a public-spirited and climate-informed design, planning, and architecture in the public interest might look like in the years ahead retains its importance, and the Green New Deal conversation may have something to offer those fighting for a just transition.

The Many Lives of the Green New Deal

As we outlined in Chapter 1, struggles for just transitions have long been intertwined with conversations around the need for a programmatic agenda centered around climate action, jobs, and equity. However, the specific formulation of a "Green New Deal," with its explicit referencing of the Roosevelt-era New Deal, has its own lineage and has advanced through several iterations. The first substantial iteration of this discussion in 2007–8 fleshed out a range of potential pathways that a GND could take. The *New York Times* columnist Thomas Friedman outlined in a 2007 article (Friedman, 2007) and then in his 2008 book (Friedman, 2008) a vision of an "energy New Deal." Writing that "The New Deal was not built on a magic bullet," Friedman suggested what could be learnt from that 1930s precedent was the need for "a broad range of programs and industrial projects to revitalize America." The GND, framed as a national (and indeed nationalist) project for rejuvenation, could be accomplished according to Friedman by the United States embracing an "all of the above" energy mix: "If we are to turn the tide on climate change and end our oil addiction, we need more of everything: solar, wind, hydro, ethanol, biodiesel, clean coal and nuclear power"—an enormous industrial project. In addition, Friedman argued for a mixture of government regulation (such as enforcing efficiency standards for buildings and appliances), encouraging energy conservation (by working with utilities to get consumers to lower energy use), and price reform (such as forcing congressional action on a carbon tax or "cap and trade" scheme (Friedman, 2007)).

Friedman's vision of the GND circa 2007/8 essentially sought to target the moderate wing of the Democratic Party. Describing his version of the Green New Deal as an attempt to make "Green" geostrategic, geoeconomic, capitalistic, and patriotic, his proposal combined faith in the innovative power of markets while nevertheless cautiously seeking to nudge Democrats toward affirming the virtues of a technocratic ecomodernism. All of this was firmly wrapped up by a vision of threatened US dominance and ultimately presented as a project to enhance "American economic power, American moral power, American geopolitical power."

A more internationalist Green New Deal was proposed by the European Green New Deal Group and the New Economics Foundation (Simms et al., 2008), focused on the financial crisis, climate change, and peak oil in 2008. This intervention argued for a Green New Deal adequate for addressing this "triple crunch" required a series of Green Keynesian/social democratic policies that focused on green job creation through massive government investment in renewables, energy efficiency, micropower generation, and the building of low-carbon infrastructure. It was argued that this should be funded by progressive taxation of the rich, international regulation of finance, capital controls, a clampdown on tax avoidance, a windfall tax on fossil fuel companies, and breaking up large banks to facilitate the development of green banking.

If both proposals had a moment in the sun, with some elements of this discussion finding their way into investment in green jobs and renewable energy as part of national and EU-level recovery packages, fossil fuel interests coupled with US congressional gridlock largely derailed this agenda. Obama-era climate policy moved forward much more incrementally through executive actions and the Clean Power Plan, which did manage to support some renewable energy expansion through tax incentives and large leases of public lands, resulting in a four-fold expansion of wind energy and a 40-fold expansion of solar between 2008 and 2016 (Simon and Hayes, 2017). As the financial crisis abated, talk of Green New Deals largely disappeared as most G20 polities returned to business as usual, and after the 2016 election the administration of Donald Trump set about weakening those clean energy incentives that did exist and propping up fossil fuel energy.

The second wave of the GND emerged in the wake of the 2018 midterm elections in the United States. In many respects this moment was politically foregrounded by a surge of energy on the progressive Left carried over from democratic socialist Bernie Sanders's attempt to obtain the Democratic Party nomination in the run-up to the 2016 Presidential election, the defeat of Hillary Clinton, and the collective recoil from Trump's first victory in 2016. The rise of youth-led climate activism represented by the emergence of the Sunrise Movement and a renewed surge of indigenous environmental activism, which perhaps most dramatically came to public consciousness during the Standing Rock protests against the Dakota Access Pipeline in 2016, helped pave the way for new openings and new coalitions. With the 2018 midterms marked by a wave of opposition to the climate rollbacks of the first two years of the Trump administration, the balance of energy, confidence, and fight in the Democratic party briefly shifted from moderates to a range of currents that took the view that climate action paired with labor and social justice priorities provided the strongest position to mobilize against Trump. The 2018 midterms swept a wave of new progressive politicians into Congress, including Alexandria Ocasio-Cortez who at twenty-nine became the youngest woman to be elected to the House of Representatives and gave the Democratic Party back majority control of the House. The Sunrise movement chose this moment to stage a sit-in protest in the office of Representative Nancy Pelosi, who was about to retake the helm of the chamber as incoming Speaker of the House. The protesters called for Pelosi to support a Green New Deal, a demand further amplified when the freshly elected Ocasio-Cortez broke with decorum and stopped by the sit-in to speak with the protesters and lend her support to their cause. The ambitions of the 2018 version of the Green New Deal found perhaps their clearest legislative expression in House Resolution 109, a nonbinding resolution that Ocasio-Cortez introduced in the US House of Representatives upon taking her seat in Congress; the resolution was quickly seconded in the Senate by veteran Massachusetts Senator Ed Markey.

H.R.109—“Recognizing the Duty of the Federal Government to Create a Green New Deal”

Working with a title that recognized “the duty of the Federal Government to create a Green New Deal,” H.R. 109 articulated the GND’s core objectives: “to achieve net-zero greenhouse gas emission through a fair and just transition,” “to create millions of good, high-wage jobs,” and “to promote justice and equity” by stopping and repairing oppression against “frontline and vulnerable communities.” H.R. 109 offered a laundry list of possible goals and strategies of how this could proceed. These included a national target of 100 percent “clean, renewable, and zero-emission” energy generation; creation of a national “smart” grid; aggressive building upgrades for energy efficiency; decarbonization of the manufacturing, agriculture, and transportation sectors; increased investment in carbon capture technologies; the establishment of the United States as a global exporter of green technology—as well as other progressive priorities, such as a universal employment guarantee ensuring jobs with decent wages, paid vacations and adequate leave, retirement security for all people in the United States, and affordable housing and high-quality health care for all.

While H.R. 109 never moved forward in Congress, it did catapult the Green New Deal into the public conversation in the United States and elsewhere—and alongside several follow-up GND bills that targeted specific sectors, such as the Green New Deal for Public Housing Act (S.1218) and Green New Deal for Public Schools Act (H.R.4442), it arguably represented the first US policy platform to propose anywhere close to the scale of economic transition that would be required to approach the decarbonization targets agreed to in the 2015 Paris Agreement.

The arrival of this second-wave GND upended several orthodoxies that had previously dominated the discourse over climate policy in the United States. It significantly accelerated the timeline of proposed climate action from the 2050 targets that had previously been the dominant formulation, emphasizing a ten-year time frame for a “national, social, industrial, and economic mobilization,” in line with the urgency in IPCC’s SR15 (Special Report on Global Warming of 1.5°C), which had just come out a month before the 2018 election. Correspondingly, H.R. 109 demanded the new mobilization take place “on a scale not seen since World War II and the New Deal era.” Green New Dealers argued that only the public sector had the ability to deploy the kind and scale of investments needed to rapidly decarbonize the US economy—by using the procurement power of government to act as a first mover and buyer for emerging transition technologies, acting to create a market for them and jumpstart their mass production, and taking the bulk of the risk when investing in the kinds of promising but not-yet proven emerging technologies that typically spook private investors (Hanauer and Farrell, 2019). Influenced by the environmental justice and climate justice movements, it also sought to fuse the third leg of "justice" to the more established climate and labor provisions that had dominated "blue-green" coalitions and Friedmanesque versions of previous GND framings. The focus on the "millions" of good green jobs that stood to be created was a reframing that shifted the discourse away from the daunting financial cost of attempting to carry out a national energy transition, toward the equally large economic opportunity to be had. But beyond the potential employment numbers alone, GND advocates believed that direct public investment and its Keynesian repercussions could be steered to provide additional economic opportunities to workers in frontline and vulnerable communities and address the unequal vulnerability of communities disproportionately affected by both climate harm and the energy transition.

Perhaps the most notable change that the GND brought to the climate and energy conversation was a shift away from austerity-orientated climate arguments, a focus on market mechanisms such as carbon taxes or cap and trade schemes. Rather, the central frame became a focus on the need for direct public investment in larger systemic changes that could drive decarbonization more quickly than the private sector would, a focus on the need to put people to work on a shared collective project of addressing climate change, and an invitation, in the words of Ocasio-Cortez, to "rediscover the power of public imagination." In a climate and energy discourse long dominated by technical details and policy minutiae, the Green New Deal, in contrast, purposely sought to generate an aspirational discourse, and here there was an opening for design. Moreover, and as Billy Fleming has observed, Ocasio-Cortez spoke the language of the architecture and design professions when she described H.R. 109 as a "Request for Proposals" in search of projects (Fleming, 2019). Implicitly, the GND was an invitation to reimagine the design of all sorts of public infrastructure that would have an effect on emissions, workers, and equity. But why anchor this project in the Rooseveltian New Deal, and what exactly does this moment do to enlighten our attempt to think about how contemporary design, architecture and planning could inform a just transition?

Design, Architecture, Planning, and FDR's New Deal

The original New Deal of 1933–9 is mostly explored as a political economic project focused on pulling the United States out of the Great Depression of the 1930s, through its major investments in public works, direct hiring of unemployed workers for their construction, and government-led reforms of industry, agriculture, energy, finance, and labor relations. It was not a single bill or piece of legislation, but rather encompassed a whole array of agencies and programs, jokingly referred to in the moment as Roosevelt's "alphabet soup" due to their many three-letter acronyms, which ranged from the Agricultural Adjustment Administration (AAA) and Farm Security Administration (FSA), which both sought to financially rescue farmers, to the Public

Figure 13 Fontana Dam North Carolina under construction, March 1944. Photo: Billy Glenn, TVA. (public domain).

Works Administration (PWA) and Works Progress Administration (WPA), which constructed large-scale infrastructure and local public works projects, respectively, to the Civilian Conservation Corps (CCC) that made myriad improvements in national forests and state parks and rural lands—over twenty agencies, administrations, corporations, and programs in all—along with dozens of Acts that reorganized various parts of the federal government and changed existing laws and policies, all created by Congress or executive order, and all captured under the New Deal moniker. It was precisely this multifaceted nature that resonated with Green New Dealers from Friedman to Ocasio-Cortez when they looked for historical precedents for a new national policy platform that could match the scale of the needed energy transition. Yet the New Deal can also be read as an exercise in political coalition-building, political communication, and design-politics. The New Deal generated a constant stream of posters, pamphlets, photos, and newsreels alongside Roosevelt's speeches and ribbon cuttings to reinforce the local and national benefits being conferred by New Deal programs. Workers and projects were frequently chosen on the basis of electoral geography to build and maintain support for Roosevelt's agenda. The New Deal was also a political moment that articulated a vision of what design, architecture, and planning "for the public good" could look like, and it offered a way in which it might be implemented and moved forward.

The New Deal's support for design extended across the spectrum of design fields, from architecture to landscape architecture to graphic design to planning. It ranged from the design of the iconic bridges and dams of the PWA, to the thousands of courthouses, libraries, municipal parks, public pools, public schools, and street improvement projects of the WPA, which employed thousands of architects and landscape architects to design all these projects, alongside the millions of workers hired to build them.

Planners and architects were employed by the Resettlement Administration (RA), which drew up plans for New Greenbelt Towns that could absorb the mass migration of poor farmers forced off their land by the Dust Bowl (though only a few of these New Greenbelt Towns were ultimately built). The WPA's Federal Project Number One extended a lifeline to graphic designers by commissioning poster designs promoting everything from national park visits to the ongoing work of federal and local public agencies to all kinds of public service announcements. Its Federal Art Project also had a design unit, which employed industrial designers who worked on furniture, lighting, murals, and public sculpture informed by the mission that industrial design could improve the quality of life of ordinary working Americans.

The CCC, meanwhile, hired and trained millions of unemployed young men to work outdoors, collecting seeds, establishing tree nurseries, planting seedlings, improving forests, and doing erosion control, transforming vast swaths of forest and agricultural landscapes, which left an indelible legacy on the country's natural and recreational spaces.

Even functional engineering works and energy infrastructure built in this period were imbued with a strong design sensibility. The Tennessee Valley Authority (TVA) and the Bonneville Power Administration (BPA), for example, employed architects, planners, and engineers to redesign entire watersheds and river valleys, and they lavished architectural expression on the design of dams' powerhouses, turbine halls, visitors' centers, and dams. Architectural details on the dams heightened their sense of scale and their gleaming modernity, and carefully crafted signage proudly proclaimed in bold all-caps typeface that this facility was "BUILT FOR THE PEOPLE OF THE UNITED STATES" (Wolff 2007; Creese 1990).

What we might call use of "the political power of persuasion through design" was often at work in the various built works of the New Deal, incorporating design at multiple scales to build a case for its new infrastructure and promoting a modernist optimism that acted as an advertisement for the New Deal itself; this strategy was especially clear in the case of the TVA. Writing about the agency's architectural agenda amid the Great Depression and the general poverty of the Tennessee Valley region, architectural historian Walter Creese has argued that the "TVA shot up, shone, and spread like a fireworks display against a dark and unsettled firmament. Envisioned as a noble gesture, TVA spread its architecture, site planning, and reorganization of the land as a total public reaffirmation at a time when all other, lesser signs of economic well-being were shrinking and disappearing" (Creese, 1990: 1). The TVA also went beyond just hydroelectricity by considering dams for navigation, flood control, industrial production, and rural electricity delivery within a holistic system of infrastructure and urbanization that included roads, worker settlements, and even new towns and land management, expanding the scope of design and planning across an entire watershed. In short, it engaged in master planning at the regional scale.

The New Deal grew out of the Progressive Movement's agenda of "rational planning for the public good," promoted by planners and foresters like Benton MacKaye and

Figure 14(a, b) New Deal dams exhibiting their cutting-edge modernist designs, from the architecture of the powerhouse to the exterior detailing. Pickwick Dam Powerhouse (left); Hiwassee Dam (right). Photos: Charles Krutch, TVA (CC by 2.0 by Tennessee Valley Authority).

Gifford Pinchot in the several decades leading up to the New Deal. Technocratic in nature, this Progressive agenda fit well with the generally apolitical orientation of many design professionals, who were happy to work for the New Deal's vast public bureaucracy and lend their expertise to the execution of these public projects. The American Society of Landscape Architects (ASLA) put together an organizational chart in 1935 of all the various federal agencies with New Deal funding that might conceivably be used to hire its members to do professional work (Cutler, 1985).

With the wealth of work opportunities for designers on New Deal public projects, many joined the ranks of federal agencies or as public employees for municipalities that suddenly had the funding to support the design of public works. This optimism of designing in the public interest, often as public employees, is one of the legacies of the New Deal era that was particularly salient to designers, architects, and planners in the GND moment, who were interested in pushing back against decades of privatization and neoliberal austerity, with the Progressive era–inspired public works of the original New Deal offering a counter to private real estate as the primary or even only mode by which to practice design.

Some Problematic and More Promising Design Legacies of the New Deal

While the New Deal's vast array of built projects cemented a clear political identity and a clear argument for the power of government to transform an economy and society through public projects, it has to be recognized that the impact of some of its programs on vulnerable communities and ecosystems was not without issues.

On the ecological front, the New Deal's speed and scale made the literal and metaphorical bulldozing of locally valued places and ecosystems difficult to avoid. While the CCC was one of the most popular of the New Deal programs, and while its feats of revegetation introduced the public to a conservation ethos of wisely stewarding natural resources, from an ecological perspective, its record was more mixed. The CCC's work on reforestation, fire protection, and forest thinning initially drew widespread support from prominent conservationists, but it soon began to be seen as operating counter to emerging understandings of ecological land management. The forester and environmentalist Aldo Leopold, while initially a CCC booster, later criticized the Corps for planting monocultures and for eliminating rare and threatened habitats in its zeal to replant the forest. The Audubon Society and National Wildlife Federation railed against the use of poison for "predator control" and the draining of wetlands for mosquito control. Advocates for wilderness protection, like wilderness and labor activist Bob Marshall, who originally worked for the CCC, turned against the agency's plans for proposed skyline roads along the ridgeline of Great Smoky Mountains National Park, whose forests the CCC had recently helped revegetate. Wilderness conservation organizations fought against proposed CCC roads in the Sierra Nevada mountains in California, the Green Mountains in Vermont, and the White Mountains in New Hampshire, effectively halting these roadway projects and preserving the character of these mountain forests.

Responding to such criticism from conservation groups and wilderness societies, the New Deal agencies did eventually embrace a new kind of "ecological planning" that led to the TVA's approach of trying to balance land use decisions so as to benefit both forest restoration and local economies—a balance of "natural and human development." FDR's conservation advisor (and REA director) Morris Llewellyn Cooke began urging New Dealers to "plan the whole job" and "get the whole thing mapped out" so as to avoid missed opportunities between different agencies and different CCC efforts—arguing for a kind of proto-regional planning (Maher, 2008: 203). Mindful of the growing tensions between the various factions in the New Deal coalition, Cooke urged groups otherwise politically pitted against each other—farmers and stockmen, nature lovers concerned with wilderness protection and sportsmen concerned with game habitat—to team up under a unified banner of "total conservation" (Maher, 2008: 204).

In terms of its social effects, the New Deal landed unevenly across the United States. On the matter of race, the Roosevelt Administration often deferred to the "traditional differentials," (Katznelson, 2013: 242) as FDR put it, of the nation's various regions. In practice, this meant accommodating white supremacy in the South at the same time that it was empowering progressive public works–minded "sewer socialists" in the Midwest. In the October 1934 issue of the NAACP's official journal, *The Crisis*, under the editorship of W.E.B. Du Bois, the TVA was slammed as a form of "Lily-White Reconstruction," which excluded black workers from equal pay and worker housing and Black residents from the benefits of the new cheap electricity (Houston and Davis, 1934: 290). The CCC, while race-neutral on paper, in practice segregated its camps and assigned lower-status work to black enrollees, keeping them out of the public spotlight. Nationally, New Deal–era policies of redlining via the Homeowner's Loan Corporation of America targeted minority neighborhoods for disinvestment while showering public largess on white suburban neighborhoods. Such policies ultimately helped undermine Black homeownership and laid the groundwork for the ongoing racial wealth gap in the United States.

The New Deal's impact on indigenous Tribal nations was also uneven. The 1934 Indian Reorganization Act halted the devastating fragmentation and loss of Tribal reservation lands by repealing the allotment policy, reversed a ban on indigenous spiritual practices, and created some economic benefits for indigenous communities via employment on reservations through the CCC Indian Division. But at the same time, other New Deal projects limited access to traditional indigenous landscapes, including treaty-protected hunting and fishing grounds. This damage was most pronounced with the flooding and destruction of vast tracts of Tribal reservation lands as a result of dam construction, as indigenous scholar Nick Estes has chronicled (Estes, 2019), including the Bonneville, John Day, and Dalles dams on the Columbia River that "flooded Nez Perce, Umatilla, Yakama, and Warm Springs lands" and the Fort Peck Dam on the Missouri River that "removed 350 Dakota, Nakota, and Assiniboine families from the Fort Peck Indian Reservation" in the 1930s. These New Deal–era dams inaugurated a process that would continue after the New Deal proper well into the 1960s, when the Kinzua Dam in New York flooded out over 600 Seneca (The Red Nation, 2021: 10).

The Green New Deal's New Public Landscapes

As we have seen, the original New Deal cannot be romanticized—it comes with significant limitations and flaws. Nevertheless, for all these limits, the New Deal was a transformative era for federal employment of design professionals, fundamentally reshaping the infrastructure of the United States while creating unprecedented public-sector opportunities for architects, planners, and designers. It was a moment when the energy system was radically expanded and transformed in the United States. It was a period of expanded public support for environmental conservation done in the public benefit, with the rational management of public lands and landscapes, almost obtaining the status of a "civic calling" in significant sectors of the public. It also presents probably the singular example of what a program of public architecture and design and planning focused around public works, conservation, the build out of ambitious energy infrastructure, and design for the public interest might look like in the face of an ecological crisis (the Dust Bowl) and an economic crisis (the Great Depression). Hence its appeal to designers and activities of a very different era.

Between 2018 and 2022, discussions of the Green New Deal generated a wave of essays, editorials, books, reports, and policy memos attempting to flesh out what kinds of transformations, in which sectors, could be imagined under the Green New Deal's banner of "decarbonization, jobs, and justice," with many of these interventions nodding toward the importance of reviving design, planning, and architecture as a key part of the project. In terms of the climate-labor agenda, the proposition was fairly straightforward: there were the jobs that would be created for solar and wind workers, battery plant workers, and powerline workers involved in building, upgrading, and operating the new low-carbon grid. But there were also, perhaps less well understood in the popular imagination, all of the jobs that would be needed to carry out the shift to electrification in the building and transportation sectors: the HVAC technicians who would be installing and servicing the heat pumps that could replace gas heaters and inefficient air conditioners, and the electricians that would be needed to wire EV chargers, connect solar farms to the grid, and make upgrades to electrical panels in millions of homes and businesses to make them electrification ready. There were the energy efficiency and weatherization jobs more closely tied to the traditional construction trades: insulating walls and attics, replacing windows, and so on. Indeed, Saul Griffith and his team at Rewiring America estimated that a policy of intensive deployment of decarbonized infrastructure could produce as many as 25 million net new jobs (Griffith et al., 2020).

For design, architecture, and planning, even considering what the employment implications of the production and construction of their designs might be represented an expansion of interest, reaching much further down the supply chain than most designers were used to thinking about, and as such was a useful exercise in building design-construction labor solidarity. *The Green New Deal Working Group of the Architecture Lobby* offered some of the most interesting sector-specific elaborations of how the building sector more generally could be moved toward supporting a just transition.

A Just Transition for the Building Sector: Toward a Radical Transformation of Housing and the Built Environment (Anklesaria et al., 2021) is grounded in two framing propositions: (i) that design workers—frequently exploited and undervalued—are also fossil workers and (ii) that design workers find themselves embedded in a broader carbon-intensive and exploitative design and construction economy which requires structural change to achieve a just transition. The document goes on to argue that such changes will need to be supported by legislative action and the establishment of new building codes, but it also makes the argument for new incentive structures to reshape the future direction of the industry. To that end, *A Just Transition for the Building Sector*, among other things, argues for new metrics to move the building industry beyond energy efficiency to consider "how the spatial configuration of entire built landscapes shapes energy demand." It is argued that this entails increasing density, tracking whole building life cycle carbon and establishing assessment metrics that address "full site, material, and labor landscapes." The document also argues for legislation that enforces transparent sourcing of materials, addresses labor rights, and directly addresses the social and ecological impacts of mining and the use of forced labor in the global building and construction industries. It suggests the virtues of a federally funded design sector that would direct design activities toward public decarbonization projects. It also calls for new forms of ownership that "can resist the commodification of housing" and fair labor clauses that can protect workers in the building and construction sector. Further steps require engaging in the very complicated work of *building coalitions* between different kinds of design labor (from architects and designers to construction workers in the building trades), which can develop common agendas to move forward.

If such interventions by the Green New Deal Groups of the Architecture Lobby helped stimulate a range of discussions in the United States and beyond about how urban designers, architecture, and the building industry could coalesce around labor and climate-friendly projects, in terms of landscape and rural labor issues, the agriculture and forestry side of the GND also opened up space for new conversations between landscape designers, critical environmental and social scientists, and public policy advocates (Fleming, 2019; Patel and Goodman, 2019; Hernandez et al., 2019; Mulvaney, 2020; Szumilas-Kowalzyk and Pevzner, 2019; Pevzner et al., 2021). GND advocates exploring this terrain argued here that both rural employment as well as the climate could benefit from creating jobs in regenerative farming, sustainable forest management, hazardous site cleanup, and ecological restoration, and a host of activities tied to managing carbon in the landscape. In terms of agriculture, H.R. 109 had emphasized the importance of "working collaboratively with farmers and ranchers" to reduce their greenhouse gas emissions as well as foregrounding the need to build "a more sustainable food system" that increased soil health. This suggested a potential range of strategies. It could entail helping reduce the methane produced by cattle herds and transitioning farms away from fertilizer overuse, which results in emissions of nitrous oxide. But it might also entail increasing carbon sequestration and storage in agricultural soils—with techniques ranging from no-till farming, to the spreading of compost and biochar on fields, to combining trees with crops (agroforestry) or trees with cattle (silvopasture), or practicing any number of other "carbon farming" techniques (Toensmeier, 2016). Much like the Soil Conservation Service engaged with farmers during the New Deal to help educate, incentivize, and change farming practices to reduce erosion, Green New Dealers argued that the federal government could play a constructive role supporting these practices. The Biden administration's Department of Agriculture did indeed issue some initial grants for trials of some of these practices. But as Raj Patel and Jim Goodman observed in 2019, in order for the larger transformation of agriculture from an extractive and polluting practice to a regenerative and "sustainable" one to actually take root in farm country across the United States, there would have needed to be a coalitional realignment bringing urban workers together with rural farmers, with better pay for one group allowing for fairer pricing for the other (Patel and Goodman, 2019). Incentives in the form of direct payments to farmers and ranchers might have incentivized more regenerative agriculture, but a GND agenda also imagined that reform of the current system of monocultural crop insurance would be needed, along with increased R&D funding and increased capacity at federal soil conservation programs to help farmers make whole farm plans with soil carbon in mind (Feldman et al., 2020).

If questions of how a Green New Deal could revitalize rural economies and ecologies invited and opened one set of conversations between design, planning, and public policy, the physical and spatial implications of the energy transition imagined by the GND with its potential to generate dramatically different landscapes throughout the country invited further points of engagement—especially for landscape architects. Here too the New Deal provided precedent. The TVA's dams, employed both architectural and landscape design to heighten benefit and drama. Aside from the attention to detail and massing in the dams' architecture, many of the New Deal dams were carefully integrated into their surrounding landscape, with the reservoirs and their surrounding forests treated as public recreation areas with visitors' centers and park-like amenities. Approach roads were meticulously designed to dramatically reveal the dams in an almost cinematic sequence, such as at Norris or Fontana Dam. Such precedents informed a range of Green New Deal conversation concerning how conflicts around the land use, landscape impacts, and aesthetic of space hungry renewables might be ameliorated through public-spirited landscape design. For example, Szumilas-Kowalzyk and Pevzner argued that Green New Dealers could learn a great deal from Dutch and Danish attempts to integrate renewable energy into the landscape through the design of broader multifunctional landscapes (Szumilas-Kowalzyk and Pevzner, 2019; Pevzner et al., 2021). Drawing inspiration from Sijmons, who suggested that the design of energy landscapes could benefit from approaches that treated energy as but one more part in an already-layered cultural landscape (Sijmons, 2014), it was suggested that paying attention to the site, layout, scale, configuration, detailing, colocation potential, and possibility of interpretive public access to renewable energy projects could all help generate better social and ecological outcomes. Such interventions suggested for example that wind turbines of large or medium scale, rarely seen inside of US cities, could be much more widely deployed in industrial and port lands, as they are in Northern Europe. Agriculture–energy colocation and multifunctional arrangements such as agrivoltaics combining solar panels with crops, grazing land, or pollinator habitat could expand such multifunctionality benefits further into rural lands: rather than forcing farmers to choose between growing food or leasing land for energy. In similar fashion, multifunctional design of energy infrastructure could configure solar along sound barriers next to roadways and rail lines, integrated into shade structures and awnings, above canals and rooftops, and in myriad other arrangements.

This growing body of work that is now occurring across the fields of landscape design, environmental science and engineering (see Mulvaney, 2020; Hernandez et al.,

Figure 15 The "Solar Strand" at the University at Buffalo, by Hood Design Studio—an example of a designed multifunctional energy landscape that doubles as both an educational art installation and as campus public space. Photo courtesy of: Douglas Levere | University at Buffalo (Permission to reproduce).

2019), and architecture and planning (Brown, 2014) has suggested that the variety of potential designs that plausibly address tensions between agricultural and energy land uses could open up much more interesting public discussions about how to design and plan multifunctional low-carbon and regenerative energy landscapes that are welcomed rather than opposed by rural publics.

Alongside its climate mitigation agenda, Green New Deal proposals routinely advocated for just climate adaptation strategies as a key component of a just transition. In thinking about coherent public policies informed by effective and legible public design strategies that could support just climate adaptation, Pevzner (2019) argued that New Deal precedents of public mobilization in the face of disaster offered inspiration (such as how the WPA responded to the catastrophic floods of 1937 in Kentucky and 1938 in New England or how the CCC, along with the WPA, Soil Conservation Service, and Soil Erosion Service collectively responded to the Dust Bowl of 1934–5 through contour farming demonstration projects and massive tree planting campaigns). Following these examples, he suggested that design interventions to facilitate just adaptions should include: (i) the designing and building of coastal seawalls and coastal marsh restorations to defend against sea-level rise and storm surge, (ii) the design and construction of all manner of cooling infrastructures to provide respite against heat waves, and (iii) building levees and green infrastructure to protect against riverine flooding, or facilitate managed retreat to move communities away from areas no longer habitable because of climate threats such as flooding or fire. It was also argued that rather than simply armoring existing infrastructure, a public design agenda seeking to bolster public support for climate adaptation could pair demands for funding public

Figure 16 Designs for the Green New Deal: Imagining large-scale public mobilizations, including a new CCC-inspired Climate Corps initiative for energy workers, and AAA- and WPA-inspired climate-resilience efforts, with support from allied labor organizations like the UAW. Student work by Ruth Penberthy, produced as part of the Post-Carbon Futures seminar at the University of Pennsylvania School of Design, taught by Nicholas Pevzner. Reproduced with permission.

agencies focused on climate adaptation infrastructures with design proposals for new public facilities or public landscapes, so that cooling centers are built atop libraries or school retrofits, or the levee or coastal marsh restoration gets paired with a public park and trail (Pevzner, 2019). For architects and urban planners designing multifunctional resilience hubs, making such facilities legible in residents' minds through design and giving such hubs a central place in daily neighborhood life (as a community center as well as a resilience center, for example) could attune a public-facing civic architecture to addressing both a new set of public needs and climate demands. In addition to these proposals, Green New Deal calls to activate a Civilian *Climate* Corps running alongside a revitalization of design in the public interest suggested avenues in which the public could be directly involved in designs for participatory adaptation.

Energy Transition and the Public Power Debate

The energy transition arguably stood at the center of the Green New Deal approach to decarbonization with House Resolution 109 arguing for "clean, renewable, and zero-emission energy sources." However, H.R. 109 did not commit to any particular set of low-carbon energy technologies with the bill's sponsors. Representative Ocasio-Cortez and Sen. Ed Markey adopted a position of "technological neutrality" as they rolled out their proposal to the press. Nevertheless, the question of which path the low-carbon energy transition ought to take exposed differences in thinking about questions of scale, key technologies, and labor within the coalition of forces mobilizing around the Green New Deal. Should policy seek to reward only renewable energy technologies like wind and solar (along with batteries, which help to store these variable renewable resources)? This was the position of the "100% renewable" camp, with influential environmental NGOs such as the Sierra Club setting up campaigns like "Ready For 100." Other groups and Democratic politicians such as Cory Booker and Elizabeth Warren argued nuclear power had to be part of the package. The influential research of Jesse Jenkins and his colleagues in the Princeton Net Zero Lab recommended a pragmatic approach to these questions. Their *Net-Zero America Report* maintained that a decarbonized grid in the United States would have to draw from: (i) fuel-saving variable renewables (which includes solar PV, solar thermal, wind energy, run of river hydro, solar thermal with hydro and reservoir hydro); (ii) fast-bursting balancing resources (demand response, price curtailment, flexible demand, battery storage, and long duration storage), and (iii) some further mix of clean firm, low-carbon resources (geothermal, nuclear, gas or coal with carbon capture and storage, biomass or hydrogen) (Jenkins et al., 2018; Jenkins et al., 2021). Yet, critical social scientists suggested that such exercises in technocratic modeling did not address a range of further issues such as the labor, human rights, and environmental injustices of the renewables industry (Mulvaney, 2019; Huber, 2022), the impact on public lands, biodiversity, and indigenous lands that solar deployment can generate (Mulvaney, 2019), concerns about the exploitative financing of community solar on communities of color in the United States (Lennon, 2017, 2021, 2025), and broader worries that a neoliberal energy transition might simply repeat extractive relations between Global North and Global South (Riofrancos, 2025a).

If most Green New Deal energy advocates did generally agree upon a need for increased public ownership over energy to speed up the energy transition—"public power" for short—further questions emerged concerning whether the bulk of energy generation, regardless of technology, would be primarily centralized and held by large utilities, or decentralized and distributed throughout the grid? Many "big public power" advocates looked to the large state- and federally owned utilities in the United States, dating back to the Progressive Era push, to regulate and municipalize and regulate utilities in the first two decades of the twentieth century and the creation of large federally owned hydroelectric projects like the Wilson Dam at Muscle Shoals, Alabama (begun by the US Army Corps of Engineers to produce explosives during the First World War) and the TVA and BPA systems that soon followed. Where the private market has been slow to decarbonize, they see public ownership as holding potential for expedited action. As energy reporter Robinson Meyer has suggested, "states like New York with publicly owned power authorities are able to intervene more forcefully in their own power markets than states that lack such capability. That's because the state itself can act to build its own large-scale power plants."

On these lines, Matt Bruenig's *People's Policy Project* (Bruenig, 2019) suggested that the quickest path to building decarbonized public power in the United States would be to "green" the TVA, which remains the largest publicly owned utility in the United States, providing not-for-profit power to 10 million customers, or 5 percent of utility-scale electricity on the US grid. Bruenig argued that changing the existing federal legislation to expand the authority of the TVA to build even more infrastructure across the country could facilitate this. Advocates of "big public power" further suggested that with unionization rates of close to 60 percent (compared to around an average of 13 percent unionization in the wind and solar industries) and its history of providing cheap rural power, something like the TVA's institutional framework could counteract the worst practices of the residential solar industry.

With the TVA being one of the few large-scale public utilities in the United States that has expertise in nuclear operations and megaproject management (the TVA's current energy mix draws around 41–43 percent of its energy from nuclear, with the rest made up of natural gas (23 percent), coal (14–24 percent), hydroelectric power (9–13 percent), and renewables (3–4 percent)), Matt Huber and Fred Stafford also argued that the TVA should be a vehicle for building cheap decarbonized public nuclear power as well as public renewables (Huber 2022; Huber and Stafford, 2023). They note that nuclear plants in particular produce vast amounts of virtually zero-emission 24/7 baseload power, that there are good precedents for nuclear facilitating rapid decarbonization in France and Sweden, and that whatever problems are generated by nuclear waste, such problems pale in comparison to the broader environmental impacts of continuing fossil fuel production. A Nuclear Green New Deal, as such, which focused the TVA's public sector expertise on building conventional large nuclear plants and small modular reactors (SMRs), they argued, would lock in the support of labor unions for decarbonization and bring more well-paid and much-needed union jobs to the locality.

Historically, both nuclear energy and large hydroelectric dams have been opposed by many conservation and environmental justice NGOs, that have argued it is indigenous communities that have borne the brunt of uranium mining pollution and dam-caused dislocation. Looking for less top-down precedents than the big river dams of the New Deal, some early proponents of a Green New Deal such as Kate Aronoff pointed to the system of rural electric cooperatives set up under the New Deal's Rural Electrification Administration (REA) as a different potential source of insights. Like the TVA, the REA was designed to provide electricity for rural Americans, 90 percent of whose homes lacked electricity at the start of the 1930s, and private electricity providers had deemed rural areas not profitable enough to electrify. But unlike the TVA, the REA was not a unified regional authority but rather a decentralized program. As Aronoff explains, "the REA set up a slew of cooperatives, requiring that groups of town residents apply together for federal loans to erect power lines and associated infrastructure. After an initial burst of federal funding, the Rural Electric Cooperatives (RECs), would be entirely owned, operated, and funded by their members, setting rates so as to ensure both affordability and financial solvency" (Aronoff, 2016). The REA still supplies somewhere around 20 percent of total US energy use, and Aronoff has suggested that the "centralized decentralization" that was pivotal to the REA's initial success could provide one vehicle for enabling more democratic and locally sensitive versions of the energy transition that keep the public interest at its center. Expanding upon this model of many smaller electric cooperatives could yield public ownership at the scale of the community, fine-tuned to the members' energy demands and local landscape character.

Pier LaFarge (LaFarge and Roberts, 2024) offered a third possible model centered on public–private collaborations to greatly expand the role of distributed energy sources (DERs)—rooftop solar panels, ground mounted solar arrays, batteries, and "virtual power plants" composed of building energy management software handling demand response. LaFarge has argued that the fastest path to new clean energy in the United States is not going to be achieved by building large projects that take a decade or more to permit, but by allowing utilities to install many DERs on their customers' rooftops and in their buildings. He argues that phase one of the proliferation of DERs has occurred in the United States through a consumer-driven, unplanned private market context. This has had the benefits of driving technological innovation; it has built green industries and generated economies of scale in the manufacturing of DERs, which has facilitated extraordinary cost declines. However, he acknowledges that this model has also generated considerable equity and justice issues and suboptimal placements, made grid planning and demand management increasingly difficult for utilities, and created a boom-and-bust model for DER manufacturers. A second, public stage of DER institutionalization as such is needed, and the model that LaFarge argues for would have public regulators (such as the Public Utility Commissions in many states) allow utilities to procure the hardware (which would still be made by the private marketplace) and deploy it on customers' property in a decentralized manner while retaining ownership. LaFarge suggests this strategy would enable utilities to engage in more effective and strategic investment, deployment, and central planning at a gigawatt scale to better meet load growth, grid stabilization, and decarbonization targets. It would also provide guaranteed purchasers for manufactures of DER hardware and software and allow such systems to be more efficiently incorporated into utility planning and management of decarbonized energy systems. A public approach to DERs would reconceptualize DERs as public goods functioning more like the pipes of a sewer system or the wires in an electric grid, with the distributed energy infrastructure on many residential rooftops operating more as a public utility rather than privatized energy infrastructure.

Transition pathways for the United States that endeavor to meaningfully decarbonize the grid also rely on rapid electrification of the transportation and industrial sectors and of the built environment (Griffith and NREL, 2017). These electrification trends, along with reshoring of manufacturing and the retirement of fossil fuel power plants, all drive demand for electricity. The energy transition will need much more clean electricity, and this suggests that there will be demand for all of the low-carbon technologies at all the scales: more utility-scale renewables, more renewable DERs, more nuclear, more energy conservation, more energy storage, more virtual power plants and demand response. The observation that renewables has much lower rates of unionization than the nuclear and fossil industries (Huber 2022) is concerning. The path forward though clearly suggests the importance of organizing union drives and enhancing labor protections across all sectors of the clean energy industries. Offshore wind has offered a good example of a renewable industry working closely with unions, with public procurement of offshore wind by multiple states along the Eastern US seaboard incorporating Project Labor Agreements and workforce training into the design of wind farm projects, and in turn winning support from coalitions of existing unions. With solar energy rapidly growing ever cheaper as it benefits from technological advances and economies of scale, a near-term future of solar dominance is in sight (as is already the case in Australia where residential solar costs less than $1 per watt). Demanding that some of these cost savings be redirected toward worker pay and worker protections seems like a good position for just transition advocates to take as they advocate for an expanded role for renewables.

If the particular energy mix of a decarbonized future has remained a matter of contestation for Green New Deal public power advocates, a core point of agreement would center around the proposition that many of the pathologies of an energy transition dominated by the long legacy of energy deregulation are only going to be resolved if we return to understanding energy as a public good and provide a leading role for ownership of these energy infrastructures by the public sector, rather than the dominant investor-owned utility model. Public power does not offer a magical road to a just energy transition. As Brett Christopher (2024) and Sandeep Vaheesan (2024) have pointed out, historically, both decentralized and more centralized public pathways have been subject to oligarchic capture. In the United States, both the Rural Cooperative Model and the TVA public power models have been criticized for failing to maintain appropriate systems of public oversight. As Aronoff has reported, "many RECs have either embraced or fallen prey to old boys networks of board members … often featuring little to no minority representation on their boards and staff," and "the overwhelming majority of coops are inordinately dependent on coal, deriving some 70 percent of their electricity from coal-fired power plants" (Aronoff 2017). Globally, there are many examples of state-run energy companies around the world from Saudi Aramco to Petrobras that are locked into furthering fossil fuel production. Nevertheless, as Schwarz et al. (2023) note, state-owned firms do potentially have some advantages in decarbonization due to their ability to access public funding and implement policies without the same profit-driven pressures faced by private firms. This can make them potentially more capable of transitioning than privately owned companies if aligned with government policies. Vaheesan's (2024) assertion is that it is important that public power campaigns today attend not just to energy questions but also how renewed institutions of public power structure their governance to prioritize not just decarbonization but building democratic norms into the culture and legal frameworks that direct their operations. Developments in New York State offer some possibilities here.

New York State has the largest state public power organization in the United States and has had a hand in public electricity generation since 1931, when FDR, then serving as governor of New York, created the Power Authority of the State of New York, which would later become the New York Power Authority (NYPA). NYPA went on to build two large hydropower dams on the US–Canada border in upstate New York, just as Canadian provinces Ontario and Quebec were building their own dams, and today operates 17 generating facilities and over 1,500 miles of transmission line.

In 2023, the New York State legislature passed the Build Public Renewables Act (BPRA), which enabled NYPA to plan, construct, and operate utility-scale renewable projects—onshore and offshore wind, solar, and hydropower—to meet the state's climate goals. It requires that all public renewable projects be subject to project labor agreements. Funds in the BRPA have also been allocated to workforce training and to issuing credits for reducing the electric bills of low- and moderate-income customers. In the same way that in 2025 New York's governor directed NYPA to build a new nuclear facility to increase the amount of zero-carbon power on its grid, Meyer has noted that "using their state own power authorities, governors in other states—or even the federal government, with an entity like the TVA—could take a similar step" (Meyer, 2025).

While the energy mix and optimal institutional forms remain a matter of contention among advocates and critics of Build Public Renewables Act (BPRA), we can also identify a growing pragmatism in many public power discussions following the realization that the second Trump administration is effectively attempting to undercut the solar and wind industry in its entirety. Globally, it has become increasingly clear that the heated but sometimes parochial debates that dominated US energy circles around the "optimal" decarbonizing energy pathways are having little impact in China, India, Korea, or the UK to take a range of examples where different resource endowments, engineering cultures, security, and geopolitical concerns and different levels of risk tolerance are already generating a plurality of possible pathways to decarbonization. More generally, such debates circulating around US energy transition discussions can often miss the fact that all lower-carbon energy systems come with both benefits and drawbacks, and our critiques of the failings of any one particular low-carbon energy system must recognize that maintaining the existing fossil fuel system involves orders of magnitude more mining and extraction, more pollution, more greenhouse gas emissions, and more human suffering (see Wang et al., 2022; Mirletz et al., 2023; Eisenson et al., 2024; Ritchie, 2025).

Housing, Urban Form, and the Green New Deal

While the transitioning of the energy system justifiably occupied a lot of room at the heart of many decarbonization discussions, housing represents another major area of conversation at the confluence of climate impact, climate adaptation, equity, labor, and the design of the built environment. The Green New Deal discussion opened up a range of propositions about what new decarbonized and affordable social and public housing might look like, and with the growing convergence of the climate crisis and an acute housing crisis, this topic has since leapfrogged to the core of the progressive policy agenda. Housing only got a passing mention in 2019 in the H.R. 109 resolution, coming just after clean energy and smart grids and sandwiched between the other project areas. H.R. 109 was additionally short on detail about how the grandly ambitious task of "upgrading all existing buildings in the United States and building new buildings to achieve maximum energy efficiency, water efficiency, safety, affordability, comfort, and durability" would be achieved. Nevertheless, the sociologist and housing scholar Daniel Aldana Cohen has persistently argued that the battle for affordable housing and climate action should run together because as important as it is to focus on green jobs, "The crushing cost of housing is just as central—if not more so—to class struggle and workers' economic pain than stagnating wages" (Cohen, 2019).

If the first rollout of the Green New Deal initially largely missed the housing issue, subsequent congressional resolutions by the Green New Deal's congressional supporters did seek to elaborate what a GND for housing could entail. In 2021, Rep. Alexandria Ocasio-Cortez and Sen. Bernie Sanders introduced the Green New Deal for Public Housing Act (S.1218), which proposed $172 billion over ten years to retrofit all 950,000 of the country's existing public housing units. The legislation proposed to create two new grant programs designed to cut an estimated 5.6 million metric tons of carbon emissions through a combination of deep energy retrofits with upgraded insulation, windows, and energy-efficient appliances and fixtures; building electrification; and on-site renewable energy generation—with the bonus benefits of reducing energy and water bills for residents and improved climate adaptation and emergency disaster response—while creating 240,000 union jobs per year. The Homes for All Act of 2019 (H.R.5244), introduced by Rep. Ilhan Omar, proposed to repeal the Faircloth Amendment, which would permit the federal government to build public housing and invest $800 billion over ten years with the ambition of building 8.5 million new units. The Green New Deals for Cities Act of 2021 (H.R.2644) introduced by Representatives Cori Bush, Gerry Connolly, and Alexandria Ocasio-Cortez proposed a $1 trillion investment by the federal government that would authorize the Department of Housing and Urban Development to provide funds to states, local governments, or Tribal Nations "to carry out Green New Deal projects, such as projects to clean up pollution, climate adaptation measures, or conservation projects on family farms."

Such legislative interventions are interesting for the ways in which the importance of design in the public interest is increasingly acknowledged. For example, the GND for Public Housing Act makes an explicit argument for design quality, under a section titled "Improved Architecture Design in Government Housing Programs." It states that "even within the necessary budget limitations on housing for low- and moderate-income families, architectural design and environmental performance can be improved not only to make the housing more livable, but also to better suit the needs of occupants," stating furthermore that emphasis in providing public housing should be given to "encouraging good architectural design" that offers "maximal environmental performance" and reflects "the highest international architectural standards and the architectural standards of the neighborhood and the community in which it is situated, consistent with prudent budgeting." In this act, we thus see a pushback against the trends of minimum passable quality and total disregard for neighborhood context that marked the last 50+ years of American public housing.

Daniel Aldana Cohen has further argued that housing and a housing guarantee should be central to future climate policy for three reasons:

> First, exploding costs have made the housing crisis as big a threat to basic well-being as low wages and under- and unemployment. Second, millions and millions of people will need new homes as extreme weather makes swathes of the country unlivable. And third, building a ton of new housing to low-carbon standards can be a massive lever for decarbonizing the building sector, which is responsible for 39 percent of US energy consumption. (Cohen, 2019)

The second point is critical for Cohen, in that homelessness and climate chaos need to be connected directly by the climate movement, representing as they do two of the century's main existential threats for poor and working people. He has argued that to work at scale, such a housing justice movement would need to overcome not only the high cost of land in already-dense urban centers, but also resistance from suburban residents against the construction of new housing for their poorer, more racially diverse new neighbors—which it could do with a muscular enforcement of the Fair Housing Act and a public sector response pioneered in 1920s Vienna: the levying of "punishing real estate taxes" in order to dampen speculation, and then a massive construction push. Cohen and his colleagues at *The Climate and Community Institute* have argued that over the coming decade, 11.3 million new homes will be required just to keep up with housing demand, which doesn't count the number that will face damage from extreme weather and the number that will need extensive energy upgrades and retrofits (Wagner et al., 2025).

A focus on housing affordability would mark a pivot for many designers and architects in the United States, many of whom more often finds itself working for developers as clients, with a focus on higher-end markets. But this topic is emerging as perhaps the most central need for bringing down the runaway cost of living. Housing is a wicked problem, connected as it is to the imbalance of demand to supply in the housing market, zoning limits and federal housing policy, the elevated cost of land, the costs of permitting, labor shortfalls, and elevated cost of materials. Additionally, the construction industry in the United States has historically struggled to produce economies of scale that might bring down unit costs, due to relatively small numbers of modularized and mass-produced homes compared to site-specific construction. The climate impacts and potential carbon savings of the sector, the potential of improving the pay and labor conditions for workers in construction-related, energy retrofit, and climate-hardening jobs, and the centrality of housing in addressing systemic injustice place this conversation squarely at the center of the decarbonization, jobs, and justice puzzle that characterizes GND-aligned design.

While a GND for housing remains unrealized at the national level, the discussion has had an impact. At the municipal level in the United States, cities like Los Angeles and Boston have embraced the GND moniker for their municipal-level climate plans, with L.A. focusing on electrification and emissions reduction targets and goals of increasing the construction of transit-connected housing. Boston's mayor, Michelle Wu, similarly borrowed the GND mantle for her program of building electrification and energy efficiency retrofits, municipal electric vehicles procurement, and green infrastructure deployment.

From the Green New Deal to Build Back Better and the Inflation Reduction Act

At its high point during the 2018–2020 period, the Green New Deal polled well, winning support from the US public well outside the progressive camp (Carlock et al., 2018). The broader Green New Deal era of climate policy, moreover (perhaps for the first time in decades), galvanized a focused wave of concrete policy work emerging out of think tanks and policy groups from New Consensus, The Roosevelt Institute, Data for Progress, to the Climate and Community Project. It opened up spaces of intellectual and political engagement on climate-labor and climate justice issues, for a range of voices that spanned from technocratic energy modelers and policy experts, to groups in the progressive Elizabeth Warren and Bernie Sanders factions of the Democratic Party to even more progressive activist groups outside the big tent (such as Sunrise and the environmental/climate justice movement) as well as indigenous activists and scholars located in the Indigenous Environmental Network and Honor the Earth. Other radical currents on the Left during this period also felt the need to frame their debates in proximity to the Green New Deal. As such we saw the elaborations of calls for a Feminist New Deal, indigenous demands for a Red Deal, and racial justice advocates proposing a Red, Black and Green New Deal.[1] Indeed, it could be observed that for a brief moment, even centrist technocratic ecomodern groups such as The Breakthrough Institute (see McBride, 2018) and some degrowthers felt the need to indicate some possible points of common ground with the idea of a Green New Deal (Mastini et al., 2021).

The conversation coincided with a high point of public concern over climate change, at least among Democrats, 85 percent of whom in 2021 said that climate change was a *very serious* problem (Monmouth Polls[2])—the number among all Americans regardless of party was 56 percent, still a majority. However, in terms of the legislative career of the Green New Deal in the United States, we might track here a story of slow whittling, which led to an increasingly fragmented Green New Deal coalition. H.R. 109 did not move forward in Congress, but by the 2020 election, the GND did play a role in pushing many of the Democratic party nominees for the 2020 presidential election to adopt more ambitious climate commitments, which linked decarbonization with job creation and social justice. The Biden campaign did not explicitly embrace the Green New Deal during the Democratic primary, but upon taking office (and convening a Biden–Sanders Unity Task Force with allies of Bernie Sanders, Biden's main competitor from the Left, which included Green New Deal advocates Ocasio-Cortez and Varshini Prakash of the Sunrise Movement), many policy ambitions of the GND were recognizable in the first push climate proposals of the Biden administration's American Jobs Plan. Popularly known as "Build Back Better," this plan proposed over $100 billion in renewable energy investments and carbon capture; $621 billion in transportation infrastructure funding; $418 billion for affordable housing, public housing, public schools, child care facilities, community colleges, and federal building upgrades; and over $580 billion for boosting manufacturing, supply chains, research and development, and labor protections. Moreover, attentive to the input of the Democratic Party's GND-aligned progressive flank as well as that of Environmental Justice scholars and activists, the Biden administration rolled out early in its term the "Justice40" (or J40) initiative, which explicitly foregrounded environmental justice concerns which it implemented across all agencies, a whole-of-government approach to distributing investments with an eye on helping "disadvantaged communities that are marginalized by underinvestment and overburdened by pollution." It also pushed for labor protections, issuing executive orders that mandated Project Labor Agreements

(PLAs) on federal construction projects and crafting incentives that rewarded projects that employed unionized workforces and used PLAs.

Over the course of almost a year of intense Congressional wrangling, a significant piece ($108b) of Build Back Better's transportation infrastructure funding got repackaged as the Bipartisan Infrastructure Law and passed with the support of centrist Republicans, while much of the renewable energy and electric vehicle incentives made their way into the Inflation Reduction Act of 2022, which passed by a single tie-breaking vote in a 50–50 split Senate over vehement Republican objections. The universal provision of affordable housing, health care, and education—some of the more aspirational progressive "social infrastructure" agenda items included in the GND vision, some of which had made it into the early drafts of Biden's Build Back Better bill—did not survive, stripped out by the more centrist members of the Democratic caucus like Senator Joe Manchin of West Virginia, along with the proposed emissions controls of power plants or any other "sticks" that could have punished companies for their carbon emissions.

The majority of the "carrots" in the IRA, for clean energy and domestic green manufacturing, improbably survived the legislative process. In terms of overall funds for emissions reduction and clean energy deployment, the IRA directed as much as $295 billion to solar, wind, and battery manufacturing, the largest such infusion ever. Since the tax credits for clean electricity production and energy storage paid out based on how much energy was produced, the total amount would have depended on the appetite of developers to invest in projects, with estimates made prior to the reelection of Donald Trump suggested these investments could have amounted to somewhere between $400 billion or more over a ten-year period. The IRA was only passed by deference to Senator Manchin's demands for expanding fossil fuel leases, fast tracking the Mountain Valley natural gas pipeline through his state, removing any emissions controls on fossil power plants, adding domestic manufacturing requirements on EVs and batteries in order for them to qualify for tax credits, and a back-room promise to introduce infrastructure permitting reform in a follow-up legislative push in the face of complete opposition from many environmental justice groups (Bozuwa and Mulvaney, 2023a). The fact that the IRA passed at all, despite the compromises, after seeming to have died by Manchin's hand weeks before, seemed miraculous to climate activists, but left many environmental justice groups feeling betrayed, sacrificed yet again while clean energy backers celebrated. The second half of the Biden term was dedicated to implementation of the IRA, staffing up newly bolstered agencies, awarding grants through the Environmental Protection Agency and the Department of Energy, and developing the tax rules to get the private sector's investments into clean energy and manufacturing flowing.

The GND's embrace of public sector–led procurement helped open the door to a more muscular industrial policy in the United States, to which the Biden Administration responded with a variety of stimuli for domestic manufacturing of clean technology and low-carbon materials, from incentives to onshore EV and battery supply chains, to Buy Clean commitments that directed federal agencies to procure domestically produced low-carbon concrete and steel for federally funded infrastructure projects. In terms of geopolitics, the broader hope was that the IRA would further spark a global "race to the top" in clean energy investment, perhaps most notably reflected by the EU responding with the European Green Deal and its Green Deal Industrial Plan. At the same time the IRA became rather more sharply repositioned in the last years of the Biden administration's term as a geopolitical tool in ways that were surprising to many progressive Green New Dealers but in some respects reflected the nationalistic nature of Tom Friedman's original GND proposal in 2007. Indeed, it is notable that as the 2024 election loomed, decarbonization and industrial policy started to become increasingly reframed by both parties as a geopolitical counter to the looming threat of China, and moreover as merely one moment of a broader "all of the above" energy agenda, as Democrats increasingly emphasized the record levels of oil and gas production under the Biden administration to appeal to voters in fossil fuel–producing states.

Other critics have observed that while the IRA subsidized private interests to incentivize decarbonized technologies and reshore green manufacturing with the theory that this would ultimately crowd out bad actors, as Kate Aronoff (2024) observed the IRA had no "sticks" to force a public winding down of the fossil fuel economy nor did it fully develop mechanisms such as equity stakes, income-contingent loans, or profit sharing to ensure public investment delivers public value (Mazzucato and Rodrik, 2023).

In any case, the IRA experiment turned out to be short-lived. Trump's return to the Presidency following his victory in the 2024 election inaugurated an abrupt about-face on virtually all climate-focused federal initiatives in the United States, along with a clawing back of federal grants to justice-focused projects and initiatives handed out by federal agencies during the Biden administration. Trump's particular animus against solar and wind energy upended leases of public lands for those technologies and chilled the IRA-initiated appetite for green investment among the financial set. Despite hopes that the geography of new IRA-supported renewable energy projects, which skewed heavily toward Republican districts, might spare its clean energy tax credits, the Republicans' passage of the "One Big Beautiful Bill Act" in Congress in 2025 essentially eliminated the incentives for solar and wind projects going forward, though it left in place through 2033 the IRA's support for emerging clean energy technologies, such as advanced geothermal, nuclear, and energy storage. On the world stage, the Trump administration pulled the US out of the Paris Agreement for the second time and set about pressuring other nations to scale back their climate commitments and purchase more American oil and gas.

What are the forces that ultimately brought both the IRA and the second act of the Green New Deal to a halt, and what can be learned from this moment?

Postmortems on the Green New Deal and the IRA

Everyone has a plan until they get punched in the face. Mike Tyson, 1987.

While a number of factors shape the outcomes of US presidential elections and any one thread is difficult to separate, a number of forces worked against the Democrats and helped propel Trump back to the White House, chief among them being inflation and the increase in Americans' cost of living. While the global COVID pandemic of 2020 to 2022 initially opened up a brief window for a temporary expansion of the welfare state of the kind that Green New Dealers had been advocating through the care agenda, the macroeconomic context also changed considerably in all high-income countries between 2022 and 2024, with a "double whammy" of pandemic-induced supply chain shocks colliding with a global energy crisis produced by Russia's invasion of Ukraine in 2022, fueling global inflationary pressures that intensified a preexisting cost-of-living crisis. Such conditions created new opportunities for deficit hawks to argue that the era of large public expenditures underwritten by low-interest debt had now come to a close and for the fossil fuel industry to argue that the United States needed to ramp up oil and gas production to guarantee energy security. Deficit orthodoxies may well have been rejected by critical economists as well as the Biden administration, which remained committed to the view that a return to economic prosperity, coupled with a decarbonizing agenda, was only possible through a public investment-led agenda focused on green industrial policy. However, as US oil and natural gas production surged from 2023 onward, so did public worries about energy prices and inflation. Certainly, going into the 2024 presidential campaign, most US voters had not in fact heard about the IRA or CHIPS and Science Acts (Campbell 2024). The hoped-for hiring benefits remained years away while voters' immediate concerns were overwhelmingly focused on the rising cost of living.

While some of the jobs-plus-decarbonization goals of the Biden administration's industrial policy were aligned with GND movement priorities, it could be argued that the manner in which many IRA policies ended up being designed around often quite opaque systems of tax credits, subsidies, and rebates generated a technocratic policy program beloved by policy wonks, energy modelers, and upper-income suburban green liberals but whose direct material benefits were often quite unclear to most working people. Notably, in striking contrast to the New Deal of 1933–9, where the communication strategy was relentlessly effective and a political language existed around public initiatives aimed at the public good, in part enabled by design, somewhere between the launch of the Green New Deal and its transformation and whittling down into the IRA, the thread was lost. While Biden got to do a handful of ribbon cuttings for announcements of IRA- and CHIPS Act–funded factories, there were few actual built projects to point to, and working-class voters were unlikely to know about the IRA or its programs for aiding the purchase of electric vehicles, heat pumps, rooftop solar, and home and energy efficiency measures. Indeed, rather than proceeding with a view that high-quality public goods and infrastructure funded from the public purse through redistributive taxation might make life easier and better for working Americans, a number of the benefits of the IRA's more visible climate- and energy-focused programs went to sectors of the upper-income professional-managerial classes who both owned their homes and had large sums of disposable income to make such investments. A $7,500 tax break for purchasing a $60,000 electric vehicle was a policy measure largely meaningless for middle- and lower-income working families, renters, or young people struggling to pay spiraling housing, healthcare, or educational expenses. The removal of the housing, care, and other social infrastructure provisions from the IRA made it much less immediately helpful for working people. The process of making compromises to bring along the final centrist members of the Democrats' caucus, like Manchin, on this single mammoth piece of legislation was a far cry from the sweeping Congressional supermajorities that FDR had at his back during the entire New Deal era, allowing for experimentation with a flurry of various programs.

A further challenge for the IRA was the long-time lag of redirecting supply chains, building factories, upskilling and retraining workers, and starting up whole new green industries. Policies that rewarded large private companies for embracing domestically situated green manufacturing may have been premised on the idea that these near-term subsidies would eventually trickle down to workers via job creation. But once again, without a coherent vision and language, such public investments can be thoroughly opaque to the public, take a long time to get going, and are not felt as material relief by everyday working people. Unlike the New Deal WPA's or CCC's immediate creation of widespread job opportunities for low-skilled, high-paying work during a time of rampant unemployment, the IRA landed during a period of low unemployment, the COVID dip notwithstanding. The IRA's jobs in new battery factories or building new wind farms would only come on line after those factories could be built and only be available to qualified and trained workers. Where there were not domestic workers ready to fill the new jobs, such as at Hyundai's and LG's brand new EV and battery plants in Georgia and Alabama, the jobs would go to skilled foreign workers or to robots in increasingly automated workplaces. Moreover, while there is some evidence that policies such as the IRA were popular once Americans learned of them (Campbell et al., 2024; Leiserowitz et al., 2023), they barely factored into voters' assessment of economic performance or helped motivate them to turn out to vote.

A victory in 2024 for the Democrats could have perhaps allowed some of these issues to have been worked through more, for the jobs benefits to materialize, and coalitions to solidify. Minimally, the investment gains of the IRA and CHIPs Act would have had more time to be made tangible, which might have solidified support for its provisions in the geographies that benefited from its investments, and opened up opportunities for an IRA 2.0 to address the limits of the 2022 legislation. Emily Grubert and Sarah Hasting-Simon's observation that post-carbon design for the "mid transition" simply faces an enormously uphill struggle not just because decarbonization is politically and technically difficult to do but because of how it additionally impacts all manner of industries, private powerful interests but also workers and communities across a hugely complex division of labor remains insightful (Grubert and Hasting-Simon, 2022).

A Third Act for the Green New Deal?

So, might there be a third act for the Green New Deal, or something like it, in the years ahead? With the Trump Administration's comprehensive dismantling of virtually the whole Biden-era climate and energy policy framework and its turn to "energy dominance" based on oil, gas, and nuclear, many would argue that the Green New Deal era of climate policy resides firmly in the past. If the shifting sands of the contemporary moment are symptomatic of more general class dealignments, nativist isolationism, and a political consolidation of the Populist Right, the Green New Deal could indeed

Figure 17 Transmission line corridors have been envisioned as multipurpose infrastructures, with the help of design. Image by Nicholas Pevzner.

be viewed as the last gasp of climate justice–oriented social democracy in the United States rather than marking its renewal. The term itself, while widely vilified in right-wing media from 2018 onward, is now mostly missing from climate policy discussions. The words "Green New Deal" most frequently appear in the public mainstream at the invocation of Trump administration staffers to deride various aspects of the Biden administration's energy or climate legacy, which in any case only ever contained no more than a faint echo of the GND's original agenda. However, if the current right-wing coalitions of economic nationalists and climate skeptics, tech-libertarian billionaires, religious conservatives, and the general aggrieved prove to be an unstable coalition of forces, other opportunities may emerge.

Despite the current backlash to climate issues (which we will explore further in Chapter 8), the challenges posed by climate change are not going to go away by calling it a hoax, and calls for the state or public agencies to step in to address the range of issues posed by the Green New Deal—from coordinating the energy transition to facilitating climate adaption of urban and rural areas to life on a hotter and more turbulent planet—are not going to disappear. The drumbeat of extreme weather events supercharged by climate change is already having impacts on the cost of house insurance in climate vulnerable areas, from Florida to California, and proving to be one of many areas that are not going to be resolved by market actors. An increase of heatwaves, wildfires, or flooding could well drive public calls for effective public agencies and interventions that can coordinate and consolidate preparation for such events, coordinate cleanup after the events, and build public capacities for maximizing just climate adaptation that leaves no one behind. Such outcomes could revitalize the role of the social insurance and climate-resilience state.

Energy dominance achieved through doubling down on fossil fuel production may sound boldly populist, but such policies contain their own risks and vulnerabilities. Domestic oil and gas prices are subject to volatile market price swings; coal is now economically unviable in the United States and without significant state support, planning and public acceptance, and major improvements in price controls and project management, the renaissance of nuclear power in the United States will continue to struggle. Sabotaging a surging domestic solar industry and handing over green manufacturing and the development of "next generation" low-carbon technologies (e.g., carbon capture and storage, "green" hydrogen, small modular nuclear reactors, advanced geothermal, long duration energy storage) to China will have economic and geopolitical consequences for the United States.

With cost-of-living concerns, and particularly the cost of housing, groceries, power, and childcare dominating the political conversation of virtually all high-income countries, the social infrastructure provisions of the GND and Build Back Better seem newly relevant. With the federal bureaucracy that previously might have overseen loans to promising pilot projects, or used the procurement process to help scale up emerging low-carbon technologies in disarray, momentum on climate policy in the United States has shifted back to states, cities, and private industries to advance decarbonization wherever they can. The project for states and cities that have adopted aggressive targets for building new housing now face the task of reigning in their NIMBY voices, ensuring that construction can actually proceed by opening up publicly controlled parcels for affordable development and through careful reform of permitting red tape, dampening speculation through policies that let working people actually enter the housing market, and keeping construction costs low by rewarding mass production and replicability. The same is true for energy infrastructure. In case electricity prices begin to spike, as Robinson Meyer has predicted (Meyer, 2025), then the states that "can find cheap, zero-emissions electrons, as soon as possible," perhaps via public power at the state level, will be in a position to keep energy bills affordable for their residents.

Despite the near-term uncertainty at the federal level in the United States, electrification will proceed across the sectors, creating disruptions and contestations. There will be a need for holistic visions of how this transition could be steered, at a variety of scales, and many opportunities to experiment in how to do this, from smoother EV charging integration into city planning, more livable and survivable climate-resilient neighborhoods, more energy production integrated into all aspects of the built environment, and more interaction with energy-producing landscapes. As clean energy continues becoming cheaper, the need for large-scale planning at the grid scale to help more projects come online will not abate. There will be further openings for design to engage with climate and energy questions, and whenever it does, labor and justice implications won't be far behind, and a repository of GND proposals and visions on hand to offer a reference, and if the context fits, perhaps a template.

7

Planetary Designs and Speculative Climate Design Futures

The just transition is, by definition, a project that proposes a future and requires us to imagine possible futures beyond collapse and apocalypse. This future, moreover, will have to be designed, planned, and materialized by labor, communities, and working people across the planet for the struggles for just transitions to have any purchase in the real world. How is this going to play out over the long run? We know the physics of the climate system gives us some insight into future warming pathways, which currently suggests we are heading in the general direction of 1.7 to 3 degrees warming. Nevertheless, while the climate system provides a bounded array of possible future pathways, innumerable scenarios within this frame can be entertained.

In this chapter, we seek to explore the diverse ways in which design-informed imaginaries, as well as design-adjacent interventions in the critical social sciences and cli-fi (climate fiction), might usefully inform discussions of just and unjust transitions. Our interest here in climate-designed futures is less *predictive*—none of us have crystal balls to determine precisely how the intersecting forces of climate, ecology, demographics, geopolitics, technology, culture, crisis, and events are going to play out over the next century—and many fields that confidently proclaim otherwise have a fairly dismal record (see White and Roberts, 2019 for a comprehensive review). We are more interested here in the *performative force* of climate-design imaginaries. Rather than explore climate design futures with a view to bet on the future that is coming, the aim of this chapter is more to explore what climate futures that currently circulate within design may tell us about the political balance of forces that are seeking to shape futures, their desires and aspirations, and the question of whose agencies and whose labor such imaginaries recognize and whose they suppress.

As we have stressed through this book, there is no intrinsic reason why design, architecture, and planning are going to ally with the just transition. Indeed, we shall see in this chapter that there are many currents of architecture, design, and planning that are already actively stepping up to provide design-futuring to embed unjust transitions. However, this is not the only story to be told. We also mark out in this chapter important "counter-hegemonic" imaginaries of possible design-futuring imaginaries to business as usual which might keep open other possibilities and help other, more emancipatory visions of design-politics to emerge in the future.

Design Futuring as Green-Theming and Decorative Ecomodernism for the 1 Percent

In Chapter 3, we saw that numerous *éminences grises* of the architecture and design avant-garde in the United States have played a significant role in resisting any engagement with questions of ecology, sustainability, and climate or environmental justice. While outright and public resistance to a technicized version of sustainability or low-carbon design in the past decade has become more rare (but see Schumacher, 2024 for an exception), design theorists such as Ester Choi (2020), Joanna Boehnert (Boehnert, 2018), and Sasha Costanza-Chock (2020) have all observed that the problem of inaction has often been replaced with new modes of design-futuring that are largely embedded in asocial and apolitical technicisms, green-theming, decorative ecology, and futuring as escapism.

The move from resistance to green-theming has a long history in the intersecting spaces between the design industry, "lifestyle environmentalism," and green consumerism. As Joanna Boehnert has noted, communications and graphic design have long played a central role here since "a primary role for design is in camouflaging the specific ecological impacts of particular extraction, production, and disposal processes" and the wider systemic dynamics of human exploitation. Design-futuring in this mode "conceals the social and ecological unintended consequences of individual products and consumerism" (Boehnert 2018). As Sasha Costanza-Chock has argued, "Most design processes today … are structured in ways that make it impossible to see, engage with, account for, or attempt to remedy the unequal distribution of benefits and burdens that they reproduce" (Costanza-Chock, 2020: 19).

"Green theming" has been standard practice for designers working in sustainable packaging, marketing, and advertising in the food industry, fast fashion, and consumer goods—where the whole raison d'être of such ventures is to ensure that spectacular imagery draws attention away from the exploitation of human labor and nonhuman suffering that is embedded in the factory, the abattoir, the call center, the construction site, the mine, the waste dump, and so on. Esther Choi argues that the rise of landform buildings, as well as the interventions of the *Heatherwick* Studios in London and Bjarke Ingels Group (BIG) in Denmark, have become particularly effective at practicing a kind of "*horticultural dilettantism*," where "the use of 'urban greenery' as 'an outfit to appear characteristic of an ecological agenda" is typically a cover to minimize the visual impact of commercial developments and urban growth funded by corporations and real estate developers" (Choi, 2020: 4). Choi makes the additional observation that the obsessive preoccupation in contemporary architecture with virtuosity and the fetishization of building facades, the ongoing blurring of boundaries between the architecture and real estate industries, and the expanded capacity of digital tools used for image making has further generated modes of design-futuring that can promote forms of design escapism entirely untethered from questions of justice, labor, impact, carbon emissions, and ecological impacts. As she notes:

> In the realms of imaging software and computer-aided design—the architects toolkits for shaping the world—gravitational forces are mere variables, and decisions

> can be erased with a few clicks and undos. With three-dimensional modeling, it's easy to scale the design of a building or a city upwards from zero to infinity like the supposed freedom of global capital. Software directs designers to create with an additive mindset, mirroring capitalism's ever-expanding surface. One never has to answer responsibly to pre-existing conditions, communities, limited material resources, socio-ecologically damaging construction materials, and methods, or the growing need for common things. Practicing worldmaking in an imaginal space with zero consequence is an invitation to perpetuate a capitalist and colonial legacy which treats nature and society as pure abstractions. Reliance on this design method encourages architects to bypass the inequalities of labor, colonialism, resource extraction and expropriation—processes which capitalism necessitates—that are embedded in the act of architectural construction, and to relinquish accountability for their designs beyond a project's ability to generate revenue. (Choi, 2020: 8)

Choi observes that the modes of design-futuring in which significant currents of commercial architecture are trapped is best described as "logotecture." Logotecture is "an approach to architecture concerned with the accrual of what Arantes refers to as 'the rent of form': the profit or value culled from "the extractive exercise of designing buildings and city plans as symbolic capital." In this respect the circulation of powerful future imaginaries and "*idyllic renderings*" of what are claimed to be green buildings, landscapes, or green cities of the future act as "a conduit for the expression of power and perceived value" (Choi, 2020: 8–9). This operates in turn as "symbolic capital in service of global finance" (Choi, 2020: 7).

Of course, distinguishing between purely decorative green-themed technocentric and apolitical eco-designs and modes of techno-utopian design solutionism, which may have genuine recuperative insights, is the critical issue here. As our review of LEED and decarbonizing discussions outlined in Chapter 3, the line between technically interesting-but-apolitical-interventions, decorative, experimental, incomplete, and just bad ideas in design is not as easy to discern as some armchair critics imply. Design for climate futures will involve ongoing experimental practice, some of which will fail, but other interventions might offer insights for other developments. How much the ecological buildings and cities proposed by Stefano Boeri (Bosco Verticale and Wonderwoods) or the sustainable floating cities designed by the Bjarke Ingels Group and OCEANIX designed (and are now to be built by UN-Habitat) might be recuperated or co-opted, retrofitted, or repurposed and reprogrammed by more radical and politically literate modes of design is a complicated question. Nevertheless, design solutionism and greenwashed design utopianism clearly remain serious problems for all design future interventions.

Designs for Climate Colonialism, Climate Haussmannization, and Dreamworlds of Eco-Apartheid

If green-theming offers one future for design, there are, of course, more direct ways in which architecture, design, and planning might more proactively and brutally

Figure 18 Oceanix City. Copywright Bjarke Ingels Group.

further unjust transitions. The architectural historian Felicity Scott has observed that if architectural utopianism is often associated with revolutionary (Left) politics, these fields also have a long history of designing structures, buildings, services, and processes that actively and purposely "maintain dominance," "quell dissent," and facilitate the "monitoring and managing" of environments and populations. Such spaces include "refugee, detention and resettlement camps, arenas of civil war and guerrilla insurgencies, occupied territories, areas under martial law, so-called urban 'ghettos', squatter settlements, and other residues of social and environmental injustice, including company headquarters and social scientific laboratories" (Scott, 2016: 16). Scott reminds us here, for example, that historically, the politically flexible Le Corbusier, after his brief flirtation with syndicalism, was happy to promote his "radiant city" vision as a "counterrevolutionary" strategy to industrialists, bankers, and fascist governments (Scott, 2016: 19). Ijlal Muzaffar (2024) has further suggested that for much of the postwar period, many manifestations of architectural modernism and planning played a critical role in the Global South through grand modernizing projects which promised much but invariably delivered less and displaced political questions of colonial debt. Such undercurrents are not limited to the past, as architecture and planning continue to be embroiled in neocolonial projects and land contestations. For example, Eyal Weizman, the founder and director of *Forensic Architecture*, has done much to document the ongoing role that architectures of control have played in colonizing, ordering, and containing the Palestinian people in the West Bank and Gaza and systematically dispossessing them of land, water, and life (Weizman, 2017). Mike Davis, Daniel Bertrand Monk, and their collaborators have argued that more and more elite green design and architecture are now involved in producing a range of "dreamworlds of neoliberalism." Such low-carbon dreamworlds would include multiple architectural, design, and developer proposals to build low-carbon smart green cities that present themselves as solutions for eco-living but are largely focused on consolidating climate-resilience futures for the 1 percent.

Examples of such developments would include the proposed Crystal Island in Moscow (run aground during the financial crisis), the improbably billed "eco city" of Masdar, Abu Dhabi, both designed by Norman Foster and Partners, and King Abdullah Economic City in Saudi Arabia, whose master planning was completed by Danish architectural firm Henning Larsen Architects, and multiple high-profile architectural firms further commissioned to design its sixty-one towers, including HOK, Omrania, SOM, CallisonRTKL, and Foster + Partners. Projects like Eko Atlantic near Lagos, Nigeria, that are elevated against the sea level rise due to problematic sand mining nearby wear a green identity, but are ultimately examples of speculative real estate development. Such ventures seek to actively sustain the lives of the superrich through the construction of temperature-controlled gated eco-communities that can offer green consumption, "sustainable" golf courses and skiing opportunities in the desert, "low-carbon" super yachts, pet cloning facilities, all supported by digitally regulated workers living in either residential segregated dorms, or company towns further afield, with networks of detention centers, walls, and other devices to keep unwanted immigrants out.

Perhaps the real-world project that has attracted attention most recently is NEOM (also known as The Line), which has been conceived as a 170-km long linear city around a high-speed train line featuring a full array of AI-powered sensors, robots, and autonomous vehicles, which somehow also promises to be zero-carbon. NEOM is particularly interesting here in terms of the human cost of such ventures. The ITV 2024 documentary "Kingdom Uncovered: Inside Saudi Arabia" has alleged that the construction site of NEOM has been marked by harsh working and living conditions and severe labor rights violations with construction workers expected to labor for anything up to eighty-four hours a week and some workers having to wait months for payment. The low-carbon green branding, meanwhile, might be well-intentioned but will quickly run into the energy intensity realities of cooling glassy mega-buildings in the desert, not to mention the embodied carbon cost of construction materials.

Urban planners Zachary Lamb and Luna Khirfan (2022) have argued that if we consider that the history of urban design is littered with examples of where planners have sided with the powerful and the megaproject has flattened all opposition, we need to be vigilant that demands for climate-resilient urban futures do not unleash a new wave of "climate Haussmannization." Climate Haussmannization is a term they use to describe "a new generation of heavy-handed inequality-exacerbating interventions undertaken in the name of climate action" (Lamb and Khirfan, 2022: 2). What role could design-futuring play in shoring up this kind of regime? Joshua Long, Jennifer L. Rice, and Anthony Levenda have argued that climate planning in urban design and policymaking has moved from "speculative conversations" to a "nearly ubiquitous planning priority among leaders in the world's largest cities" (Long et al., 2020: 39), climate resilience has become a key means through which the unequal urbanscapes of neoliberal urbanism are secured and reproduced. Notably, under the frame of market-friendly climate resilience, we have seen a massive shift of financial and material resources in urban policy to projects such as "defensive sea walls, new digital infrastructures and platforms, exclusive self-sustaining eco-districts, and the physical reinforcement of institutions, transportation networks, utility infrastructures, and industries" (Long et al., 2020: 39). Long et al. argue that some of these projects may well be necessary. However, without a robust account of public power and forces to assert the importance of design for the public good in these plans we can often end up with market-led and private developer–shaped climate-resilient infrastructures that "prioritize projects offering the highest return on investment," which are focused on "creative experimentation in exclusive housing projects focused on resiliency," and expand the "retreat opportunities for the super-rich, whose transnational search for insulated, climate-hardy, and 'heavily secured environments' has disrupted housing markets on a global scale" (Long et al., 2020: 39).

Design-Futuring for Green Social Democracy: Green New Deal Utopias

The core future imaginary behind a revived vision of green social democracy or eco-socialism in the form of the Green New Deal is that the climate emergency necessitates that the state, public agencies, and public institutions will need to take the lead on driving complex multi-decade iterative projects to decarbonize the whole economy and adapt as best we can to a warming world. As we have seen in Chapter 6, a great deal of Green New Deal discussions have pragmatically pursued this agenda through policy proposals. Nevertheless, Green New Deal imaginaries have inspired various design and design-adjacent forays in more speculative futuring.

Alexandria Ocasio-Cortez explicitly embraced a critical utopian standpoint in a video she made in collaboration with Avi Lewis, Naomi Klein, and the artist Molly Crabapple.[1] "A Message from the Future with AOC" offers an illustrated narrative of a Green New Deal unfolding across the 2020s foregrounded by Congressional action. The narrative focuses on the life of Ileana, a college graduate who moves through "AmeriCorps Climate" working on union jobs in climate restoration, as a solar plant engineer, working as a care worker, and ultimately taking a seat in Congress to move the Green New Deal project forward. While this account of a climate future presents a US transformed by high-speed rail, a national smart grid, conservation, climate remediation work, and building retrofits, this social democratic account of the future accents public works, the establishment of universal care (Medicare for all), and a federal jobs guarantee with union protection that puts people to work building this future.

A more internationalist and comprehensive account of a Green New Deal can be found in Kate Aronoff, Alyssa Battistoni, Daniel Aldana Cohen, and Thea Riofrancos's *A Planet to Win* (2019). While this intervention is largely a work of political analysis and policy, it has distinct utopian socialist and futuristic features, linking a publicly driven post-carbon energy transition to a broader aspirational vision of a post-carbon future that is underpinned by good green jobs and affordable low-carbon housing, transport, and health care. Yet, it expands this imaginary further by drawing together an eco-socialist vision of future low-carbon energy infrastructures, sustainable urban worlds, and regenerative rural worlds rebuilt, restored, repaired, and repurposed by public action and well-paid union labor. *A Planet to Win* argues for public recognition for

Figure 19 Anastasiia Raina, Poster Design. Climate Futures, Design and the Just Transition RISD 2019.

"feminized labor" and public support for the care economy. It argues for a reimagining of public goods and a reorganization of work that could support public luxuries and forms of low-carbon communal leisure that undercut the narrow limitations of an individualized, isolated consumerism. And perhaps most notably, it seeks to ground the Green New Deal in a freedom-focused internationalism.

For a more specific design-focused intervention, which has near-term speculative future dimensions, *The Green New Deal Superstudio* co-organized by Billy Fleming in 2021[2] provided a remarkable catalogue of place-specific plans and ideas that could "spatially manifest the principles of the Green New Deal"[3] in ways that complement the politics of *A Planet to Win*.

In terms of more speculative projections, it could be argued that the writings of Kim Stanley Robinson have offered perhaps the most sustained inquiry into the processes that could give to societies informed by something like a governance structure and culture envisaged by the most utopian versions of the Green New Deal. The trilogy *Red Mars* (1992), *Green Mars* (1993), and *Blue Mars* (1996) sit at the most fantastic end of his oeuvre here. This trilogy imagines how processes of political-, cultural-, and design-steered worldbuilding toward a kind of eco-socialist utopia might occur across several centuries instigated by the colonialization and terraforming of Mars and the political contestations, conflicts, revolutions, and postrevolutionary reformulations that occur between assorted radicals, reformists, and reactionaries as a new, postcapitalist society on Mars is slowly brought into being. *The Ministry for the Future* (2020), in contrast, offers a more Earthbound, technical, darker, and somewhat conspiratorial account of how a devastating heat dome in India triggers a series of cascading further developments—unilateral geo-engineering, the rise of climate terrorism, international conflict, and so on—which ultimately creates the space for international diplomacy, smart finance, and politics to create other pathways to global decarbonization. Robinson's *New York 2140* offers us something in-between.

As the title suggests, *New York 2140* is set in a world unraveled by "two climate induced great deluges" that occur across a forty-year period, produced by the melting of significant parts of the Arctic and Antarctic ice sheets. Following such disasters and the concurrent massive sea-level rise that follows, Robinson sketches a world that has gone through various climate and ecological crises as well as moments of economic and political turmoil. These systemic crises generate frantic efforts at technical decarbonization through the transformation of energy systems, agriculture, and transportation and massive ongoing urban adaptation; yet after these decades of such technical transformations, social relations remain intact. Green capitalism of a kind has been forced into being after the "four horsemen have already bolted," but social and political relations remain unequal and unhappy.

Perhaps the most interesting feature of *New York 2140* in terms of climate fiction is that in contrast to the foreclosing imaginaries of a great deal of the genre, which often bring together climate determinism and capitalist realism, passive human subjects and overfamiliar, old end-of-the-world tropes, premised on Hobbesian struggles against all (Paolo Bacigalupi's *The Water Knife* comes to mind), Robinson's depiction of green capitalism after the crisis is interesting for its mapping of partial victories and more mundane struggles. The world of *New York 2140* is neither an ecotopia or fully fledged dystopia. Rather, the drama of the book unfolds in a world that has been scarred by climate turmoil, the mass suffering of many, and the retreat of global techno-elites to gated green communities, higher ground, and more temperate and safer geographies. But it is also a world marked by resistance and struggle, solidarity, and organizing by a majority that may still work for these elites but who are progressively looking for ways out, experimenting with all manner of strategies, tactics, and design interventions to not only sustain their own futures but build better lives.

The waterlogged coastal cities of "paranoidly poldered" London and Washington DC, "swampy Bangkok," or "nearly abandoned Buneos Aires and Copenhagen" are past their best. But New Copenhagen is now a "spectacular city" in Greenland. The Russians have built innumerable new cities in a warmed Siberia, which sees herds of de-extinct "pseudo-Mammoths" wandering the Tundra. There are townships on the oceans and sky villages. Intercontinental air travel is conducted by carbon-neutral and solar-powered blimps that dot the skies. Rapid developments in materials science, new waterproofing and reinforcing materials from graphene to diamond sheeting, new pumping systems, and Dutch engineering have created room for urban mitigation and adaptation to a kind of unequal and uneven but functional blue urbanism. If elites have mostly departed to higher ground, the rest find themselves negotiating a precarious "mélange of old and new" a hotch-potch blue urban landscape. New York is now "a super Venice." It is a city that has in part technically adapted to a warmer world via canals and sky bridges, water taxis and renovated berths. But within the contradictions of the city and the spaces where diverse working people are to be found, we find pockets emerging of collective ownership, mutuals, co-ops, solidarity economies, self-built alternative structures, systems, services, and designs. It is here where Robinson suggests the social and political battle for something like a Green New Deal emerges. *New York 2140* is a reflection on what might be considered a half-finished transition. Amidst all manner of upheavals and struggles, life goes on as does collective political and class struggle to make things otherwise.

Eco-Communitarian and Solar Punk Maker Worlds

If some social democratic, eco-socialist and utopian socialist Green New Dealers have tended to focus their post-carbon imaginaries on what we might call a "statist garden vision" of a reconstructed public realm, driven by public institutions, and an eco-democratic state that enables future citizens to live well, accessing "communal luxuries" in a post-carbon future that has been carefully designed and planned *for them* by public-spirited architects, designers, planners, and policymakers, we can trace a rather more dynamic, shambolic, decentralized, participatory, and anarchic vision of post-carbon futures discernable through a range of currents we cluster together as enthusiasts of eco-communitarian and solar punk "maker-worlds." Here, the designing is done by many, if not all, planning is sporadic, cacophonous, and a hit or miss affair, marked by a good deal of participatory chaos and reconfiguring, and engaging in redirective practice (Fry, 2009) across a warming world never stops.

Points of design inspiration can found for solar punk–inspired futures in the "the tools for social transformation" traditions of design that stretch from the Whole Earth Catalogue and Ivan Illich's "tools for conviviality" to contemporary currents of

Figure 20(a, b) Post-carbon Future. Images. Mitchell Joachim/Terreform 1. Image used with permission.

Open-Source Ecology, to attempts to reclaim the informal spaces of design innovation that runs from Colin Ward's writings on DIY innovation by the rural and urban working class making their own worlds, to Manzini's proposals for design where everyone designs. Solar punk with its origins in Brazilian science fiction draws further inspiration from liberation theology and eco-anarchism, Latin traditions of communist, communalist, and anarcho-communist experiments, aspects of Bookchin's social ecology, designs for degrowth, and enthusiasts of bottom-up design. Aesthetically, solar punk maker worlds often depict future worlds as a mash-up of the kinds of art nouveau and retro-futuring to be found in many Studio Ghibli films, mixed up with walkable cities, Stefani Boeri's style lush vegetations on buildings, communal gardens, with further blended high-tech solar, indigenous and Afro futurist motifs, (depending on the authors preferences and choices).

Within contemporary design advocacy, glimpses of such solar punk sensibilities can be found in the fascination that transition design demonstrates for attempts to build low-carbon solidarities and opportunities for collective, pleasurable downscaling through community gardening, permaculture, regenerative agriculture, and different forms of community provisioning, care, and mutual aid; attempts to build cultures of community recycling, maintenance, and repair; low economic trading schemes, community-based solar and wind, energy democracy, and community-supported agriculture, tool and lending libraries; and climate-conscious public art or low-carbon streets, restaurants, and festivals (Irwin et al., 2015; Kossoff, 2015; Tonkinwise, 2003). We can also see traces of such a vision in proponents of a new commons vision of the future sketched by the P2P Foundation (2015) who argue for a decentralized-designed post-carbon future based on participatory democracy and open-source circular economies, cooperative and sustainable peer-to-peer production, expanded by local makerspaces, fab labs, local agro-food production, additive manufacturing, and digital currencies.

In many solar punk accounts of post-carbon and ecological futures, the state has either retreated or acts as an enabler to "cosmopolitian localism" (Kossoff, 2015) and networked resilient communities that are focused on sustainable living, ecological regeneration and adaptation, building diverse/solidarity economies (Gibson-Graham, 2006), and stewarding common goods and resources that are managed collectively for the benefit of all. Some elements of this world are sketched in Juliet Schor's *Plenitude* (2010) that aggregate up subcultures of self-provisioning and downshifting, DIY cultures, and various attempts to build low-carbon communitarian ventures, which it is argued could open up very different pathways for low-carbon living for the future. Slow food, sustainable pleasure, and leisure, sustained by institutional supports like a universal basic income, Schor argues, might allow us to power down, decelerate, shift our life/work balance, and ultimately live better.

In some respects, the general sensibilities of solar punk are anticipated by Octavia Butler's vision of "Earthseed" elaborated in the *Parable of the Sower/Parable of the Talents* "All that you touch/You Change/All that you Change/Changes you/ The only lasting truth is Change." Perhaps the most elaborate far future account of such developments, though, can be found in Cory Doctorow's science fiction novels. *Walkway* (2017), for example, imagines a transition to an egalitarian post-carbon future occurring in

North America through maker movements equipped with all manner of bottom-up technologies from 3D printers to defense drone technologies, leading a collective "walk away" from polluted corporate cities to establish new sustainable egalitarian maker communities in territories opened up by climate change.

Walkway could be seen as a twenty-first century speculation on what the world might look like if the kind of project envisaged by the *Whole Earth Catalogue* had gained majority support. Doctorow's more anarcho-social democratic novel, *The Lost Cause* (2023) offers something of a pragmatic fusion of Green New Deal and eco-communitarian maker visions. A Green New Deal has been partially brought into existence in a near future California, but it is endangered by both right-wing fanatics and social stresses and strains produced by people movement and other forms of climate-amplified crises. The core of the book explores both ideological and what we might call socio-technical battles occurring between neofascists who seek to respond to a worsening social and environmental crisis by organizing militias, disruptions, and terrorism and a Green New Deal–inspired movement that is seeking to expand the social capacities for mutual aid and solidarity. If Rightist militias are constantly trying to develop technologies of surveillance and destruction, Doctorow's counters this with clusters of hardy young Green New Dealers, politically savvy but also equipped with a range of engineering and design skills, who are attempting to constantly outflank the Right through retrofitting, repurposing, designing and building climate-resilient and adaptive infrastructures, housing, transportation through making use of publicly available distributed technologies, robotics, drones, and 3D printing. The premise of *The Lost Cause* is that the forces of reaction can be circumvented by building systems of mutual aid that improve the lives of the locals while having enough slack in the system to accommodate newcomers and climate-displaced people.

Design-Futuring for the Pluriverse

If a great deal of Global North climate-design-future speculation has pivoted around the Green New Deal/ecomodernism/degrowth and climate pragmatism space, visions of transition influenced by decolonial and postcolonial currents, anti-imperialist approaches to global environmental justice, and indigenous movements have increasingly argued for other starting points and future imaginaries. Notably, all these currents have suggested that one of the central problems with many transition discussions that emanate from the affluent world is that they are often embedded in ways of knowing (or forgetting) that purposely obscure the global power relations at play in all designs for transition and the flows of energy, food, resources, materials, and racialized, classed, and gendered labor from the Global South to the affluent North that have been intimately entangled with systems of colonialism and imperialism, slavery, genocide, and plunder for five hundred years (see Escobar 2018; Muzaffar, 2024). As such, attempts to theorize transition design futures in terms of US energy choices for decarbonization that render invisible these exploitative geographies of resource and energy extraction between minority/majority worlds, that render invisible the masses of migrant labor from care workers, construction workers, agricultural and domestic workers that are central to maintaining the existing "extractive circuits" of global capital (Dharia, 2022), and that ignore the social and ecologically uneven terms of global trade that continue to govern these relationships or the role that militarism plays in policing and sustaining carbon- and energy-intensive lifestyles of the top tier of the global population must be recognized as deficient.

What follows from this, then? First, no progress on just transitions will be made unless there is acknowledgment of the ecological carbon debt and colonial debt that is owed by the Global North to the Global South (Muzaffar, 2024), an imperial drain that itself has been premised on centuries of colonial extraction of wealth from periphery to metropole. Second, and as we have seen in Chapter 6, there must be full recognition that many technologies underpinning proposals for achieving an energy transition in high-income countries raise significant issues around resource extraction (particularly for rare earths and lithium, land use impacts for solar and wind) and disposal and could contribute to the ongoing displacement of environmental pollutants from North to South (Riofrancos, 2025a). As such the environmental impacts of affluent nations must be internalized rather than displaced, and many argue this will now require that the affluent world rapidly shrinks its vast overextended ecological and carbon footprint to allow for sustainable growth of the "survival emissions" of the Global South (Agarwal and Narain, 1991). Third, transition strategies must not only stop the use of the Global South as "sink" and "tap" for the Global North but also provide reparations to allow communities in the Global South to mitigate and adapt to climate change and build systems that can offer sustainable prosperity for the majority world (Abdulla et al., 2019; Táíwò, 2022).

Within urban design, the indigenous urban planner Ted Jojola and his colleagues (see Walker et al., 2013) have long argued that the temporal horizons of indigenous "seven generation planning" are central for the kind of long-term planning required for climate change. Design scholarship on "Maroon Societies" have recovered African American, Afro-Caribbean, and Afro-Brazilian attempts to establish intentional communities and build societies premised on mutual aid/communalism beyond slavery and white supremacist violence (Price, 1973). Jessica Gordon Nembhard (2014) and Kali Akuno of *Cooperation Jackson* have variously sought to ground contemporary attempts to build cooperative design economies in much older late-nineteenth- and early-twentieth-century African American experiments with economic democracy and cooperative systems of ownership (Gordon Nembhard, 2014; Akuno and Nangwaya, 2017).

Landscape architect and urban designer Julia Watson offers an aggregated compendium of possibilities here in *Lo-TEK: Design by Radical Indigenism. Lo-TEK* argues that the design, architectural, and engineering knowledge of indigenous, peasant, and agricultural peoples has often been entirely dismissed by decades of high-modernist design and architecture, but we are increasingly coming to appreciate that such communities tested and refined multiple ecologically coherent solutions to their problems over millennia. From cooperative land and water/flood management systems, which have also facilitated complicated systems of resource sharing, to building practices that use sustainable materials and deploy innovative systems of shading and cooling, to agricultural practices that enhance biodiversity, to forest management practices that are effective at controlling wildfires, Watson suggests we can find all manner of remarkable innovations that have often been independently but simultaneously discovered across geographically diverse communities and societies. Such systems are not necessarily small scale—rice terracing with complex hydrological, irrigation, and aquaponic systems can transform vast landscapes. The water management system of *Qanat* underground aqueducts of Persia, which built a subterranean infrastructure of tunnels transporting water from mountain aquifers to low-lying cities, were "monumental Earthenworks" (Watson, 2019: 244). But they are often multifunctional, use local skills, and are embedded with symbiotic systems of traditional ecological knowledge that are often remarkably sophisticated. In a world marked by rapid climate change, she has argued that what is needed is a design movement *Lo-TEK*, which can catalogue and amplify these practices so as to "rebuild an understanding of indigenous philosophy and vernacular architecture that generates sustainable, climate resilient infrastructures" (Watson, 2019: 9). *Lo-TEK* does not offer simple solutions to environmental or climate issues that can be directly translated regardless of place or space. Watson argues that some examples of Lo-TEK, if "scaled and hybridized," could offer alternatives to business as usual. They are systems that in some cases are time and labor intensive or embedded in a particular communal or ontological worldview. Many cases, most recently, have suffered disaggregation by the expansion of urban modernities. But Lo-TEK does point to a range of successfully designed forms that significant human populations have adapted and incrementally improved for millennia, often in the most inhospitable contexts. It serves as a reminder that human societies have lived in many different contexts in the past and have to do so again in dramatically climate-changed futures.

The post-development orientation of *Lo-TEK* largely counterposes modernity/coloniality with communal cocreation, low-tech interventions, and traditional ways of knowing/being. In terms of indigenous design politics, though, it would increasingly seem that there are many different potential relations between modern and amodern/non-modern pathways evolving. To take one example here, Sen̓áḵw is a large high-density housing and urban project proposed and developed by the Squamish First Nation people on Squamish territory in Vancouver. The Squamish First Nation worked with Revery Architecture/Westbank to develop a proposed sustainable development of 11 towers with 6,000 units, 1,000 of which are affordable. The end result, though, could be viewed as offering more of a synthesis between green growth, modernist impulses, and indigenous design that the literature on *LoTEK* anticipates.

The Nigerian-born and Brooklyn-based visual artist and architect Olalekan Jeyifous offers a similar hybrid fusion of perspectives in his climate-future inspired speculative designs, which provide collages of retrofitting, adaptive reuse, contemporary African urbanism, and Afro-futurist ecotopian speculation to reimagine future possible neighborhoods of Brooklyn or Lagos. Here, we are presented with a mode of design-futuring that increasingly seems to be interested in intermingling and blending the traditional, modern, indigenous, and hypermodern into new amalgams in far more promiscuous ways.

Figure 21 2021 artistic rendering of Senakw. (Revery Architecture/Westbank/Squamish First Nation).

Designs for a Carbon Removal Society

If designs for a decentralized future marked by downscaling, energy descent, and degrowth have become the focus of a good deal of the climate Left in their attempts to push back against a designed future defined by a technocratic ecomodernism, other currents suggest that this account of a post-carbon design-future is missing a good deal. Holly Jean Buck (2019, 2021) has argued that we should focus on the hard realities of where we are in the climate crisis. Notably, for all the current attention paid to surging renewables, fossil fuel still accounts for over 80 percent of global primary energy use. The renewables roll out is already experiencing disenchantment and political resistance in the US (Buck, 2021: 12–15), with concerns about land use and material impacts growing (2021:44). Attending to the IPCC models demonstrates that limiting emissions to 2 degrees is very difficult without negative emission technologies, and 1.5 degrees is probably unobtainable (Buck, 2019: 7). Decarbonization of many sectors—from industrial decarbonization to aviation—remains "possible but really challenging" (2021: 40). If hopes for fast emissions reduction fail and rapid warming continues, it is very plausible to believe that social and political pressure will build to "try something else." Simply avoiding a conversation about these issues does not prepare us well to manage them when that time comes. Approaches to decarbonization that focus on mitigation and adaptation remain critical, but grappling with climate restoration, carbon removal technologies, and practices is not something that can be avoided.

Buck argues that this predicament suggests the need to look beyond "the tired binary" between degrowth and ecomodernism (Buck, 2019:161). Clearly, we will need some degree of behavior change, demand reduction, and reduced consumption (Buck, 2021: 45–8), but we will also need "more of certain kinds of industry and less of others"—simultaneous building and unbuilding. More generally, there is little reason to believe that "zombified neoliberalism" is going to drive these processes. Successful forms of carbon drawdown will involve industrial restructuring as well as the building of new essential industries for carbon removal on a scale that will rival today's infrastructure of fossil fuels. It may also have to involve a great deal of community reinvestments and plans for rural economic development with communal buy-in because not only renewables but also many possible carbon drawdown approaches will occur in rural contexts. However, Buck suggests that because of the uncertainties surrounding their profitability and because capitalism doesn't have the capacity to act and engage in "macro level, strategic long-term thinking" (2019: 30), the state and the international systems working with communities that could benefit from carbon removal and entrepreneurs that could open up new pathways for innovations in the field are going to be central (Buck, 2019: 33).

What would a carbon removal society look like? Buck acknowledges that despite the centrality of these technologies limiting emissions, what we have at present are a series of propositions, proposals, and technologies that are at varying stages of the development to deployment pipeline. A good deal of geo-engineering discussions have revolved around the extent to which solar radiation can be managed, which could variously involve turning down incoming sunlight via space mirrors, cloud albedo modification (brightening clouds to increase their reflective properties), stratospheric aerosol injection, or modification of the surface albedo (such as painting urban areas white to increase their reflectivity). It is argued that many of these projects may never be financially viable or actually work. However, we are going to need ongoing research to figure out what might be possible, and here, and beyond solar radiation management, the spectrum of possibilities could range considerably from what are often referred to as "nature based" solutions—ecosystem conservation, restoration and land management that increase carbon storage or avoiding greenhouse gas emissions (Buck, 2019: 95)—to engineering-based approaches involving direct air capture of CO_2 or bioenergy with carbon capture and storage.

On the ecosystems side of the ledger, Buck suggests some of the most promising options might involve cultivating carbon sequestration through regenerative agriculture and afforestation, and new approaches to soil management using biochar, silvopasture, and other kinds of regenerative agriculture. Ocean farming and marine cultivation approaches to carbon capture could range from seaweed cultivation, marine microalgae, to kelp cultivation. These approaches could be affordable and have the support of many conservation NGOs and industrial groups (Buck, 2019: 149). On the engineering side of the ledger are biomass energy with carbon capture and storage (BECCS) or direct air capture (DAC) and storage, which come with many more challenges. Other possibilities might also emerge. We may be able to store carbon in building materials, soil, rocks, or products (Buck, 2019: 7). Expanding opportunities for cultivated meat could replace grazing land.

Buck argues that a shift to a carbon removal society is certainly compatible with an intensification of existing forms of exploitation, land grabs, and projects to expand new frontiers of accumulation, which reinforce existing inequalities and colonial relations. But this is not given. A focus on planning, building, and designing a carbon removal society might also open up strategic opportunities to rethink how we approach design, planning, and building for the just transition. Foregrounding regeneration, removal, and restoration could involve fighting for a very different model of the just transition than one focused only on harm mitigation. For example, such a frame could ask questions about how carbon removal might be done in ways that facilitate the reinvigoration of rural economies, address the need for climate reparations for marginalized or indigenous communities, and/or open spaces for revaluing restorative labor. Carbon removal could facilitate new opportunities to think about new modes of community, public, and collective ownership. It might entail the development of a broader carbon removal culture that is focused on the labor of care. Interestingly, Buck speculates at one point that a carbon removal society will require a transformation of education that among other things would involve the foregrounding of critical design skills across the population. This is because we need to visualize what we can't see—carbon, microbial life, labor (2019: 194) but also a cultivation of design skills more broadly focused on sustainability and carbon management, which could facilitate a disposition to redesign.

Perhaps one of Buck's boldest propositions is that focusing on a managed decline of the fossil fuel industry could open up a viable path to transition in ways that are not currently on the table. Winding down the fossil fuel industry is hugely complicated not just because of its enormous political power and capacity to defer legislation but because of the extent to which all manner of workers, communities, and countries are dependent on fossil fuel and ancillary industries for jobs and tax revenue. Once again and as we have seen with the closing down of the coal industry in the UK or deindustrialization across the rust belt of the United States, an unplanned and inequitable transition can be done very badly leaving workers without jobs and communities without income or hope, and with regions or developing countries sitting on stranded assets or facing huge bailouts or cleanup costs if all things remain equal. Another approach to the fossil fuel industry could be to put it "under public ownership and run it in reverse" (Buck, 2021: 176). The critical issue here regarding fossil fuel companies and carbon drawdown is that "if we don't need them for fossil fuel extraction, we will need them for removal services, since they have the very technical capacity that's needed" (Buck, 2019: 203). Drilling into bedrock and pumping CO_2 in looks a whole lot like drilling into oil reservoirs and pumping oil out. Effective and scaled-up carbon removal in short is going to entail building up a whole new industry. This industry could variously entail building facilities for direct air capture, building renewable energy plants to run these facilities, developing drilling and seismic expertise to inject the CO_2 underground, building pipelines to transport CO_2, and developing a whole series of carbon-utilizing support industries in green cement, chemical, equipment manufacturing and construction to move this forward (Buck, 2021: 176–7). In short, a range of carbon removal activities could fit with the existing skill sets of oil and gas workers and allow pathways to open up where transitions occur that maintain good industrial union jobs. More generally, creating this industry "according to a progressive vision, grounded in worker's demands and public ownership" (Buck, 2021: 177) and structured so that impacted communities gain clear community benefits could open up many more potential alliances between organized labor and climate advocates. Ending fossil fuels and the fossil fuel industry then is not just a technical project, but about challenging the whole system of exploitation and the social relations embedded in this mode of energy production (Buck, 2021: 179).

Clearly, Buck's attempt to sketch a future pathway for a just transition by following the pathway of a carbon removal society has many risks (as she fully acknowledges). There are always legitimate concerns that carbon removal takes the focus off the urgent need to focus on mitigation and adaptation. We can see some of these moves occurring now as the fossil fuel industry touts its abilities to produce "clean fossil fuels" (Buck, 2021: 16). The uncertainties around all carbon removal approaches are enormous. There are limits to how much carbon can be stored in ecological systems and conditions where carbon stored in ecosystems, such as soils, can reach upper limits or be released back into the system through, for example, fires that potentially undercut gains. Carbon removal done badly can have all kinds of knock-on effects. Scaling up biomass production for bioenergy with carbon capture and storage will likely impact food security and increase deforestation. Monocultural afforestation projects could destroy biodiversity (Buck, 2021: 28). Carbon sequestration through direct air capture is expensive, relies on cheap and abundant renewable power, and deep drilling could lead to leaks or be disturbed by seismic activity (Buck, 2021: 50). There are already plans by fossil fuel companies to attempt to combine direct air capture with enhanced oil recovery, which could ensure that carbon is treated "as something to be cleaned up and reused rather than eliminated" (Buck, 2021: 16–17). Without a price on carbon, most of these approaches are not at present cost competitive. More generally, speculative accounts of the ways in which geo-engineering could spur or amplify more catastrophic futures proliferate.

At the same time, Buck's point is that the potential for disaster is why we need to continue research on all possible routes to carbon drawdown and why we need to think about how a fight for a just transition might institutionalize these forces and possibilities in other ways.

Designs for Planetary-Scale Computing and Stack-World

In some respects, the recent writings of the design theorist Benjamin Bratton can be seen as elaborating on the contours of a carbon removal society while foregrounding the central role that planetary-scale computing will take in shaping this world to come. *The Stack* (Bratton, 2015) is Bratton's proposal that we view the various types of *planetary-scale computations* such as cloud computing, smart grids, robotics, universal addressing systems, mobile and urban-scale software, ubiquitous computing, and so on "not as isolated, unrelated types of computation but as forming a larger, coherent whole" or an "accidental megastructure." Bratton suggests that design is now fundamentally embedded in and reliant on this digital infrastructure. In some fields of design, software has replaced theory not only as a theory but as a tool for thinking. More broadly, Bratton has argued that the Stack as planetary-scale infrastructure "is changing not only how governments govern, but also what governance even is in the first place." He argues, though, that responding to the challenge of the Stack, with dystopian or utopian responses, will get us nowhere. Rather, the Stack model "is global but it is not immutable. To the contrary, it is intrinsically modular, and so this megastructure is also a platform, and an interface even, for the redesign and replacement of the Stack-we-have with a Stack-we-want (or perhaps with the Stack-we-want-the-least)."

One of the core themes running through *The Terraforming* (Bratton, 2019) is that speculations on the future of our planet in the light of climate change or the pandemic are poorly served by a technophobic and romantic environmentalism. Bratton returns us here to the observation that the "age of ecology" pronounced by the 1960s counterculture we explored in Chapter 2 was always meshed with cybernetics. Notably, our very engagement with the image of the blue planet from space, our discovery of climate change, and our ongoing attempts to model, understand, adapt, and influence future climates and other ecological disruptions were and are fundamentally dependent on the Stack. It is the space program and later planetary-scale computing that are in part responsible for one of the iconic images of the Earth that generated the iconography of the modern environmental movement. The general circulation models that inform our understanding of how increased greenhouse gas concentrations are impacting physical processes in the atmosphere, ocean, cryosphere, and land surface are entirely reliant on supercomputers. Moreover, Bratton argues that addressing and designing post-carbon and post-COVID futures is irreducibly going to drive us further into "Stack World." Whether we consider the kinds of digital infrastructures that will have to be constructed to contain outbreaks of zoonotic diseases or the increasingly elaborate digital infrastructures that Bratton argues will be essential for monitoring carbon emissions and protecting biodiversity, this is going to give rise not to a retreat back into "nature" but ever more elaborate methods of digital monitoring, surveillance, tracking, and testing. To think otherwise, for Bratton, is to lapse into a naïve technophobic purism or a kind of conspiratorial politics that is politically useless.

Bratton argues, moreover, that the scales of our design interventions moving forward are going to have to be commensurate with the scales of the problem. Different kinds of design interventions, from climate geo-engineering to the expansion of nuclear power, the construction of new modes of quarantine urbanism to the reprogramming and governance of the smart city, all are going to require an array of governance structures, both horizontal and vertical. The Terraforming then, which is very much a reflection written during the COVID-19 pandemic, is a moment where we will need to embrace a "positive biopolitics" that would allow more to see "the suppressed potential of such technologies."

Bratton's elaboration of Stack-World clearly has its insights. The proposition that the "age of ecology" and the "age of cybernetics" could be intimately related is a theme that hovers in a range of thinkers we have engaged with in this book from Bookchin's 1960s and 1970s left-libertarian writings on technology or Stewart Brand and the general trajectory of the Whole Earth Catalogue project to Donna Haraway's writings. In foregrounding the centrality of AI, robotics, and planetary-scale computing to any plausible design future points to all manner of ways in which design and struggles for just transitions will have to be addressed.

Some aspects of the implications of Stackworld on climate change and biodiversity are elaborated in the final works of the digital socialist feminist and geographer

Karen Bakker. *Gaia's Web* (2024) for example provides an elaboration of the ways in which digital technologies have become central to biodiversity conservation, enabling smart grids, buildings, and cities. At the same time, it could be noted that Bratton's various design projects and interventions tend to have very little to say about labor or potentially environmental/climate impacts of Stackworld, the detrimental political impacts that neo-Feudal Silicon Valley has had on enshittification of political discourse, the corrosion of democracy or the energy demand that is anticipated by data center growth. Bratton argues that existing forms of liberal governance demonstrated by the failure to address climate change or the handling of a global pandemic demonstrate that the existing political system is marked by deep structural failings. What is required is a "geopolitics based on a deliberate plan for the coordination of the planet." Bratton does not provide any clear account of who the political forces might be that might steer this plan, or what this plan might entail. More generally, his enthusiasms for the many ways through which Stack-world dissipates human agency shades toward an account of a transhumanist designed future that seems to have little account of how a democratic (human) citizenry could have some control over the Stack. Indeed, it could be suggested that as Bratton's project has evolved into ever more ambitious collaborations and think tank intervention, the nature of his project seems to bear more and more resemblance to Buckminster Fuller and his proposition that we might ultimately replace the irrationalities of mere human politics with a new "design science revolution" part two. The critical question going forward for advocates of design for just transition in contrast will surely be how much the infrastructure of Stack-world can be recuperated by labor-environmental forces for other ends, how much will need to be hacked and redirected, and how it can be stopped in its tracks and resisted.

Design-Futuring in Uncertain Worlds

In this chapter, we have tried to sketch out a range of possible pathways that could emerge as design, architecture, and planning are forced to address the challenge of climate change and broader ecological questions and further forced to engage in designing post-carbon sustainable futures. As we have seen, there are some fairly clear trajectories that the design world is already taking and could move further toward in future decades, which will reinforce the unjust transition pathway that we are presently on. The forces driving design as decorative ecology, design as climate colonialism, climate Haussmannization, and design that orientates toward building dreamworlds of eco-apartheid are of course much bigger than the design industry. However, we might say that—all things remaining equal—there are good reasons to believe that the future of architecture, design, and planning is firmly heading in this direction. As we have seen in this book, while there have been many notable forces within design that have drawn attention to these developments and proposed other ways forward, the capacity of the world of design to collaborate with eco-authoritarian, climate destabilizing, and exploitative design projects should not be underestimated. The defining features of professional design ideology under neoliberalism—*market realism* (it's a business opportunity), *design narcissism* (I'm an artist and this is a great opportunity to make a mark), *design innocent* (what is colonialism?), and *design helplessness* (I'm only one cog in the system, what can I do?)—all move this project forward as does a broader design industry that has happily served fossil capitalism, racial capitalism, and imperialism for centuries and more. If such authoritarian currents further incorporate the more repressive aspects of geo-engineering or digital surveillance, we could be heading for some very dark futures *by design* to be sure. It will take significant forces to make things otherwise.

However, we have also tried in this chapter to reinforce the observation that our possible designed futures are not given, that professional ideologies and political circumstances can change, and that there are many possible "counter hegemonic" pathways that design could take to facilitate just transitions. And here, rather than assert the value of the one true pathway, we have tried in this chapter to map and appraise some of the most plausible counter-hegemonic paths that could bring design, architecture, and planning closer to supporting just transitions. All these design-future imaginaries we have mapped here are partial and incomplete. None are predictive or definitive. They all have insights, drawbacks, and limitations. As we have seen in our discussions of geo-engineering, carbon removal, and "Stack-world," some of these scenarios contain possibilities and the seeds of disaster. Academic life is full of isolated individuals staking out absolutist positions that must be defended at all costs. Real-world emancipatory design futures are most likely to end up as blends of our scenarios or take entirely unexpected trajectories that we have failed to anticipate. Designs for labor-friendly post-carbon futures may well involve approaches that mix degrowth and green growth, that incorporate moments from high modernism and designs for the pluriverse, and that incorporate new blends of digital technology and indigenous wisdom to achieve just carbon mitigation, adaptations, and yes, carbon drawdown. We don't know the future ahead of us, but the purpose of design-futuring is to clarify what is at stake in possible pathways forward and be clear about the costs and benefits of courses of action.

8

The Politics of the Just Transition in an Age of Disenchantment

While the elites are talking about the end of the world, we are talking about the end of the month.

"Gilets Jaunes" (Yellow Vests) protest slogan 2019

Transition Impossible?

In this book, we have made the argument that a world that is negotiating a climate crisis and a broader socio-ecological crisis is also now immersed in a range of deep-seated conflicts over socio-ecological transitions. These conflicts over transitions raise questions about the classed, raced, and gendered *labors* of transition. They also raise questions of *justice* such as, who will pay for such transitions? who will be disrupted? who might prosper under new conditions and who is going to be left behind, eliminated or discarded? We have also argued that this is a battle over the *design* of our future. The just transition irreducibly raises questions about the extent to which we can plan, design, redesign, imagine, and materialize worlds that are very different to business as usual or the extent to which the forces of the status quo will lock us into fossil capitalism and life on a much warmer, more turbulent planet.

It has to be soberly recognized that this project of transition is not going according to plan at the moment. *Designing the Just Transition* may have demonstrated that both credentialed and non-credentialed designers have developed enormous capacities, resources, schemes, dreams, and potentialities for thinking about how we might design and plan more equitable, democratic, and ecologically sane futures and that these currents could, in theory, align with and reinforce movements pushing for just transitions. Nevertheless, there have always been bleaker voices, perhaps most famously those of the energy historian Vaclav Smil, that have long sought to temper "over-excited transition talk" by observing that at the macroscale, technological-energy transitions of the past have been slow and hugely disruptive, drawn-out affairs that have stimulated enormous amounts of social and political conflict within societies and geopolitical tensions across societies (Smil 2019).

Whether past dictates future remains a significant point of contention for energy scholars focused on the question of decarbonization. In terms of global trends, transition pessimists (or realists depending on your perspective) observe that despite the accelerating addition of low-carbon energy sources, global carbon emissions are at an all-time high, and 80 percent of global energy production remains stubbornly locked into fossil fuels. Jean-Baptiste Fressoz (2024), most recently, has maintained that the proposition that we can understand historical energy transitions through a stagist model of wood → coal → oil → gas → renewables, which assumes that the era of one energy source ends and is replaced by another, obscures the fact that at a global level we are using more and more energy from all kinds of energy sources. The current renewable surge at a global level, as York and Bell and more recently Daniel Yergin argue, is *additive* rather than displacing fossil fuel (York and Bell, 2019; Yergin, 2025). Fressoz (2024) has further observed that much energy focused analysis of energy transitions do not adequately capture the broader material impacts or "material symbiosis" that energy transition can generate. Talk of transitions occurring from wood to coal ignore the fact that the coal age increased the consumption of wood, the oil age was underpinned by the continued use of wood and coal. Shifts in energy forms can still generate an overall increase in materials used. And such observations offer sober lessons for current claims surrounding decarbonization.

As Thea Riofrancos and others emphasize, the renewables revolution entails significant material, land, and biodiversity impacts that extend far beyond carbon metrics (Riofrancos, 2025a). Lithium, coltan, and the rare earth elements essential for renewable energy technologies and electric vehicles must be extracted from somewhere, and within a neocolonial global order "green extractivism" ensures that the burdens of this intensified mining are disproportionately shifted onto the "green sacrifice zones" of the Global South (Zografos and Robbins, 2020).

A rather different view of the current state of the transition to Smil and Fressoz emerges from the scholarship on socio-technical transition theory that has long maintained transitions are best understood *as non-linear processes*, often characterized by what Frank Geels refers to as "hype-disillusionment cycles" as well as "regime inertia and unexpected developments," which can give rise to temporary reversals or slowdowns, even as the long-term trajectory is ultimately moving toward systemic change (see Geels, 2002, 2011, 2020). In this vein Nafeez Ahmed (2025) argues that *transition pessimists* miss that, regionally, we have seen absolute declines in the use of biomass, wood, and coal and the UK, Denmark, Costa Rica, Uruguay, Germany, and France have all reduced their fossil fuel reliance. "Green extractivism" may well be a problem, but energy analysts such as Seaver Wang (Wang, 2021; Wang et al., 2024) and Hannah Ritchie (2025) argue, we need to have a clear-eyed view of the scale of this problem. Notably, the pollution, human rights, biodiversity and material impacts of all low carbon sources of energy, are real, concerning and need to be carefully attended to, but this should not make us lose sight of the observation that maintaining the existing fossil fuel system generates pollution, human rights, biodiversity and material impacts that are orders of magnitude more consequential than green extractivism.

As Seaver Wang (2021) reiterates, perfectly safe energy, or energy with no material impacts, does not exist. This is an important point, particularly in light of the endless misinformation surrounding the social and ecological impacts of low-carbon technologies which continually circulates through right wing media (Brulle et al.,

2024). At the same time, an intellectually and ethically consistent commitment to a just transition cannot accept utilitarian arguments positing that some communities or peoples—whether in Chile or the Congo—possess lesser value than the global need to decarbonize. As Riofrancos underlines, enforcing labor, human rights, and environmental standards in mining; ensuring impacted communities are properly consulted and recompensed; and making judicious evaluations of extraction volumes (how much is truly necessary, or whether we could do things differently) must all be central to just transition advocacy (Riofrancos et al., 2023; Riofrancos, 2025a).

Michael Liebreich has further argued that transition pessimists are often in danger of repeating fossil fuel industry talking points and more generally guilty of committing *the primary energy fallacy* (Liebreich, 2023, 2024). Notably, he suggests that models of global energy use that focus on the stability of primary energy demand over the past 200 years in the global economy and project forward from this, to fix future primary energy demand based off fossil fuel usage does not take into account significant differences in efficiency and utility between fossil fuel and renewable sources. Notably, traditional fossil fuel–powered systems are marked by significant loss of waste heat during combustion. In contrast renewable systems produce electricity directly, and low-carbon technologies such as electric vehicles or heat pumps use electrical energy directly. In essence, this means that the energy transition from fossil fuel sources to renewables does not require a 1:1 replacement of primary energy; the same energy services can be provided with more efficient technologies (Liebreich, 2023, 2024 also see Rosenow, 2024). Others further argue that the speed of transition could be impacted by the fact that unlike fossil fuels, which are commodities that rely on significant transportation and infrastructure to function, solar, wind, and battery technologies are modular technologies that can be manufactured, are subject to learning curves, and tend to get cheaper over time. Hence, they can be built and installed much more quickly (Ritchie, 2025: 48).

The extent to which the current speed of renewable deployments and the price drops of solar and wind power will continue is a further point of uncertainty. If veteran climate activist Bill McKibben's *Here Comes the Sun* has sought to buoy activist energies in dark times by foregrounding the unprecedented exponential global growth of solar energy (see McKibben, 2025)—which McKibben maintains is unstoppable—more skeptical voices, such as Brett Christopher's (2024), have cautioned that while the costs of renewables may have dramatically fallen, private capital (at least in the West) is not shifting away from fossil fuels to renewables because the expected returns on investment are too low. Tens of billions of dollars a year continue to be invested in financing continued oil and gas exploration (Christopher, 2024). Of course, the critical issue here though is how much Christopher's analysis remains true for the rest of the world.

How much the macro-energy transition will be driven by the simple physics of climate change and its material impacts or further derailed by active sabotage by fossil fuel interests and ongoing culture wars, high interest rates, the AI/data center energy boom, or other dramatic social, political economic, demographic, and technological developments over the next few decades is perilous to forecast. The truth of the matter is that we simply do not know how the many dynamics forces shaping our socio-ecological and climate relations are going to play out in future decades. In the rest of this chapter though, we try and get some further grip on this contingent and chaotic moment (Wallace, 2026) by focusing on a range of "headwinds" and "tailwinds" impacting all our attempts to think about designs for the just transition: from the impact of China's emergence as green manufacturing superpower (Wang 2025) and perhaps the first "electrostate" (Tooze, 2025c) to the climate backlash (Patterson, 2023). We consider surging currents of climate nihilism circling around much of the climate movements and the ways in which liberal discourse in the West has recently pivoted between currents of so-called climate realism and debates over abundance. Having mapped these discussions, we conclude this chapter by thinking about *the impossible project* of the just transition and reflect on the modest but still important contributions that labor-environmentalisms and a design politics of transition might contribute to opening other possibilities, and other horizons.

China Shock/Electro-State?

If the core lesson of Jean-Baptiste Fressoz's work is that a decarbonizing energy transition that reduced material impacts would be historically unprecedented, the economic historian Adam Tooze has argued that China's dramatic rise, over the past four decades, as the fastest urbanizing society in history, the center of global manufacturing, and more recently, as clean tech superpower and geopolitical rival to the United States, may well be offering us a world historic rupture which fits the bill. Indeed, Tooze has suggested, China's rise is possibly an event of such historical consequence that it may well consign the eruption of Euro-American modernity, the foundational event of Western social theory, to a mere preface in the history of modernization (Tooze, 2024, 2025a, 2025b). Whether this claim proves correct, the view that China (and India's) energy and green manufacturing transition is now the critical axis shaping the future trajectory of decarbonization and energy transition pathways is compelling.

The basic facts here—now repeated by the business press on a weekly basis and received by Western elites with a degree of shock, awe, and fear—are remarkable. China began to invest heavily in solar manufacturing, wind, and battery technologies in the 2000s. However, even by 2010, European, Japanese, Korean, and US manufacturers still played a significant role in this marketplace, while Chinese firms were only beginning to localize their supply chains, remaining largely reliant on foreign technology for key components (Helveston and Nahm, 2019). However, between 2010 and 2025, this landscape underwent a complete transformation, which has enabled Chinese companies to achieve global market dominance almost across the board in renewable energy and green manufacturing.

As a number of commentators have observed (Xu, 2022; Hilton, 2024; Tooze, 2024; Wang, 2025), such circumstances have been significantly facilitated by a series of factors, such as (i) comprehensive and proactive modes of long-term state planning by the central government that have strategically prioritized the "new quality productive forces," placing renewable energy, green manufacturing, and EV production as core elements of a new growth model; (ii) national policies mandating market creation through subsidies, procurement quotas, the use of green industrial policy, and targeted funding to achieve rapid green technology transfer and integrated supply chains; (iii) extensive encouragement of public–private partnerships between domestic and overseas firms that has helped upskill the domestic labor force and provide spillover scaling and clustering advantages for domestic manufacturers; (iv) proactive local governments that have often played a critical role in providing land, infrastructure, tax subsidies and investments in human capital for domestic and overseas investors; (v) a state-owned banking sector that has channeled very large amounts of low-cost credit into renewables, EVs, and grid infrastructure, cushioning firms through losses in their scaling-up phase; (vi) a business culture of intensive competition between domestic manufacturers, which can generate dramatic economies of scale in production and, ultimately, battle-hardened regional and national champions that can evolve into globally competitive companies (see Xu, 2022; Hilton, 2024; Wang, 2025).

Analysts may differ over the relative significance and contribution of these factors to facilitating current conditions; however, it's clear that the cumulative impact of these developments have led to staggering cost declines in green technologies that are now reshaping the global manufacturing and energy sectors. Notably, it is China that now makes 80 percent of the world's solar panels (compared to 2 percent in the United States), approximately two-thirds of the world's electric vehicles, lithium-ion batteries, and wind turbines and 40 percent of the world's heat pumps. With twenty-seven nuclear reactors under construction, China is increasingly recognized as a world leader in nuclear technology (Ezell, 2024). China is also gaining an edge in energy efficiency and energy-management systems. It dominates the mining of many of the critical minerals essential for all manner of decarbonization technologies, electric vehicles, battery production, and so on. Perhaps most striking though is how these developments have led to a massive acceleration of overseas investment by Chinese clean tech manufacturers in the Global South spanning batteries, solar, wind, and green hydrogen. Xue and Larsen (2025), for example, have argued that pledged investments by Chinese manufactures now reach over fifty-four countries, and the $220 billion dollars invested surpasses US investments over the four years of the Marshall plan. Such developments are helping to expand sites of green manufacturing beyond China to Southeast Asia, the Middle East, Africa, and Latin America but also help underpin a global solar/battery storage boom in such regions as well.

If China's rise as a global "Green tech superpower" is one part of this story, the internal transformation of the country is no less striking. Domestically, China has not only built more high-speed rail, urban metro systems, and long-distance ultra-high-voltage transmission development than any other country over the past decade but deployed more solar, wind, and low-carbon nuclear power than any other part of the world. Stylianou, Tauschinski, and White argue that China is on course "to source 50 per cent of its power from low-carbon energy including hydro, solar, wind, nuclear and battery storage systems by 2028," which has led to the suggestion that it is on its way to becoming the world's first "electro-state" (Stylianou, Tauschinski and White, 2025). This is a state that is both dramatically electrifying its energy system and, as Tooze notes, "the first large economy in which clean energy investment has become the principal driving force of overall investment and economic growth" (Tooze, 2024).

In terms of labor-environmental and climate issues, these developments pose all manner of challenges for just transition frameworks—frameworks that it has to be acknowledged have been heavily influenced by the attempt to think about the challenges posed by transitions in the context of liberal democracies. Chinese modernization has had staggering social and environmental impacts over the past four decades. Hydrological projects, such as the Three Gorges Dam, have induced major land and ecosystem changes, including mass displacements of rural communities, landslides, and altered water systems. The material impacts of Chinese modernization and ongoing industrialization remain substantial (Wang, Nordhaus, and Ramachandran, 2025). As Daniel Wang has observed, the urbanization process in China more generally has proceeded at breakneck speed since 1978. A program of "compulsive construction" means that "in effect, that the state built a new city the size of greater New York City and greater Boston combined every year for thirty-five years" (Wang, 2025). To be sure, highways and superblocks built in haste and often with little regard for citizen input has produced very mixed results in many places, with plenty of high modernist urban blight running alongside the explosion of new hyper-futurist cityscapes of Chongqing, Shenzhen, or Shanghai. The material impacts of this urban transformation are captured by the much-repeated observation that in three years of the twenty-first century, China used more cement than the US did in the entire twentieth century. Chronic air and waste pollution, water scarcity, and desertification all remain significant challenges. As of 2025, China is also the largest greenhouse gas emitter in the world, with 75 percent of its emissions coming from coal. Energy policy analysts such as Gong and Lewis (2024) note that while China may have met its economy's carbon intensity reduction targets early and is committed to becoming carbon neutral by 2060, current government policies remain primarily focused on *energy security* and economic growth and are ambiguous about the actual transition plans to wind down the coal industry.

If China has the potential to be an electro-state, we may have to acknowledge that this capacity was built through decades of fossil-fuel-intensive industrialization (Wang, Nordhaus, and Ramachandran, 2025). There have been periodic claims made by the climate policy community that the growth and deployment of renewables in China may be large enough to soon lead to actual declines in coal use and carbon emissions. One scenario here may be that recent coal additions will increasingly become less frequently used "peaker plants" or backup power for what will become a renewables- and nuclear-dominated grid (see Ritchie, 2024; Tooze, 2025a; 2025b). Another plausible view is that there are a range of conflicting forcing agents trying to shape China's energy future and how these forces play out remains radically uncertain. Forcing agents that could further propel breakneck electro-state transitions would include geopolitical and economic advantages gained from the consolidation of China's global green tech dominance as well as recognition that the existence of chronic environmental problem can threaten the stability of the regime—air pollution generated significant public protest in the 2010s. Forces that may slow this process down could include chronic overproduction of solar, wind, and batteries, which could generate an oversupply glut, and the manner in which influential provincial governments in China remain supportive of local coal producers. China's constrained access to oil and natural gas and fears that this opens up geopolitical vulnerabilities may well ensure that it keeps its domestic energy options open (with a solar/nuclear, wind, and coal mix) but ends up as the principal provider of all low-carbon technologies for Africa and Asia.

Additional complexities emerge when we have to recognize that these achievements with regard to poverty reduction and the build out of renewables in China have occurred in broader labor-environmental contexts where the government is entirely hostile to independent worker organizing, environmental and labor protests are regularly met with repression, and the country's green manufacturing and mining sectors is marked by a range of ongoing human rights, environmental and labor concerns. The International Labor Organization (ILO) estimates that 45 percent of the existing production of solar panel and electric vehicle batteries in Xinjiang province remains reliant on Uyghur forced labor (ILO, 2025). As the ILO further points out, many of the materials required for green manufacturing and energy transition in Europe and the United States also originate from global supply chains in the Global South ranging from the Democratic Republic of the Congo, to the lithium triangle of Chile, Argentina, and Bolivia, and many other spaces in between which come with their own labor and environmental issues (Riofrancos, 2025a).

Of course, China, the BRICS, or Global South more generally are not going to take labor, environmental, or human rights lessons from the United States or the European Union, power blocs that have been involved in the colonialization and wholesale plundering of Southern ecologies, land, resources, and raw materials and the super-exploitation of Southern labor for centuries while in contemporary times failing to meet their own emissions cuts, let alone address questions of climate debt, ecological uneven exchange, uneven terms of trade and so on. The fact that China has lifted more people out of poverty in the past fifty years than any other country in recorded history and managed to move up the value chain from a semi-peripheral economy to a world leading market-state green manufacturer are achievements that have no parallel for countries in the Global South following the Washington Consensus. It is no wonder then that many other BRICs countries and other countries in the Global South but also the United States and the EU are looking carefully at what might be learned from China's hybrid market-state model. This model may well leave little to be desired in terms of democratic accountability. Nevertheless, and as Isabelle Weber and Thea Riofrancos have argued, this should not foreclose exploration over what could be learnt from China's successes with long-term planning, its developmentalist industrial policy, its history of joint ventures to build domestic technological and design capacities, and its administrative capacities to direct public investment and market activities to specific strategic ends (Weber, 2021; Riofrancos, 2025b).

In terms of how these developments impact design, planning, and architecture, up until recently, conventional wisdom within Western design culture had it that while China has established significant technological and engineering leads in electric car engine design and manufacturing, battery chemistry, solar and wind manufacturing and demonstrated capacity to build low-carbon nuclear power plants at speed and scale that had exceeded the capacities of the United States or the EU, in terms of *design* and *urban planning*, Euro-American design culture remained hegemonic, with the direction of traffic of the most influential urban designs and propositions still largely flowing from West to East. The I-phone may be manufactured in China but was designed in California, so the well-worn cliché goes. More recent developments suggest these rather complacent assumptions (the creative West/versus the routine low-skilled assembly work of the East) may need to be revised.

For example, Chinese firms have significantly advanced the use of prefabricated construction methods and emerged as a leader in integrating robotics and 3D printing within the building sector. Greater attention is also being paid to China's innovative use of drone and robotic technologies for transmission lines and solar farms in difficult to access areas (Temple, 2018). We have seen the extensive rollout of "smart city" digital urban planning technologies to facilitate traffic management, pollution control, *and* citizen surveillance in Shenzhen and Hangzhou, notable design innovations in urban service design and transport design planning spurred by the rapid electrification of Chinese cities. In landscape design and planning, we have seen the development of "sponge city" urban water management systems—first formally proposed by Chinese landscape architect Kongjian Yu—which have now been exported to Copenhagen. More generally, in areas such as car manufacturing there is growing recognition that Chinese car companies such as BYD are, in many contexts, not just producing better electric vehicles but higher-quality designed vehicles at much cheaper prices to the once mighty German auto-industry. Hence, the growing interest by German car companies such as Volkswagen and BMW in developing joint ventures with Chinese car companies, so they can learn from China.

How much the rise of China as clean-tech/electro-tech superpower is going to open up new pathways globally for designs for the just transition or merely facilitate a vast experiment in authoritarian social engineering and designing that sidelines democratic input and displace environmental impacts onto new "green sacrifice zones" (Wang, 2025) remains to be seen. Nevertheless, confronting China's "breakneck" process of hyper-modernization and its potentially critical role in decarbonization further highlights the extent to which a great deal of design-transition-ecology discussions in the second half of the twentieth century were heavily shaped by environmental movements and environmental design discussions emerging disproportionately out of the United States, the UK, Germany, Northern Europe, and Australia, drawing particular influence from the pivotal 1960s/1970s counter-culture moment and the concerns of the post-materialist sections of the middle classes of these countries. How much of the guiding assumptions, environmental aesthetics, political frames, favored designs, and taboo technologies that preoccupied such discussions are going to remain hegemonic as climate impacts intensify—particularly in the Global South, popular demands for development and shared prosperity also intensify and Southern governments map their own distinct climate mitigation and adaptation strategies—is worth reflecting on.

More generally, how design culture and the flow of critical design knowledge, ideas, and practices around just transitions is going to be impacted by a fragmenting world system marked by growing tensions between "petro-states" (perhaps the United States, Russia, and the Gulf States) and emerging "electro-states" (led by China and its allies) (Tooze, 2025c), declining influence of the "environmentalisms of the rich" (Fraser,

2022), and geopolitical re-orderings, as China, India, Brazil, Indonesia, Turkey, South Africa, and Nigeria (among others) emerge as *regional hegemons*—with their own geopolitical ambitions, their own economic, resource, demographic, material, and energy endowments/constraints and socio-ecological challenges, their own complex relations with their own minorities, subordinated populations, and regional neighbors, and their own distinct approaches to optimal forms of design, planning, and architecture—remains a critical issue.

For example, some of the most influential analyses of shifting North/South design dynamics over recent years have emerged from the decolonizing design movement. Such currents have not only produced important interventions highlighting the parochialism and extractivism of Euro-American design history but provided compelling scholarship highlighting the ongoing colonial entanglements of much of the design industry and design education (Abdulla et al., 2019; Escobar, 2018). Influential voices in the decolonizing design discussion such as Escobar have further argued that a reversal of the imperial flows of design knowledge from North to South could give new impetus to revitalizing pluriversal political ecologies, emancipatory peasant and indigenous knowledge practices, and a revaluation of rural craft practices, and facilitate rebuilding connections between design and agrarian populism through reclaiming discussions around tools for conviviality, agroecology, alternative cosmologies of making, and so on (Abdulla et al., 2019; Escobar, 2018). Such interventions have been heavily influenced by Latin American currents of decolonial thinking and the explosion of radical agrarian social movements and Left parties that occurred in the region during the "pink tide" moment, as well as resurgent indigenous intellectual and cultural contributions from North America and Oceania and currents of post-development theory premised on the claim that we should understand development, industrialization, and modernization as Eurocentric colonial impositions, and impositions that had essentially failed and run out of steam.

This project clearly has had significant and possibly transformative impacts on the design academy in many places (most strikingly at the moment in Canada and Australia to Latin America). Such currents in particular have powerfully made the argument that credible designs for climate mitigation have to attend to ongoing processes of colonial extraction, land dispossession, and resource exploitation that remain key dynamics of green transitions in the Global North. As we have seen in Chapter 7, such currents have drawn attention to the vast stores of design innovation that is produced by indigenous, subaltern, and rural people on an ongoing basis that has facilitated climate adaptative strategies for millennia. The current moment of upheavals and geopolitical turbulence in the whole relations between North and South also poses new challenges. We can point to five particular challenges that emerge for such frameworks, notably: (i) the consolidation of diverse Southern industrial *modernization* projects from China and India to the BRICS that are not adequately read as Eurocentric impositions and are clearly seeking to challenge US hegemony (possibly successfully) and are often involved in projects to assert regional hegemony and subordinate their own peripheries; (ii) ongoing retreat or containment of many agrarian populist movements from Bolivia to Brazil under multiple pressures from state violence to ongoing rural to urban migration: (iii) the rise of nationalist and supremacist movements from Hindutva to Bolsonarism in the periphery to resurgent white supremacist or white identity politics movements in the core that use the language of nativism and even decolonial theory (see Ansari, 2024) to assert the claims of their exclusionary politics; (iv) the rise of diverse indigenous futurisms, Sino-futurisms, Afro-futurisms, and Latino/a/x futurisms, among others, whose cultural and political projects seem less focused on directly counterposing cosmopolitical knowledge to modernist frames but rather are often engaging in generating cultural meshings of pluriversal *and* hybrid multiple modernities as points of inspiration and departure; (v) in the light of the successes of Chinese hyper-modernization, a return to questions of role that the developmental state, South–South technology transfers (such as Belt and Road), and green industrial policy might play in contesting US hegemony and opening new development paths for industrial development (see Arboleda, 2020, 2023; Estevez and Riofrancos, 2025).

Such altered geopolitical and geo-cultural contexts suggest that while a significant rebalancing of the respective influence of the design cultures of the high income and majority world is likely to proceed, what might replace it in the future could generate very different majority world environmental-design conversations to those that currently predominate. It is already evident that many BRICS countries have very different views to many Northern environmental NGOs (or even Northern Left-Green radical currents in design) about the extent to which their populations should have access to air conditioning, biotechnology, nuclear power, AI, and water desalinization technologies or how much fossil fuel infrastructures should be retired, maintained, or even expanded. We have also seen fractures and fissures emerging between indigenous social movements seeking to defend their autonomy, land, and ecologies from mining, logging, and other extractive industries coming up against (sometimes) Leftist developmental projects in Bolivia, Brazil, Ecuador, and Chile keen to prioritize the exploitation of natural resources to fund social programs and/or improve the national accounts (Riofrancos, 2020). Further battles over water management, land use, conservation, carbon capture, and removal technologies can all be anticipated in the future.

The Populist Right, Climate Denial, the Climate Backlash, and Net-Zero Meltdowns

If China presents one set of deep challenges to our abilities to discern a straightforward directionality to transition/design futures, the backlash to climate and environmental policies (Patterson, 2023) that has been brewing in virtually every liberal democracy since the 2010s raises a further set of issues.

An insurgent global populist Right that now ranges from assorted national conservatives and techno-feudalists to ethno-nationalists and neo-fascists fully committed to the fossil fuel system are clearly on the march in virtually every elected democracy. This is a politics that takes on different forms in different places. The manifestations of Right populisms surging in India or the Philippines, Hungary or the US are not all the same. But Right populists across the board have had strong ties with the fossil fuel industry and have commonly tended to downplay or dismiss the dangers of climate change and the broader environmental crisis, and they are increasingly at the center of a climate and environmental backlash politics. In the United States and Europe such currents have mixed with a resurgent white identity politics and attempts to frame a community of interests around ethnic identities rather than class solidarity, civic nationalism or cosmopolitan universalism. Such currents are relentlessly focused on contrasting what is presented as a detached cosmopolitan and internationalist environmentalist project that seeks to impose global climate agreements over the heads of national interests and democratically elected national governments. We have seen such politics draw from environmental aesthetics focused on wind power or solar power as a "blight on the landscape" or "destroyers of rural countryside and natural habitats." Post-COVID such arguments quickly tilt into epistemological wars against all manner of "establishment experts," where "instant social media experts" help circulate all manner of vaguely conspiratorial "alternative facts," half-truths, and misinformation that insist that decarbonizing technologies are more environmentally destructive than fossil fuels (Eisenson et al., 2024; Ritchie, 2025). As Cara Daggett has argued climate skeptics in the United States have also converged with broader "manosphere" currents, where petro-masculinity comes together with broader gender anxieties (Daggett, 2018) generating worries that advocacy of effeminate green lifestyles are "neutering" masculine agencies and "warrior dispositions."

Demographic and economic shifts are also undeniably playing a role in shaping these new political coalitions. Much of the electorate of high-income countries is aging and will be increasingly reliant on immigrant labor coming from countries that are experiencing youth bulges. Perhaps more complex still though for advocates of just transitions is that Right populist resistance to environmental and climate concerns is increasingly relentlessly focused on the costs and disruptions of "net zero" policies and seeking to mobilize sectors of the population that are either climate skeptics or fossil fuel dependent, or those workers who have already experienced unjust transitions into a broad politics of grievance that touches on a range of broader economic, cultural, and political concerns.

Here complexities emerge in that Right populist arguments against transition *as they are currently proceeding* may be opportunistic and partial but their salience comes from the fact that working people dealing with decades of austerity, neoliberal deindustrialization and fragmented union protections and safety nets have good reasons to be concerned about further jobs losses occurring in carbon-intensive industries and sectors. The current populist focus on "greenflation" where broader cost-of-living concerns are applied to any and all net-zero policies may often involve a good deal of misinformation but their resonance clearly comes from the perception that the private market deployment of numerous green technologies to date – from EVs to rooftop solar - has been focused on providing elite green goods to upper income groups rather than making life easier for working people. Such arguments also derive their potency by mixing cultural and class components, whether focused on the hypocrisy of "jet setting eco-elites," cultivating resentments against urban green middle classes imposing their "implausible, impractical, and irrational sustainability preferences" for public transportation or dense green urbanisms over the needs of "motorists," or "authentic suburbanites and rural folk."

If the general strategy of Right populist forces in opposition is to rage against "net zero lunacy" as part of near-constant culture wars propagated through corporate media channels and social media, from where an increasingly large swath of the population gets their information about climate questions (Buck, 2023), perhaps the more salient question that arises is whether Right Populists in power actually have a coherent programmatic agenda beyond a politics of resistance.

The Right in power in the United States which has staked its future on a petro-nationalist and nativist project makes for an interesting test case here. Most notably, in terms of climate and environmental questions, the second Trump administration (1) has effectively destabilized and undermined confidence in permitting and financing of solar, wind, and EV production in the United States while undercutting much federal government support and incentives for second-generation decarbonizing technologies; (2) is actively working to increase energy inefficiencies in all manner of appliances and household goods; (3) is curtailing green manufacturing and green jobs initiatives initiated by the Biden administration; and (4) is defunding climate science, federal weather forecasting capabilities, and federal capacities to address extreme weather events. And for all talk of resistance to liberal elites, it is increasingly clear from Republican economic policies—such as the One Big Beautiful Bill Act passed by 119th Congress of 2025—that the US Populist Right is as committed as traditional neoliberals to achieving dramatic transfers of wealth and power from working people to the rich.

The populist Right, in its hard denialist mode, ultimately has no programmatic or credible socio-technical answer to the looming real material threats posed by climate change. The basic tool kit followed of deny, underplay, delay, continue building fossil fuel infrastructure will ultimately make life more dangerous, communities more vulnerable, goods more expensive, and energy less reliable and more polluting for the constituencies they profess to represent. Nevertheless, the rise of the populist Right clearly presents serious challenges for any credible account of the just transition, and it requires some serious and unflinching critical reflection.

Racial animus and the search for security in patriarchal traditionalism and harmony in imagined ethnically homogeneous communities of the past provide partial insight into the rise of the populist Right. But such explanations are clearly insufficient for explaining the ongoing electoral success of Right populism in many different political contexts or its capacity to make inroads into previously social democratic working-class strongholds in Europe and multicultural urban and suburban areas in the United States, let alone explain the electoral success of populist Right currents in India, Brazil, the Philippines, and so on.

As Oatley and Blyth (2021) among others have noted, in the United States the heartlands of populist Right support are to be found in states and regions tied to fossil fuel extraction and mining that have either the most to lose from post-carbon transitions or have already experienced "transition" in terms of deindustrialization. The political ascendance of this "carbon coalition" in the United States through the Trump administration has produced a global strategy that, according to Blyth and Driscoll, is attempting to impose "carbon dominance" on the rest of the world (see Blyth and Driscoll, 2025; Blyth, 2026). Carbon dominance can be seen as a realpolitik strategy based on the view that both domestic attempts at decarbonization in the Biden era and

China's green industrial build-out constitute an existential threat to the interests of the carbon coalition and to the comparative advantage of the US as the largest exporter of fossil fuels in the world. To curtail and unravel both threats requires the deliberate obstruction of not only domestic but also global decarbonization efforts. It further requires coercive trade policies (tariffs) and militarized adventures—Greenland, Venezuela, Iran—to reinforce the carbon dominance of the US across the globe.

Blyth (2026), interestingly, has suggested that this strategy might have short-term advantages for the US economy but has considerable drawbacks in the long term. The US is a hegemon in trouble, falling behind China in many critical technologies. Climate change is real, and a strategy to subsidize US fossil fuel infrastructure indefinitely could lead to stranded assets and a whole series of technologically and economically obsolescent industries. For much of the majority world, fossil fuels are expensive and impact the balance of payments. Low-carbon technologies could delimit the current account vulnerabilities of developing economies that emerge from the need to maintain dollars to buy fossil fuels.

The reduction of European and other right-populist forces to vassal states of a US carbon dominance project could well create further cracks and fissures in the coherence of right populisms. It is also evident that populist reactions cannot simply be explained focusing only on material dependencies or white supremacism/xenophobia. As Arlie Hochschild and others have noted, the populist Right has successfully tapped into much deeply rooted feelings of anger and anxiety that pervade sectors of the population that have been impacted by decades of economic uncertainty, and unjust transitions, who feel disrespected and talked down to by cultural elites and are profoundly dissatisfied with the status quo and with the poll-tested politicians of both Left and Right (Hochschild, 2018). Sociological research has pointed out for decades that green parties and climate and environmental movements in most high-income countries have been dominated by the urban professional and post-materialist middle classes, and green movements—apart from environmental justice movements—have struggled to expand their base of support outside these strata. The dominance of neoliberal environmental policymaking in high-income countries has made it easy to dismiss green products, transportation choices, or green urbanism as green lifestyle-ism, or simply modes of conspicuous consumption for green upper-class elites. Even Biden-era green industrial policy—as we argued in Chapter 6—ended up with a byzantine array of tax credits to purchase expensive decarbonizing technologies that were skewed toward homeowners with large disposable incomes.

The political space occupied by Right populism has clearly been further enabled by the atrophy and emptying out of much of the classic Left's key organizational base—trade unions, working-class cultural associations, civic associations and clubs, political parties, and social movements with deep roots in communities, workplaces, and churches (Putnam, 2000; Skocpol, 2003). The replacement of these social forces by a very online, much smaller leftish-greenish progressive culture heavily rooted in an NGO/higher-education sector grounded in the professional middle classes has to be recognized as a problem for moving just transitions forward (see Piketty, 2020; Huber, 2022; Jäger 2026). Most notably, it could be argued that the structural position of much of the green Left in academia, the NGO sector, and middle-class downshifting cultures is detached from the material practices of making, building, constructing, designing, planning, and the many forms of labor involved in doing this work (White, 1995; Sennett, 2008; Wark, 2016). Social democratic, socialist, and labor parties bridged these gaps at their highpoint in the twentieth century by a coalition politics that brought trade unions and working-class constituencies into a broader political arena dominated by technocrats and political elites, providing mechanisms through which political, material, and technical knowledge and skills gained by working on nature could have a place in the political arena. Scholarship and organizing around "labor-environmentalism" today could be seen as an explicit attempt to bridge this gap, yet it remains difficult and fraught work.

This would suggest that forms of sustainable and green design, architecture, and planning that are stuck repeating counter-culture agendas of old, naturalized green neoliberalism or taken the post-materialist sensibilities of the liberal greenish professional middle classes as a normative standard, and are devoid of any engagement with socioeconomic inequalities are unlikely to prosper in this new environment. Indeed, such currents are very likely to become targets of Right populist aggressions and will offer few tools to counter such challenges. More generally, the proposition that working-class electorates in high-income countries living in communities that have been downsized by decades of austerity, and squeezed by an ongoing cost-of-living crisis are going to be peeled away from the populist Right by a framing of designs for the just transition structured around an ecology of scarcity, middle-class agrarian romanticism, a mixture of green welfarism and economic stasis, calls for degrowth, or a "prepare for the worst" fortress ecology would seem utterly deluded. Rather, it would seem increasingly evident that green design, architecture, and planning will have to both center the question of labor in all its complex dimensions (as we have argued throughout this book) – and offer alternatives to the status quo that are more affordable, more attractive, more convivial, and simply better than the high-carbon alternative. Without this dual focus on labor and ecology, it is going to fail as a project·

Climate Nihilism, Climate Realism and War Ecology

If misinformation and populist grievance politics mark one very considerable challenge to any possible just transition, the rise of *climate nihilism* and *climate realism* presents a range of further obstacles. Over a decade ago, the political theorist William Connolly argued that a core intellectual underpinning to the energy politics of the Right was a discourse of nihilism, a positionality of "drill baby drill and screw the consequences." It could be noted that since this time, "climate doom" has come to circulate much more broadly as a mode of contemporary common sense in many political spaces beyond the far Right (Connolly, 2017). The proposition that "we" are heading toward inevitable and self-evident climate doom informs a great deal of first-world Leftist environmental analysis and social movement mobilization and indeed has pervaded youth culture more generally. A mixture of climate doom, climate realism, and realpolitik functions as evidence of "seriousness" in a great deal of climate and ecological commentary. Moreover, climate nihilism can easily attach to all manner of unrelated reasons for melancholy, discontent, and disappointment around the failures of modernism, the neoliberal colonialization of the university, the decline of the humanities, and so on—hence its capacity to inflect a wide range of regressive modes of first-world climate politics, from survivalist anarcho-primitivisms, to fortress eco-fascisms, to celebrations of propaganda by the deed.

Climate doom has received continual pushback from climate scientists such as Michael Mann (2022) and Kate Marvel (2025), from indigenous authors such as Kyle Powys Whyte (2017), Nick Estes (2019b), and environmental justice scholars such as Mary Annaïs Heglar (Heglar, 2019). If Mann and Marvel have continually argued that doom has no basis in climate science, Heglar, Estes, and Whyte have further argued that first-world climate doom is a deeply reactionary and classist politics, a sensibility where the struggle for just transitions will go to be buried (Heglar, 2019; Estes, 2019; Estes et al., 2021; Whyte, 2017). Climate nihilism nevertheless remains a potent force – particularly in the US and Europe - and one that is likely to grow if climate inaction continues. Moreover, doom-framed arguments are difficult to argue against because we are not anywhere near making the progress we need to make, one can find endless evidence for troubling developments and both potential and real disasters in the climate science literature on a near weekly basis, and the actual energy, climate, and social policies of the populist Right in power are clearly making worst-case climate scenarios far more likely.

If climate nihilism marks out one enemy of the just transition, we might also note that the policy manifestation of doom-lite can now be found in the troubling zone of so-called "climate realism." A number of powerful establishment institutions, from *The Council on Foreign Relations* in the United States (Sivaram, 2025) to the *Tony Blair Institute for Global Change* in the UK (Fursman 2025), are now making climate realist noises. Such a program is premised on the (reasonable) proposition that the mainstream international climate process has failed, and that we should accept that we are facing a world on track for 3 degrees of warming. It moves from these arguments to the assertion that Right populists have the wind in their sails, that publics in the EU and the US are disillusioned by net-zero aspirations, and increasingly will not tolerate the economic impact of such policies. What follows from this then is not an argument for replacing green neoliberal transitions with green economic populism to link low carbon and low cost but rather the proposition that "national interest" should now shape how the first-world countries respond to the climate crisis and that last-ditch technologies such as solar radiation management should be foregrounded as "we" prepare for worst-case outcomes.

It is interesting to note that UK and US climate realists that are now emerging from US establishment liberals and former champions of the "Third Way" in Europe are now making blatant imperialist climate arguments of a "lifeboat ethic" variety. As such, the climate realist position argues that however we might want to divide up historical responsibility for climate emissions, it is not the United States or UK that will be responsible for *future* emissions, but rather China, India, and Africa. Essentially, then, climate change is "their problem" to address, since individual emissions count for little, emissions in Europe are going down, and asking citizens in the high-income countries to sacrifice anything is both politically untenable and meaningless considering that the battle has moved to the Global South. This position conveniently ignores the fact that the Global North is responsible for the vast majority of historical emissions of greenhouse gases since the start of the Industrial Revolution, eliding the temporal and spatial aspect of justice in the responsibility for climate change. It is also interesting to note how, after lots of talk of "helping" the Global South to transition, the stick is revealed, with the view that the first-world nations must both help the South transition and compel countries to do so if they resist. Varun Sivaram has ominously warned that "Advanced economies should develop aggressive trade tools, tariffs, and so-called 'climate clubs' that penalize countries with large and fast-growing emissions." Such an approach to the Global South may be "fundamentally unfair," but nevertheless, Sivaram argues, "Every tool of the U.S. and allies' arsenals, spanning diplomatic and economic coercion to military might, should be on the table" (Sivaram, 2025).

If many US- and UK-based conversations around climate realism can be seen as centrist attempts to articulate a climate-lukewarmist politics marked by fealty to a faltering Pax Americana, Pierre Charbonnier has observed, following the outbreak of the Russia-Ukraine conflict, a not dissimilar turn to discourses of "war ecology" in Europe. War ecology captures a moment in which discussions of the "green transition" have been increasingly submerged by broader debates around energy security and resilience, weaponized supply chains, anxieties around escalating energy costs, and growing desires to break or mitigate the "toxic dependencies" that petrostates hold over the energy system (Charbonnier, 2022). Since Charbonnier's insightful marking of the war ecology moment in 2022, we have seen, if anything, a violent amplification of his thesis across the political field, turbocharged by the global chaos unleashed by the Trump administration.

What follows from this new period of energy and geopolitical crisis is hugely indeterminate. The ongoing energy and supply-chain shocks produced by conflict stretching from the Russia-Ukraine war to the US invasion of Venezuela and the US/Israel/Iran war have left much of the world exposed to fossil-fuel price spikes. From India and Indonesia to Pakistan and Brazil, many rising powers may now see their long-term material and strategic interests at odds with an unreliable US and its "carbon dominance" project (Blyth, 2026). Unlike the 1973/74 global energy crisis, which occurred when alternatives to fossil fuels, with the exception of nuclear power, were still awash with as-yet untested claims for the virtues of a "soft energy path," today we have China more than willing to sell low-carbon and low-cost solar, wind, EVs, and nuclear power to the rest of the world. As such, the present geopolitical energy crisis could plausibly lend new legitimacy to the Chinese renewables-plus-nuclear pathway, and force a new moment of Sino-European/African/South American cooperation (Tooze, 2026). Equally plausible, though, is that such a moment could further empower national-conservative and red-brown neofascist populist forces that insist sovereignty, security, and energy affordability can only be restored through revanchist fossil-fuel nationalism, marked by militarism, the protection of legacy fossil-fuel industries, and a return to fortress national ecologies stocked by manly men and culturally homogeneous societies.

What is clear, though, is that the violent return of geopolitics to discussions of design for transition throws an already difficult project into an even messier and more dangerous terrain, marked by unhappy trade-offs, unreliable allies, and rapidly shifting political alliances. Thinking about what a geopolitical ecology of design might entail under conditions of war ecology does bring new challenges to the field. Notably, after a long period of being primarily preoccupied with local communities, war ecology could push design and planning upward and outward, as it is increasingly forced to grapple with questions relating to grids and logistics, supply chains and infrastructure, ports and transit networks, and consider how the design fields relate to, and can contribute to, discussions around green industrial policy, state capacities, and territorial planning for adaptation and resilience (Brenner and Schmid, 2014; Easterling, 2014; Brenner, 2016; Baxi et al., 2018; Hutton, 2019; Brenner and Katsikis, 2020; Goh, 2021).

The Design Politics of Liberal Abundance

If climate realism could be viewed as a set of political currents that are preparing to weather the consequences of unjust transitions, a rather different set of arguments has recently emerged in influential liberal US circles to reset the contours of climate and progressive politics. Ezra Klein and Derek Thompson's book *Abundance* (2025) is an intervention that has been deeply shaped by the failure of Biden-era climate policy and the defeat of Kamala Harris. It is an argument that has also been influenced by the mobilizations around the Green New Deal of 2018–22 and it reflects the shock of the US pundit class with declining US hegemony in the face of China.

Abundance (2025) begins with a brief sketch of a possible future to come that is something of a hybrid between centrist Obama-era Silicon Valley climate solutionism and some of the more progressive speculative futuring provoked by the Green New Deal (Aronoff et al., 2019). Klein and Thompson's account of the United States of 2050 presents us with a world that has now achieved the widescale deployment of an assemblage of technologies, some familiar to ecotopian speculations (abundant and cheap energy provided by solar, wind, geothermal, and nuclear power; abundant mobility achieved through electric cars and scooters), some inflected with a modernist brush (abundant water through desalinization plants; abundant food provided by vertical farming and cellular meat; transformed health care through universal access to anti-addiction and weight loss medicines; transatlantic jets using traditional and green synthetic fuels), and some more recent preoccupations (drudgery alleviated and leisure expanded thanks to automation, AI, and drones). The social and institutional features of Klein and Thompson's world of 2050 is a little less clearly articulated, with social peace secured by loosely drawn "labor rights" and "economic reforms." The core theme of the book is that the reality of our existing world, marked as it is by climate and ecological crisis, a lack of affordable housing, a "broken" innovation system, and a zero-sum politics of scarcity is largely the product of broken state capacity, which results in an inability to get big projects done.

Klein and Thompson acknowledge that this "broken state capacity" is in part the product of decades of market fundamentalism on the Right, but the main focus of the book is how such conditions can also be traced back to a transformation of liberalism and broader Left advocacy dating back to the 1960s and 1970s. Notably, they argue, we have gone from "a liberalism that could build" during the New Deal era to a liberalism from the 1970s onward that has so wrapped itself up in legalism and proceduralism that it has ended up with a politics that stops both good projects and bad projects alike. This is an "everything bagel liberalism" that produces well-meaning legislation chock full of environmental reviews, project labor agreements, requests for public input, and the like, which nevertheless results in ever more costs and cumbersome bureaucratic red tape. They also note that since the 1960s we have seen in the United States at least, a fetishization of the "the local" and "communities" in left liberal activism and politics, but this has often expanded the abilities of middle- and high-income NIMBY property owners to say "no" to everything from multifamily and affordable housing in the suburbs to renewable energy projects and transmission lines that could carry clean energy from where it is generated to where it needs to go.

Klein and Thompson acknowledge that this liberal-left politics has delivered real successes such as the growth of an environmental regulatory state since the 1970s—which they argue was essential for dealing with the environmental challenges of the postwar era—but they suggest that this is a regulatory state that is not designed to meet the housing, climate, innovation, and infrastructural challenges of the present or the future. This is evident by the ways in which the biggest champions of this regulatory framework, notably, the bluest or most Democratic states in the United States, such as California, Hawaii, and Massachusetts, are states presiding over a crushing cost-of-living crisis and an inability to build much-needed transportation, affordable housing, or clean energy infrastructure quickly or efficiently. Examples of blue state inertia offered in the book include San Francisco's failure to build affordable housing and California's failure to build high speed rail—a project signed off on by Governor Jerry Brown in 1978, of which some 50 years later (and in the time that China has built 23,000 miles of high speed rail), only 500 miles of rail have been built from Merced to Bakerfield, behind schedule and massively over budget, a project that few support and even less will use.

What is the solution to these issues? *Abundance* argues for a range of smaller and larger moves that need to be made, ranging from proper public funding of the US innovation system to zoning reform, from permitting reform to a shift from proceduralism to delivery. But the main thread of the argument, skips over questions of redistribution (Bruenig, 2025) and instead gravitates toward emphasizing the need for a reform of regulations to support the expedited construction of critical projects. Klein and Thompson argue the politics of scarcity is the common theme that informs the zero-sum thinking of both the grievance-fueled populist Right obsessed with "those immigrants that will take away your wealth" and a NIMBY degrowth Left obsessed with (selectively) preaching the gospel of less to avoid catastrophe, but that neither current can move us forward. Rather, the "big tent" ideological shift they recommend needs to move beyond the politics of scarcity to embracing the idea of the abundant society.

What can we make of this intervention from the standpoint of design, planning, and architecture for the just transition? Klein and Thompson's version of abundance is an intervention disciplined by the horizons of liberal political discourse in the United States, and those looking for some insightful analysis of the role of the United States in the world system, climate debt, or ecologically uneven exchange will have to go elsewhere. In addition, it is a discourse that has already proved to be sufficiently pliable to not only generate dismissive critiques by the degrowth Left but attracted the support of both Democratic moderates and Left economic populists seeking to claim the abundance mantle. Nevertheless, and taking these limits into consideration, this is a book that raises important questions around what might constitute the most effective governance structures and means for implementing public policy proposals in ways that are often glossed over by design-driven transition discussions, which are often heavy on "aspirational visions" and light on how exactly political coalitions might be constructed to move such projects forward in the real world.

Klein and Thompson's analysis does capture elements of political inertia in the United States that are not just related to a hyper-partisan legislative process but to the ascendance of legal liberalism, proceduralism, exclusionary zoning, and hyper-decentralized localist public consultation processes that can often end up empowering those with the time and money to engage. These are all problems that can stand in the way of and unravel social democratic struggles for just transitions, and they need to be acknowledged and addressed. The focus on institutional ossification and broader failures of vision and implementation that Klein and Thompson point to in left-liberal governance are abysmal. They are right to argue that for all the talk of organizing and gaining public power on the Left, public power that cannot negotiate around interest group sectional interests and policy gridlocks to efficiently and effectively implement a program and actually deliver on basic promises undermines public confidence in public institutions and state action.

Yet, this is also a politically amorphous text with numerous narrative maneuvers in play and numerous absences in the analysis. For example, the manner in which Klein and Thompson largely put to one side the role that both the free market Right *and* centrist progressive neoliberals both played in producing and undercutting state capacity for much of the past five decades through privatization, outsourcing, amplifying corporate power, and market fundamentalism leaves us with a one-sided analysis. The text has little to say about the ways in which the unraveling of state capacity through the general neoliberal drift of policies across much of the world—from welfare reform, which expanded public–private partnerships, to deregulation of financial services, the expansion of private prisons, charter schools, and union-busting neoliberal trade policies—all occurred under the watch of Clinton-era Democrats in the United States and Blairites in the UK. *Abundance* is underpinned by a liberal-pluralist view of US politics, which tends to see the political outcomes they critique—NIMBYs and real estate developers, rigid environmentalists and fossil fuel industries, unions and employers—all competing on a more or less equal playing field. As such, the role that financial capital and many other vested interests have deployed for decades in undercutting the provision of public goods or rendering public review bodies underfunded and dysfunctional (Bozuwa and Mulvaney 2023b) receives much less attention. If we are looking for reasons for why "we don't have nice things," the proposition that the blame resides with labor and community benefits agreements, workers demanding a living wage for installing solar, environmental justice movements trying to avoid toxic waste dumps in minority neighborhoods, or indigenous communities trying to avoid oil pipelines running through their land would seem misdirected.

To move to questions of design, it is also worth pondering whether a just transition, or indeed any transition, is necessarily going to be best served by liberal abundance agendas that are primarily framed around getting fast and cheap results through a broad deregulatory agenda. Here, we have a number of problematic past precedents in the history of postwar housing where a drive to build cheap and fast, inattentive to public desires or the broader provision of services and infrastructure, led to fast and shoddy, which didn't turn out so well. Quick and cheap public housing projects, which ripped working-class communities out of existing support networks, ended up sited in areas lacking access to transportation or jobs, and had little funds for post-build maintenance, produced urban blight across the world. In the United States, Daniel Aldana Cohen and Mark Paul have observed that public housing in the pre- and postwar United States went up quick and cheap because of constraints introduced by the real estate lobby. It was the real estate lobby that was afraid of competition from the kinds of high-quality public housing that the New Deal promised and who lobbied hard for public housing to be built at a lower standard. As Cohen and Paul document, public housing in the United States became, "an option of last resort; indeed, throughout the late 1930s and early 1940s, the industry wrote laws that limited the amount the government could spend per unit of public housing, ensuring the structures were of inferior quality" (Cohen and Paul, 2020). The second wave of suburban sprawl and McMansion buildout also did not go so well either, culminating with a financial crisis and ghost estates and saddling millions of families with debt and foreclosures. In recent years, there has been a significant increase in the construction of new housing developments in regions that are historically susceptible to natural disasters, such as fires and flooding. This would suggest abundance alone that dispenses with careful planning for climate changed futures is not going to serve us well.

After the Backlash? Thinking About Radical Abundance and Red-Green Plenty for the Just Transition

From the perspective of the just transition then, it is easy to imagine a project of abundance driven by real estate interests, mortgage lenders, and the politicians that are in the pocket of such interests fueling a "green" construction building boom. In a decade, this could leave us with an abundance of new "green" ghost estates that have no sidewalks or services and nothing to do and that no one wants to live in. We could similarly imagine an abundant green urbanism defined by "sustainable" LEED Platinum-certified luxury downtown housing units—owned by private investors and largely empty—an abundance of upper-middle-class new "sustainable" gated communities with their own local private green security and decentralized solar grids, an abundance of shoddy lower income green housing built on land vulnerable to fire and flooding and hence, now uninsurable and owned by lower-middle-class folk now under water with their mortgage payments, and an abundance of working people living in "sustainable" trailer parks and tent cities made out of "new smart" low-carbon materials with compost toilets. We could envisage an abundance of urban growth that has locked in high-carbon forms in climate vulnerable areas and all of this built with materials that are difficult to disassemble by an abundance of unhappy low-paid workers who precariously cycle between often dangerous nonunion "green jobs" (where speed is everything and worker safety nothing), periods of AI-induced unemployment, and government schemes encouraging "grow your own (organic) food" and "make do and mend" schemes, to compensate for the dismantling of public assistance.

Of course, to be fair to Klein and Thompson, this is not the world they advocate for or promote. However, it is certainly possible that the Right-leaning edges of the very broad abundance coalition that has emerged in the light of their book (which presently ranges from techno-feudalist and lukewarm-ist ecomodernists to New Democrats, economic populists, and democratic socialists (Teles, 2025)) could generate such outcomes.

The plausibility of such a scenario has inspired some to dismiss the whole abundance discussion as re-baked neoliberalism and proclaim that the old eco-scarcity politics framed around localism, frugality, and "enough" will win the day. For those doubtful that such a politics has any popular traction, a rather different approach to this discussion would be to consider what resources might abundance discussions, the radical traditions of social and ecological design we have engaged with in this book, labor-environmentalism, and an understanding of design for the just transition as involving an invitation to engage in worldmaking open up for thinking about a politics beyond the backlash.

If the intersecting climate, housing, and broader cost-of-living crisis has to address questions of design and building, as we have seen in Chapters 3 and 6, there are many potential pathways for doing this. As we have pointed out in these chapters, this will require attending to the many opportunities that exist for housing abundance and new urban forms to be in part enabled by smart retrofitting, repurposing, and adaptive reuse using the existing stock of residential and commercial buildings. As we saw in Chapter 3, we know from the work of Ellen Dunham-Jones and June Williamson (2009, 2021) who have imaginatively explored the possibilities of retrofitting suburban sites such as derelict office parks, malls, and parking lots or Lacaton and Vassal's imaginative and award winning retrofitting of public housing that when retrofitting and adaptive use is done well, and in dialogue with residents, it can deliver much improved social and environmental outcomes. It will require thinking about the forms of planning, public financing, and regulatory mechanisms that can facilitate affordable and public green housing that can disrupt the hegemony of the real estate industry, which only builds when profits can be made and often just sits on land until profitable conditions arise (Stein, 2019; Hill and Mazzucato, 2024). Policy-wise, we are going to need an abundance of diverse ownership models (private, cohousing, co-ownership, cooperatives), different housing typologies, and approaches that encourage smart, just, and convivial retrofitting of downtowns and suburbs as well as high-quality sustainable new builds. A radical vision of abundance in housing—as Zohran Mamdani has argued—could include rent control, upzoning for density, new financing for green retrofits, and public investment—because markets alone are not going to deliver affordable and sustainable housing to all.

But what are the kinds of institutional and coordinating frameworks that could bring together design, architecture, and planning with struggles for social and environmental justice in a more integrated fashion? We have argued throughout this book that the disaggregation of the fields of design, planning, and architecture with each other, with the critical social and environmental sciences, and with labor and environmental movements has curtailed their agency in the world. Dan Hill and Mariana Mazzucato (2024) have further argued that a "mission-driven whole of government" approach that can bring design, planning, and policy together under a project of providing "housing for the common good" could be one productive space for moving forward here. They suggest aligning building and retrofitting strategies with a common good policy framework that involves "transformation in practices around housing in terms of design, construction, care, economy, and more diverse forms of shared living, tenure, ownership and governance" (2024: 3). The policy goal of a common good approach should be to stimulate, direct, and shape public bodies, private developers, and housing cooperatives to embark on building high-quality low-carbon public, social, and affordable housing that is "dignified, durable, beautiful and adaptable." Here, while Hill and Mazzucato acknowledge that private sector actors, community/co-operative and even self-build ventures can play some role in furthering common good housing, they emphasize that an active "market shaping entrepreneurial state" needs to be the driving force: "governments must invest in their internal capabilities to build, or re-build, the competence and confidence to successfully codesign and deliver programs, and progressively care for and maintain places, while collaborating fluidly across business and civil society" (Hill and Mazzucato, 2024).

Abundance of green housing in and of itself, moreover, is insufficient as a guiding frame for action. We need housing retrofitting and new green construction to occur in places where people want to live and where they can have access to high-quality public transport, well-paid union jobs, good schools, rich cultural opportunities, and decent services. This involves attending not just to more housing units but thoughtful urban planning and placemaking. Such housing typologies will have to address the changing demographic profile and lifestyle choices of mid-twenty-first century urban dwellers and seek to retrofit or green build in ways that also attend to the crisis of social disconnection and loneliness, the retreat from social and ecological engagement and experiences into isolated digital cultures. This requires that demands for housing are not only linked to thoughtful urban planning but—and as we have argued in Chapter 6—a genuinely inspiring account of a reclaimed public realm anchored by multifunctional public institutions. As we have seen throughout this book, there are enormous insights that a vision of radical abundance can draw from, in terms of feminist experiments with cohousing, Black urban ecologist advocacy of inner city ecological and social retrofitting, labor-centered designers focused on public housing, transition designers focused on communing, and collective ownership models as well as lessons that can be learnt from real-world housing policy in Vienna and other places (Cohen, 2019; Cohen and Paul, 2020).

If we further consider that there will be a huge pulse of embodied carbon emissions associated with the wave of new construction that will be needed to address the housing crisis and huge material impacts, the moment to radically decarbonize construction and delimit the social and ecological impacts of material extraction needs to occur *before* these buildings go up. This would suggest that a credible vision for how we build low-carbon and just *ecological* abundance futures is not just a deregulatory challenge but one of aligning incentives, new practices, land, and labor, and new innovations in design and technology in more effective ways to achieve the outcomes we want. New regulatory frameworks and incentives will be needed that can facilitate the use of low-carbon building materials with embedded and just systems of circularity. A credible circular economy needs to address minimum standard for dismantling, recycling, and the possible reuse of materials before buildings are built. Growing evidence suggests that building affordable, energy-efficient, and resilient sustainable housing at scale could productively explore innovations in modular housing and prefabricated building construction methods, which could in theory deliver reduced construction time and costs as well as cut carbon emissions and waste (Zonta, 2024). How these changes impact labor and ecological relations, construction sites, and workers will also need to be carefully considered.

More generally, retrofitting and building abundant urban ecologies for all will need to attend not just to the social and climate crisis but to the biodiversity crisis. Radical abundance in warming times will require us to think not simply about housing units or urbanism but consider the broader social ecologies that run through and shape all moments of the build environment. This opens up discussions of how a project advancing social and ecological justice as well as carbon drawdown can integrate housing affordability and upgrade the quality of our urban lives with strategies that can protect and expand urban biodiversity and increase access to nature for working people.

Finally designs for radical abundance will have to involve not just retrofitting and green building for climate-resilient futures but, as Tony Fry has continually emphasized (Fry 2009, 2014, 2020), unbuilding and engaging in eliminative design. Building/retrofitting and unbuilding climate futures will involve many moves. On the urban resilience front, it is essential that in a world heading for 1.5°C to 3°C warming climate adaptation as well as climate mitigation has to be central to future urban design and planning. This will involve amongst other things maximizing opportunities to counteract heat island effects by sheltering everyone through affordable housing to maximizing urban cooling and shading opportunities in the home and in the broader locality and many broader strategies to futureproof urban forms as much as possible against heat and flooding, wildfires and extreme weather events. Landscape design, ecological engineering as well indigenous design and planning can all play a critical role here in revitalizing the desire for low-carbon and ecologically just *municipal splendor*. Building a municipal realm and an urban landscape open to all urban citizens, as we saw in Chapter 2, shaped some of the best aspirations of the landscape and urban planning field in the late-nineteenth and early-twentieth century. Such an opportunity to return to this project today will need to incorporate the demand for affordable housing with designs for new green public spaces that are open to all, where many humans and nonhumans can go about their business in urban parks and gardens, urban pathways, bikeways, greenways and wildlife corridors, with green roofs that encourage biodiversity, wetlands and rich spaces for biodiversity in the city, and spaces that offer respite from extreme heat or cold. Unbuilding will involve not just decommissioning fossil fuel infrastructure and engaging in the cleanup of polluted landscapes but also considering what are the parts of the city or broader urban forms that we have to retreat from, or abandon, due to sea-level rise or fire threats (Fry, 2009, 2013).

To be sure the scale of national- and regional-scale projects needed for rapid and just decarbonization, of housing and urban form we have explored here, or of electrical transmission buildout, or of an interstate rail network necessitates a larger planning framework that is capable of master planning, prioritizing projects, selecting the most critical ones from a long list of possible candidates, evaluating alternatives, and focusing attention on minimizing their impact on the communities that will inevitably be impacted. Framing a just transition around localism, as Dimitris Stevis (2023) compellingly argues, is not going to suffice here. As we have seen in Chapter 3 and Chapter 6, this necessitates a recovery of the capacity for regional planning that has not been evident in the United States since the early New Deal era. However, we could observe that while there should always be room for streamlining legislation and regulations from the past that create inertia, Klein and Thompson's account of abundance here combines a critique of current regulatory frames with something of an ambivalent engagement with labor. To be fair here, they do acknowledge and support the unionization of the workplace in general, but they also suggest that the kinds of project labor agreements, union demands to be paid the prevailing wage in construction, or community benefits agreements that the Biden Administration sought to build into the IRA were a lag on progress.

A rather different view would be that such stipulations were important for making sure that low-carbon buildouts lift all boats and that workers are not exploited. As we have previously noted in Chapter 5, we already have a great deal of evidence from the explosion of the residential solar industry in the United States, that fast and cheap without labor protections or thoughtful regulation can lead to all kinds of exploitative conditions embedded in energy transitions (Huber, 2022; Lennon, 2025), which in turn generates backlash and can lead us back to slow and shoddy.

Kate Andrias and Alexander Hertel-Fernandez (2025) have further argued that labor unions can play a productive role in delivering a working-class and labor friendly vision of ecological abundance. Notably, they suggest that unions brought into policymaking early through tripartite agreements between government and the private sector can facilitate faster construction on large-scale projects through negotiating terms and conditions to speed up building and help recruit and direct skilled labor to projects through contributing "shop floor expertise that makes production of goods and services more efficient and effective." Second, they highlight that unlike many NGOs, which may profess to speak for communities but are often just "professional staffs with no mass constituency to hold them accountable," unions by law are "democratically run organizations with memberships." As membership organizations, which at their best represent millions of workers, they have the potential to help create the broader political coalition for abundance through building labor–environmental alliances focused on leveraging those green infrastructure projects where the gains of these projects are more broadly shared. Specific unions of course can also oppose all manner of projects because they conflict with their members' interests, but sectoral bargaining could alleviate these concerns. Finally, they point out that abundance discussions are largely focused on "growing aggregate wealth and supply" and that "concerns about the distribution of income and wealth or the responsiveness of government to working-class citizens barely make an appearance in most abundance discussions." This is a major failing of abundance agendas and if not addressed will just generate bad jobs, more upward division of wealth, and populist backlash. What is required in contrast then is "an abundance agenda that serves working people—not the economic and political elites who have already gained so much power over government."

The Impossible Project of the Just Transition

> The cause of labor is the hope of the world.
> The Workers' May Pole from Walter Crane in 1894.

Designing the Just Transition is a book written in dark times that nevertheless seeks to stand against climate doom and climate nihilism, melancholic paralysis, and climate denial. Rather we have argued that life on a warming planet necessarily throws us into *a project of redesigning the future*, because however the cards fall now, we will have to mitigate and adapt to life on a warmer planet, and we will have to do so as part of a multigenerational project, *iteratively* and *inventively*, again and again and again (Tonkinwise, 2003, 2015, 2024; Fry, 2009). The many conversations we have mapped in this book have been in debt to all manner of bold activists and advocates, who have often tried to press their case in the face of a great deal of disinterest or active resistance by key forces in the design academy and design industry.

At the same time, we have also seen in this book that there are certain kinds of first-world environmental advocacy that have long engaged in a kind of upper-class "savior politics," which has not always been the friend of working people in the Global North or the Global South. Such a politics has often romanticized certain kinds of natures, landscapes, and peoples and selectively championed the virtues of certain frugal modes of living while erasing the histories of labor that have shaped and tended these landscapes and obscured the broader entanglements of ecological questions with questions of labor, class, race, and gender and empire. Such currents have additionally often pursued quite biophysically reductionist agendas entranced by the agencies of nonhuman natures or (increasingly terrified) by the agency of climate, but devoid of an equally robust account of the potential *creative agencies* of *working people* to respond to the socio-ecological challenges we collectively face. What Nancy Fraser (Fraser 2022) has referred to as the "environmentalisms of the rich" have further been singularly preoccupied with the language of limits and constraints, drawn to stories of generic humans as environmental degraders, moved on to melancholic reflections on the world "we" have lost, called for decentering "the human," and ended with prophetic claims that "we" should all brace ourselves for a great catastrophe to come. The result of this has been an often-scolding politics, which, in current times, is increasingly wrapped up in existential gloom and, unsurprisingly, has continually struggled to gain a base of political support outside small slivers of the post-materialist middle classes of high-income countries.

Sustainable and ecological design discourse is perfectly capable of amplifying these ideologies and anxieties of the environmentalism of the rich. To take an almost comical example of what can happen when savior environmentalism collides with the elite world of design, at the XXII Triennale di Milano, *Broken Nature: Design Takes on Human Survival* was unveiled by MOMA director Paola Antonelli. The exhibition offered a range of contributions that "restorative design" could make to limiting carbon emissions. It is notable though how the exhibition catalogue framed the whole event through what can only be called a rising aestheticized fatalism of the 1 percent, working with the ridiculous proposition that design should embrace the call for building "A beautiful ending." As the exhibition catalogue outlines:

> Even to those who believe that the human species is inevitably going to become extinct at some point in the (near? far?) future, design presents the means to plan a more elegant ending. It can ensure that the next dominant species will remember us with a modicum of respect: as dignified and caring, if not intelligent, beings.[1]

We might detect here a certain affinity between *Broken Nature* and the growth of first-world "ruin porn" and long-standing Euro-American white environmentalist motifs of purification through the breakdown of nature. Such currents resonate with long strands of colonial conservation thinking that has wished to save nature … without people.

Design, architecture, and planning then are fields that are clearly not going to save us and very much need to get their own houses in order. Yet, we have also seen in this book that there is much more to the history of design, architecture, and planning's engagement with ecology and labor than events like *Broken Nature* would allow us to believe.

In contrast to the environmental politics and environmental design of the rich, we have sought to demonstrate in this book that a great deal of scholarship and activism emerging out of labor-environmentalisms, critical feminist and environmental justice studies, Southern environmentalisms, and critical Indigenous political ecologies have long tried to push back against these stories. Such currents have argued that for many working people around the planet, negotiating crisis is an everyday experience and that being told to accept your limits, decenter yourself, contain your voice, and accept boundaries are the ways in which most subordinated peoples have already been addressed for much of their lives (Whyte, 2017; Estes, 2019; Estes et al., 2021; Bookchin, 1995; Heglar, 2019). More recently, we have seen elite first-world environmental discourse splinter into currents that proclaim "technology will solve all"—rendering working people passive observers in a technocratic decarbonizing project that they have nothing to do with and no stake in—versus other currents that seek to convince working people, who have experienced decades of economic stagnation and austerity and are now immersed in a cost-of-living crisis, that the containment of economic growth should be the be-all-and-end-all aspiration of environmental politics. But it is increasingly clear that *both* these currents of environmental advocacy are spent ideologies, more or less pushing an exhausted politics that has little popular resonance. Both these singular positions (which are almost entirely the products of the academy or policy think tanks in high-income countries) lead to interventions that are oversimplistic and unhelpful, drawing lines between technology versus social change, green growth versus degrowth, which do little to build support for a majoritarian popular politics that can support just transitions or help to develop positions that speak to the aspirations and concerns of working people around the planet to live better, richer, more creative lives and to contribute to climate solutions and building regenerative social ecologies, without impoverishing themselves or their children.

The Brazilian radical political and legal theorist Roberto Unger has argued across a range of works that a viable progressive politics for the twenty-first century should aspire to see economy and society reorganized and democratized, part by part and step by step, so that "ordinary men and women have a better chance to live to a larger life." And by a larger life, Unger is not talking about the capacity to shop oneself to death but rather to live "a life of greater scope, greater capability and greater intensity" (see Unger, 2005). His writings have gone on to try and sketch institutions, modes of training and education, forms of organizational and technological support, and deep democratic engagements that could all be part of a Left strategy to expand the creativity and worldmaking possibilities of working people (see Unger 2005). Unger's interventions are not without problems and have had surprisingly little to say about climate or ecological questions. This book though has argued that in a range of design-orientated discourses and alternative currents of labor environmentalists, Black ecologists, feminist critics of alternative technology, social ecologists, and environmental justice advocates, we can locate a range of important attempts to rethink a material concept of creative agency as well as a vision of democratic worldmaking that might just be useful for our current times.

We have seen many dissenting and subordinated currents that have sought to suggest ways in which credentialed experts, design-aligned workers, *and* non-credentialed lay people could play a role in creatively materializing and enacting the just transition—if connected to larger political movements. Such movements of course need to offer concrete programs that can directly address climate, ecology, and material concerns such as the cost-of-living crisis and the mechanisms and modes of mobilization that can successfully implement change in ways that keep public support open for yet more action in the future.

This book has also sought to complicate the story of the designer as world changing neoliberal subject or powerless individual artist by highlighting that over the past hundred years and more, there have been many different sites and spaces that have opened up a *design politics* that has centered on questions of labor, nature, the environment, and justice. Our relentless focus here on design in this book of course needs to come with all manner of provisos. Designers, architects, and urban planners are no more likely to bring into being "a world transformed" than isolated health care workers are going to bring about universal health care. Moreover, designing the just transition is a project that will have to encompass many more groups and skills than we have focused on in this book—from agricultural workers to engineers, computer programmers to those involved in childcare and eldercare. Nevertheless, we have traced moments when designers, planners, and architects have acted as collective subjects and/or been part of broader movements, and this has opened *institutional* pathways for change. Notably, and as we have demonstrated, designers have found employment in the state as public servants tasked to implement the physical improvements of the New Deal (Chapter 3), proposed garden cities and promoted regional planning (Chapter 2), and helped to design and plan social housing or the postwar European Welfare State (Chapter 5). They have been employed to work with unions and employers to facilitate industrial democracy and workers' participation in Scandinavia (project DEMOS and UTOPIA) and been approached to help devise ways to nationalize sectors of the economy through "cybernetic management" (as in Project Cybersyn in Chile) (Chapter 5). Designers, architects, and planners have also fought battles to form their own unions, argued for unions that address broader social and ecological concerns than just terms and conditions (important though these struggles are), and built solidarity networks across the fields of architecture, construction, and building to address common problems. They have also fought to turn their workplaces into cooperatives or develop social businesses that are driven by a social and ecological mission. Such battles continue today as architects, designers, and planners have contributed their time and expertise to further just transitions and advocate for green new deals at local, state, and federal levels.

Professional designers, planners, and architects have also built forums and networks to contribute their specific technical, spatial, and visual knowledge to questions of climate, decarbonization, labor, and spatial justice. We can see this going on today in many spaces, from the quiet advocacy work of environmental design groups working on software, educational materials, and data for tracking carbon, energy, and ethical sourcing, to the work on building codes, performance standards, or zoning that are attempting to ratchet up industry standards and cultivate best practice to advance decarbonization.

In the realm of culture and political change, designers, planners, and architects have a long history of acting in the public sphere through organizing campaigns, making visual arguments, and engaging in larger cultural projects to draw public attention to labor and ecological abuses. The history of design, planning, and architecture is also full of direct interventions such as prefigurative politics (squatting, tactical urbanism, transition design), design activism (Who Builds Your Architecture?), withdrawing labor (the strike), or refusing to work on projects that exploit nature, communities, and workers. All will be critical to force and coerce the design industry toward just transitions (see Textbox 2 for a summary of the many possible pathways forward).

We have seen in this book that any plausible history of design cannot simply tell the story of credentialed designers. If design is a form of labor, and design work includes many more people than just credentialed designers, what are the possible pathways for design politics? We have also argued in this book that designing always draws together many visible and invisible forms of labor to do worldmaking. We live in a designed world, but this is a world made for most of human history by noncredentialed designers, proximate designers, and vernacular designers with their own deep reserves of knowledge and insight. Building just post-carbon futures is irreducibly going to draw from many hands to build different futures, and established currents of design can help by ceding their grip on projects and inviting codesign collaboration. Bringing together the augmenting and amplifying of the transformative capacities of a design politics will require *new alliances* and *new coalitions* to render the labors of design more efficacious. Maximizing the potential for designing the just transition will require—as Peggy Deamer and *the Architecture Lobby* have long argued—fighting for just transitions *within* the design industry. Here fights for fair pay, labor rights, and unionization are critical for design to get its own house in order. But we will also need to see coalition building to allow for different kinds of relationships to emerge *between different credentialed design fields*. The ecological architect Pedro Gadanho captures some of the much-needed new alliances and coalitions when he notes

> the architectural field will need many other alliances if it wants to envision its own transition beyond a now narrow disciplinary heritage. It will need to partner again with technological innovators, but to let technical ideas effectively propel its spatial conceptions. It will need to confer with ecologists and environmental engineers, but so as imagine how to turn buildings into living productive entities. And it will have to learn from their close colleagues, landscape architects, not to add on some last-minute green features, but to devise how natural carbon drawing strategies can become a permanent feature of their architectural proposals. (Gadanho, 2022: 89)

Perhaps more ambitious still are arguments that have long circulated in discussions prompted by *The New Architecture Movement* and feminist groups like *Matrix Design* in the 1970s and 1980s, and more recently has resurfaced in *Production Studies* and groups such as *Building Workers Unite* and the *Who Builds Your Architecture* group, notably that outside the relatively privileged sphere of credentialed architects and designers, we are also going to need a design politics of alliance building that moves across all design workers and design adjacent workers. We have also seen in this book important currents of critical and transition design, advocates of redirective practice, and "design when everyone designs" plausibly argue that design facing a warming world is going to have to acknowledge and validate the contributions and design skills of lay, noncredentialed designers. As Tony Fry has argued, in a warming world we would do well to maximize the "design intelligence" of the general population as well as building the support institutions and skills to facilitate contexts in which design when everyone designs is possible.

Text Box 2: Possible Spaces of a Design Politics for Just Transitions

Category	Pathway	Description	Examples/ References
Design politics for institutional transformation	Public employment	Employing architects, designers, and planners in public sector roles for social and environmental projects.	US New Deal, Greater London Council, Chile Project Cybersyn
	Green workers co-ops & employee ownership	Democratically owned, sustainable businesses managed by design-workers, with shared profits and decisions.	Platform Cooperativism, Mondragon Co-op, Cooperation Jackson, Evergreen Co-op
	Unionization of design workers	Organizing workers into unions for better pay, conditions, and influence over design outcomes.	Architecture Lobby, FAECT, Sveriges Arkitekter
	Building design worker solidarity networks	Alliances across design professions for collective bargaining and advocacy on climate and labor issues.	Building Workers Unite, Who Builds Your Architecture?, UVW-SAW
	Platforming and repurposing	Building green/social businesses from existing structures; retrofitting for post-carbon practice; nationalizing fossil fuel industries for carbon removal.	Tony Fry, Holly Jean Buck, Carbon Removal Futures
Expert and technical design politics	Technical tool networks	Networks to build/share construction software, educational materials, and data for tracking carbon, energy, and ethical sourcing.	Decarbonize Design, Carbon: A Field Manual
	Technical codes & standards networks	Reforming building codes/ standards for lower-carbon construction, sustainable consumption, and circular economies.	International Living Future Institute, LEED, Passive House
	Policy literacy (designers & policymakers)	Growing designers' policy awareness, working with lawmakers, integrating design expertise into policy.	Design Council UK, AIA Advocacy

Design activism	Public campaigns	Raising awareness and mobilizing public opinion around design labor and production networks.	Fashion Revolution, Architects Declare, Low Carbon Public Power
	Propaganda and cultural production	Supporting social movements through political symbolism, protest art, cultural material (posters, films, memes, etc.).	Grapus (France); Just Seeds Collective (US); Red Womens Workshop (UK).
	Prefigurative practice and reimagining public space	Using art/design to imagine new worlds and anticipate alternative ways of living; interventions in public space.	Communes, squats, tactical urbanism; transition design.
	Refusal, sabotage, strike, moratoria	Withholding labor, refusing harmful projects, disrupting exploitative/ climate-destructive work (strikes, go-slows).	Olin College Design Refusal, Strike for Climate, Architects Climate Action Network.
The design politics of co-creation and coproduction	Codesign with labor	Codesigning with workers to center their needs and knowledge in design processes.	Worker-centered HCI design, worker-centered service design, design for industrial democracy (Lucas Plan)
	Codesign with communities	Engaging communities directly in decision-making, design, planning, and building processes.	Design Justice Network, transition design, participatory design, advocacy planning
	Capacity building for design & planning	Creating centers and programs to empower broad participation in design and planning.	Technical advisory centers, participatory design centers, community planning centers.
	Popular design education for citizenship	Educating the public in design thinking and planning for democratic participation.	Patrick Geddes and Civic Education and Participation, RPAA, Tony Fry

	Vernacular design/DIY/ Design by society	Grassroots, nonprofessional, and community-led design and making.	Maker Movements, DIY, Pro-Am Movements, Maker Faire, Instructables, Open-Source Ecology

Designing the just transition can be seen as an invitation to embark on an impossible project. It is an impossible project much like the struggles to abolish slavery, defeat European colonialism, obtain universal suffrage, unravel absolute monarchies, and challenge rule by aristocrats and priests. In their time, these were all deemed impossible projects attempting to change timeless and immutable systems by much of the received opinion of the day. Yet, the just transition is an impossible project that we need to embark on to face down fossil fascism, try to save countless species and ecosystems from extinction and collapse, and give future generations a chance to enjoy democratic, equitable, and livable post-carbon futures. It is an argument that suggests the best ways to address those that feel betrayed by the powers that be and are tempted by the dead-end politics of grievance endlessly promoted by the populist Right is to argue and demonstrate that alternative paths for rebuilding the public realm exist. Rather than embracing a politics of limits, of living with less, or a nostalgic politics of return to a world we have lost, it is an argument that maintains we should collectively embark on a politics of democratic worldmaking and of creative design, regeneration, and reconstruction. It is an open invitation that does not counterpose ecology and labor but rather a project that honors the dignity of labor and recognizes the contribution that many hands and many skills could make to remaking society. In short, it is a call for a creative labor-environmental politics that is premised on the belief that, even at this late hour, there is still everything to play for.

Text Box 3: Sites of Coalition Building for a Design Politics for Just Transitions

Fighting for just transitions *within* the design industry and design firms.	Fighting for just transitions ***across*** credentialed design & design adjacent fields.	Coalition building for just transitions via recognition, validation, and engagement of lay, noncredentialed designers' contributions to designing just transitions; alliance building for co-construction.
Examples: Struggles for rights to fair pay, labor rights, unionization attempt to build cooperatives, refusal to work on exploitative or environmentally harmful sites	**Examples:** Adaptive reuse projects: collaboration among ecologists, engineers, architects, planners Circular economy projects: joint work of industrial ecologists, urban planners, economic, community planners, ecologists Building labor alliances across architectural, design, and construction/trade fields Building labor alliances across labor-environmental fields	**Examples:** Redirective practice and "design when everyone designs" Educational initiatives maximizing public "design intelligence" Traditions of codesign, community design, participatory design, citizens' design, design justice.
Advocates: The Architect Lobby Unionization Struggles in Design Fields	**Advocates:** Building Workers Unite Who Builds Your Architecture group Green Trade Unionism + Labor Environmentalism Green Workers Alliance	**Advocates:** The Design Justice Movement Community design and planning movements and centers Technical aid centers that facilitate self-managed design such as the Brazilian collective Usina-CTAH Self-Build and DIY Design Build advocacy groups

Notes

Introduction

1. Climate reductionism and climate determinism can come in different forms, such as (i) climate discourse that follows older forms of environmental determinist modes of explanation that attempt to reduce a wide range of complex socio-ecological processes from migration to famine, war to droughts as primarily driven by climate; (ii) interventions marked by a fixation on worst-case climate scenarios that flatten the complexities of climate science; (iii) a generalized politics of catastrophism that instrumentalizes "climate" to provide the delivery mechanism for political change; (iv) tendencies to collapse the climate model of possible futures with the actual future; (v) a singular preoccupation with "carbon" as the central environmental issue, thereby ignoring or sidelining other critical issues such as biodiversity loss, deforestation, overfishing, air, water and plastics pollution, soil degradation, and land use change. Useful critiques of such practices have been made by social and natural scientists over recent times coming from various standpoints (see variously Hulme, 2011; Paprocki, Cohen, Elliott and Kozlov, 2019; Mann, 2022; Marvel, 2025). Our focus in this book is engaged with climate issues, but as we shall see, just transition approaches necessarily see climate entangled in broader environmental concerns emerging from where we "live work and play." For a broader analysis of environmental issues, see White, Rudy and Gareau (2016).
2. Diana Ürge-Vorsatz Panel comments given to 17th International Conference on Sustainability Transitions Conference June, 2026, Lisbon.
3. For example, climate scientist Zeke Hausfather's (2025) metareview of future warming projections over the past five years suggests "a median estimate of future warming under current policies of 2.7°C in 2100 (with a 5–95th percentile range of central estimates spanning 2.3–3°C). Adding in emissions uncertainties and climate system uncertainties gives a much wider range of 1.9°C (5–95th percentile range of lower bound estimates from 1.5°C to 2.2°C) at the low end to 3.7°C (3.2–4.4°C) at the high end" (Hausfather, 2025).
4. See Ken Friedman and Erik Stolterman's 2018 Foreword to the MIT Design Thinking/ Design Theory Series https://doi.org/10.7551/mitpress/11035.003.0001.
5. See https://architecture-lobby.org/.
6. Wark's critique of the limits of modes of critical theory that fail to engage with the matter of how the world is made and designed has some degree of resonance with Richard Sennett's critique of Hannah Arendt in The Craftsman (2008). Notably his observation that Arendt's attempt to distinguish between *Animal Laborens* (that sees work as an end in itself) and *Homo Faber* (making a life in common) sets up a dynamic that "slights the practice man or woman at work." As he notes "For Arendt, the mind engages once labor is done. Another more balance view is that thinking and feeling are contained within the process of making" (Sennett, 2008: 7).

7. It could be argued that at the end of the day and despite their defense of utopian imaginaries, neither Walter Benjamin nor Henri Lefebvre managed to keep process and form together in any kind of concrete way in their speculative thought.

1 The Just Transition as a Social, Political, and Design Imaginary

1. The UAW's full name is the International Union, United Automobile, Aerospace and Agricultural Implement Workers of America.
2. Mazzocchi's starting point for a critique of liberal and Malthusian environmentalism bears some resemblances to the writings of Murray Bookchin during this period—who was probably the principal critic of such currents counterposing his own post-scarcity social ecology to austerity environmentalisms. However, unlike Mazzocchi, Bookchin unfortunately gave up on labor or the workplace as a site of social and ecological transformation from the late 1960s onward (see White, 2008: 70–82).
3. Using the language of twenty-first century ecopolitical theory, we might say that there are aspects to Mazzocchi's understanding of a just transition that has Left ecomodernist elements and other features that have sympathies with a post-growth understanding of Left social democracy.
4. Additionally, see the Shell Sustainability Report, 2023, which uses similar language on just transitions. https://www.shell.com/sustainability/reporting-centre/reporting-centre-archive/_jcr_content/root/main/section_2106585602/tabs/tab/text_copy/links/item0.stream/1742906426699/4ef5cfa607e5e308ad8a68fc3ddffbe6342f8fa8/shell-sustainability-report-2023.pdf.

2 Environmental, Ecological, and "Nature-Inspired" Design Histories

1. The exchange between Gui Bonsiepe and Victor Papanek is collected in Penin's excellent volume, *The Disobedience of Design. A Gui Bonsiepe Reader*.

3 Green Design as *Threat* and *Promise* for Architecture, Design, and Planning

1. One example of this drift in the UK is the way in which pioneers in the Radical Technology Movement, such as Godfrey Boyle, ended up playing a key role in establishing renewable energy as a research strength of The Open University, the public university founded by the Labour government in the late 1960s.
2. The politics of renewable energy technologies, their relationship to the decline of New Deal liberalism, and the rise of neoliberalism have become matters of considerable dispute, particularly within US/UK left energy debates in recent years. For example, Matt Huber and Fred Stafford have argued that decentralized renewables in the US have tended to be far more compatible with neoliberal political economies than many U.S. environmentalists have acknowledged (see Huber, 2022; Huber and Stafford,

2023). Amory Lovins's position as a champion of market environmentalism and decentralized solar and wind—alongside his longstanding criticism of centralized public utilities—is offered as evidence of such an alignment in the U.S. case. Huber and Stafford further contend that this alignment has reached full fruition in the contemporary United States, as a largely privately owned solar and wind sector, remains anti-union and hostile to public utilities and advocacy of off-grid solar continues to undercut public utility provision of electricity. In contrast, they have offered a full-throated defense of centralized public utilities foregrounding the low-carbon and potentially union-friendly virtues of nuclear power, citing France's rapid build-out of its nuclear fleet in the 1970s and 1980s as a model.

By contrast, the UK based academic Simon Pirani (2024) has argued that the roles of state, community, and private interests in the development and proliferation of renewable energy have varied significantly across history and geography—and across key centers of innovation and deployment such as Denmark, Germany, the U.K., and China. From this, Pirani suggests that the relationship between renewables and neoliberalism has been far more historically contingent than something inherent to the technologies themselves. Instead, he maintains that, on grounds of speed, safety, and cost, a renewables-led rollout under public ownership is preferable to a renewed nuclear push. As we shall see in Chapter 6, underlying this debate are fundamental disagreements about the optimal/ideal institutional and technological mix and the fastest paths to decarbonizing the electricity grid. Whilst these debates are undoubted important, the extent to which many US and UK based accounts of optimal decarbonizing energy systems are now being overtaken by events is the critical issue here. As we shall explore in Chapter 8, China has followed an energy-transition pathway that has involved a record build-out of both renewables and nuclear, with the U.K. and South Korea similarly embracing wind, solar, and nuclear pathways to decarbonization. France has maintained an electricity mix in which around 70 percent of generation comes from nuclear power while also retaining wind and solar targets in its 2026 energy law. Growth in electricity supply in Pakistan and Australia has largely come from a substantial and historically unprecedented rooftop solar boom, while in Japan there has been contentious debate and partial movement toward restarting nuclear reactors. In Germany a completed nuclear phase-out has nonetheless sparked renewed public argument over whether this decision should be reconsidered. Such ongoing diversity and variability in real-world global energy policy will undoubtedly, be further upended by the consequences of the war in Iran. This does suggest that pragmatic calculations based on geopolitics and shifting security concerns, country-specific natural endowments and geography, levels of grid development, cost and ease of financing, a desire to avoid external dependencies and limit current account vulnerabilities, public acceptance, the power of incumbent actors, social movements, and state capacity, are all likely to weigh much more heavily in future majority-world energy discussions than purely ideal-typical accounts of a single optimal low-carbon pathway.

3. *2005/32/EC*; Directive 2005/32/EC of the European Parliament and of the Council of 6 July 2005 Establishing a Framework for the Setting of Ecodesign Requirements for Energy-Using Products and Amending Council Directive.
4. https://www.ellenmacarthurfoundation.org/towards-the-circular-economy-vol-1-an-economic-and-business-rationale-for-an.
5. Searchinger, Timothy, Liqing Peng, Zionts Jessica, and Richard Waite. 2023. "The global land squeeze: Managing the growing competition for land." World Resources

Institue [report]. https://www.wri.org/research/global-land-squeeze-managing-growing-competition-land.
6. BIG. 2020. Formgiving: An Architectural Future History. Taschen.
7. See https://www.houseeurope.eu/.

4 Labor-Centered and Worker-Centered Design for Just Transitions?

1. See https://michaelsinger.com/.
2. See https://just2ce.eu/about-us/.
3. See https://news.industriall-europe.eu/documents/upload/2022/12/638055883130940811_Adopted_-_More_and_fair_circular_economy_-._towards_strategic_autonomy_for_industrial_jobs_and_a_cleaner_environment_-_EN.pdf.
4. See https://rotordc.com/aboutus-1.
5. See https://aarch.dk/en/linked-bank-of-materials/.

6 The Green New Deal, Design, Planning, and the Public Imagination

1. See https://feministgreennewdeal.com/, redblackgreennewdeal.org; The Red Nation 2021.
2. See https://www.monmouth.edu/polling-institute/reports/monmouthpoll_us_050624/.

7 Planetary Designs and Speculative Climate Design Futures

1. See "A Message from the Future with AOC" https://www.youtube.com/watch?v=d9uTH0iprVQ.
2. See The Green New Deal Superstudio https://www.gndsuperstudio.com/home.

8 The Politics of the Just Transition in an Age of Disenchantment

1. See http://www.brokennature.org/exhibition/theme/.

Bibliography

Abdulla, Danah, Ahmed Ansari, Ece Canlı, Mahmoud Keshavarz, Matthew Kiem, Luiza Prado de O Martins, Tristan Schultz, and Pedro Vieira de Oliveira. 2019. "Decolonizing Design Manifesto." *Journal of Futures Studies* 23(3): 129–32.

Agarwal, Anil, and Sunita Narain. 1991. *Global Warming in an Unequal World*. N.a.

Agyeman, Julian, Robert Doyle Bullard, and Bob Evans. 2003. *Just Sustainabilities: Development in an Unequal World*. Earthscan.

Ahmed, Nafeez. 2025. "Civilisation could Leap into a New Era of Superabundance." *Club of Rome Blog*, January 9. Available at: https://www.clubofrome.org/blog-post/ahmed-civilisation-superabundance/ (accessed January 26, 2026).

Ajl, Max. 2021. *A People's Green New Deal*. Pluto Press.

Akama, Yoko, and Joyce Yee. 2019. "Embracing Plurality in Designing Social Innovation Practices." *Design and Culture* 11(1): 25–45.

Akuno, Kali, and Ajamu Nangwaya (Eds.) 2017. *Jackson Rising: The Struggle for Economic Democracy and Black Self-Determination in Jackson*. Daraja Press.

Alexander, Samuel, and Brendan Gleeson. 2019. *Degrowth in the Suburbs: A Radical Urban Imaginary*. Palgrave Macmillan.

Allen, Matthew. 2020. *Architecture Becomes Programming: Invisible Technicians, Printouts, and Situated Theories in the 1960s*. In Sebastiaan Loosen, Rajesh Heynickx, and Hilde Heynen (Eds.), *The Figure of Knowledge: Conditioning Architectural Theory, 1960s–1990s*. Leuven University Press.

Andrias, Kate and Alexander Hertel-Fernandez. 2025. "Abundance That Works for Workers—and American Democracy" published on March 31, 2025. Available at: https://rooseveltinstitute.org/blog/abundance-that-works-for-workers/.

Anguelovski, Isabelle, and James J. T. Connolly. 2021. *The Green City and Social Injustice: 21 Tales from North America and Europe*. Routledge.

Anker, Peder. 2001. *Imperial Ecology: Environmental Order in the British Empire, 1895–1945*. Harvard University Press.

Anker, Peder. 2010. *From Bauhaus to Ecohouse: A History of Ecological Design*. Louisiana State University Press.

Anklesaria, Sarosh, Joshua Barnett, Kevin Bogle, Joseph Bondi, Adare Brown, Gioia Connell, Clare Fentress, Carson Fisk-Vittori, Sean Flynn, Sam Hauserman, Niels Hoyle-Dodson, Elisa Iturbe, Valérie Lechêne, Ryan Ludwig, Mahan Shirazi, Akshita Sivakumar, Caitlin Watson, Matthew Waxman, and James Wood. 2021. "A Just Transition for the Building Sector: Toward a Radical Transformation of Housing and the Built Environment." *The Architecture Lobby*. https://architecture-lobby.org/news/a-just-transition-for-the-building-sector/.

Ansari, Ahmed. 2024. "Decolonization & Knowledge in Design." *Diseña* 25: 12–33.

Arboleda, Martín. 2020. *Planetary Mine: Territories of Extraction under Late Capitalism*. Verso Books.

Arnstein, Sherry R., 1969. "A Ladder of Citizen Participation." *Journal of the American Institute of Planners* 35(4): 216–24.

Aronoff, Kate. 2016. "How to Socialize America's Energy." *Dissent Magazine*, Spring 2016. Available at: https://www.dissentmagazine.org/article/energy-democracy-usa-socialize-renewable-public-private-cooperatives/.

Aronoff, Kate. 2017. "Bringing Power to the People: The Unlikely Case for Utility Populism." *Dissent Magazine*, Summer 2017. Available at: https://www.dissentmagazine.org/article/the-unlikely-case-for-utility-populism-rural-electric-cooperatives/.

Aronoff, Kate. 2023. "The IRA Is an Invitation to Organizers." *Dissent Magazine*, Spring.

Aronoff, Kate. 2024. *Green Industrial Policy's Unfinished Business: Managing the Fossil Fuel Transition*. New York: Roosevelt Institute. Available at: https://rooseveltinstitute.org/wp-content/uploads/2024/07/RI_Green-Industrial-Policy-Fossil-Fuels_Report_072024.pdf.

Aronoff, Kate, Alyssa Battistoni, Daniel Aldana Cohen, and Thea Riofrancos. 2019. *A Planet to Win: Why We Need a Green New Deal*. Verso Books.

Asafu-Adjaye, John, Linus Blomquist, Stewart Brand, Barry W. Brook, Ruth DeFries, Erle Ellis, Christopher Foreman, David Keith, Martin Lewis, and Mark Lynas. 2015. "An Ecomodernist Manifesto." https://www.ecomodernism.org/.

Aureli, Pier Vittorio. 2016. "History, Architecture, and Labour: A Program for Research." In Aaron Cayer, Peggy Deamer, Sben Korsh, Eric Peterson, and Manuel Shvartzberg (Eds.), *Asymmetric Labors: The Economy of Architecture in Theory and Practice*, 157–62. The Architecture Lobby.

Baker, Shalanda. 2021. *Revolutionary Power: An Activist's Guide to the Energy Transition*. Island Press.

Barber, Daniel A. 2016. *A House in the Sun: Modern Architecture and Solar Energy in the Cold War*. Princeton University Press.

Barber, Daniel A. 2019. "After Comfort." *Log* 47: 45–50.

Barber, Daniel A. 2023. *Modern Architecture and Climate: Design Before Air Conditioning*. Princeton University Press.

Barber, Daniel A. 2024. "Drawing the Line." *Places Journal*. https://placesjournal.org/article/drawing-the-line-architecture-in-the-anthropocene/

Barbrook, Richard, and Andy Cameron. 1996. "The Californian Ideology." *Science as Culture* 6(1): 44–72.

Barca, Stefania. 2012. "On Working-Class Environmentalism: A Historical and Transnational Overview." *Interface: A Journal for and About Social Movements* 4(2): 61–80.

Barca, Stefania. 2019. "The Labor (s) of Degrowth." *Capitalism Nature Socialism* 30 (2): 207–16.

Barca, Stefania. 2024. *Workers of the Earth: Labour, Ecology and Reproduction in the Age of Climate Change*. Pluto Press.

Bardhan, Ronita. 2024. Climate Adaptation: How Can Design Science Help the Transition? Working Paper.

Barry, John, and Geraint Ellis. 2014. "Beyond Consensus? Agonism, Republicanism and a Low Carbon Future." In *Renewable Energy and the Public*, 29–42. Routledge.

Bastani, Aaron. 2019. *Fully Automated Luxury Communism*. Verso Books.

Battistoni, Alyssa. 2017a. "Bringing in the Work of Nature: From Natural Capital to Hybrid Labor." *Political Theory* 45(1): 5–31.

Battistoni, Alyssa. 2017b. "Living, Not Just Surviving." *Jacobin* 26 (Summer): 65–71.

Battistoni, Alyssa. 2022. "4 A Green New Deal for Care: Revaluing the Work of Social and Ecological Reproduction." In *The Green New Deal and the Future of Work*, 103–41. Columbia University Press.

Baxi, Kadambari, Jordan Carver, Laura Diamond Dixit, Lindsey Lee, Tiffany Rattray, and Mabel O. Wilson. 2018. "Design Construction Networks. Who Builds Your Architecture?" *Ardeth. A Magazine on the Power of the Project* (3): 192–203.

Bazilian, Morgan, Gregory Clough, Juliet Akamboe, Aaron Malone, Macdonald Amoah, Ian Lange, Yasmin E. Marin, Ebenezer Manful-Sam, Brad Handler, Felix Ayaburi, and Elizabeth A. Holley. *The State of Critical Minerals Report* 2023.

Beachy, Ben. 2019. "Five Big Ideas for a Green New Deal." Sierra Club. January 1, 2019. https://www.sierraclub.org/articles/2019/01/five-big-ideas-for-green-new-deal.

Bell, Karen. 2019. *Working-Class Environmentalism: An Agenda for a Just and Fair Transition to Sustainability.* Springer.

Benjamin, Ruha. 2019. *Race after Technology: Abolitionist Tools for the New Jim Code.* John Wiley.

Benyus, Janine M. 1997. *Biomimicry: Innovation Inspired by Nature.* William Morrow.

Berger, Markus, and Kate Irvin (Eds.). 2022. *Repair: Sustainable Design Futures.* Taylor & Francis.

Black Panther Party. Fall 1974, "Landbanking." *CoEvolution Quarterly* 3: 33.

Blyth, Mark, and Driscoll, Dan. 2025. "Trump's global war on decarbonization". *Project Syndicate*, 2025. Available at: https://www.project-syndicate.org/commentary/trump-global-war-on-decarbonization-green-technologies-by-mark-blyth-and-daniel-driscoll-2025-08 (accessed January 26, 2026).

Blyth, Mark. 2026. "*Does decarbonization still have a future?*" Lecture given to the London School of Economics European Institute, 20 March. Available at: https://www.youtube.com/watch?v=09AY6X2Xppw (accessed 26 March 2026).

Boehnert, Joanna. 2018. *Design, Ecology, Politics: Towards the Ecocene.* Bloomsbury.

Bond, Kingsmill, and Sumant Sinha. 2026. "India's Electrotech Fast-Track: Where China Built on Coal, India Is Building on Sun." Ember, January 21. Available at: https://ember-energy.org/latest-insights/indias-electrotech-fast-track-where-china-built-on-coal-india-is-building-on-sun/.

Bookchin, Murray. 1971. *Post-Scarcity Anarchism.* Ramparts Press.

Bookchin, Murray. 1980. *Toward an Ecological Society.* Black Rose Books.

Bozuwa, Johanna, and Dustin Mulvaney. 2023a. "Ten Policies to Manage the Fossil Fuel Transition." *Climate and Community Institute Policy Brief.* Available at: https://climateandcommunity.org/wp-content/uploads/2023/12/23_03_14_ENG-ENERGY-3-1.pdf (accessed January 26, 2026).

Bozuwa, Johanna, and Dustin Mulvaney. 2023b. "A Progressive Take on Permitting Reform: Principles and Policies to Unleash a Faster, More Equitable Green Transition." *Climate and Community Project and Roosevelt Institute*, August 22, 2023. https://rooseveltinstitute.org/publications/a-progressive-take-on-permitting-reform-principles-and-policies-to-unleash-a-faster-more-equitable-green-transition/.

Brand, Stewart. 1995. *How Buildings Learn: What Happens After They're Built.* Penguin.

Brand, Stewart. 2009. *Whole Earth Discipline: An Ecopragmatist Manifesto.* Viking.

Bratton, Benjamin H. 2015 *The Stack: On Software and Sovereignty.* MIT Press.

Bratton, Benjamin H. 2019. *The Terraforming.* Strelka Press.

Bratton, Benjamin H. 2020. "18 Lessons of Quarantine Urbanism." *Strelka Mag, 3.*

Brenner, Neil, and Christian Schmid. 2014. *Implosions/Explosions.* Jovis.

Brenner, Neil. 2016. *Critique of Urbanization: Selected Essays.* Birkhäuser.

Brenner, Neil, and Nikos Katsikis. 2020. "Operational Landscapes: Hinterlands of the Capitalocene." *Architectural Design* 90(1): 22–31.

Brown, Ross. D. 2014. "Farming Landscapes, Energy Landscapes, or Both? Multifunctional Landscapes as a Response to Rural Energy Transitions." In *Changing Senses of Place: Migration, Place, and Identity in a Global Age*, 187–204. Cambridge University Press.

Bruenig, Matt. 2019. *Fighting Climate Change with a Green TVA*. Peopless Policy Project. Available at: https://www.peoplespolicyproject.org/project/fighting-climate-change-with-a-green-tva/ (accessed January 26, 2026).

Bruenig, Matt. 2025. "The Abundance Agenda." The People's Policy Project. https://www.peoplespolicyproject.org/2025/03/24/the-abundance-agenda/

Brulle, Robert J., J. Timmons Roberts, and Miranda C. Spencer. 2024. *Climate Obstruction across Europe*. Oxford University Press.

Buch-Hansen, Hubert. 2018. "The Prerequisites for a Degrowth Paradigm Shift: Insights from Critical Political Economy." *Ecological Economics* 146: 157–63.

Buck, Holly Jean. 2019. *After Geoengineering: Climate Tragedy, Repair, and Restoration*. Verso Books.

Buck, Holly Jean. 2021. *Ending Fossil Fuels: Why Net Zero Is Not Enough*. Verso Books.

Buck, Holly Jean. 2023. "Confronting Climate Change in Extremely Online Times." In D. W. Orr (Ed.), *Democracy in a Hotter Time*. MIT Press.

Bullard, Robert D. 1990. *Dumping in Dixie: Race, Class, and Environmental Quality*. 1st ed. Westview Press.

Bureau of Labor Statistics 2022. *Occupational Outlook Handbook: Solar Photovoltaic Installers; Wind Turbine Technicians*. Washington, DC: US Department of Labor. Available at: https://www.bls.gov/ooh/ (accessed January 25, 2026).

Campbell, Eryn, Lauren Lutzke, Heather Kostick, and Michael Mann. 2024. "Communicating Policy: Raising IRA Awareness in Philadelphia." *Kleinman Center for Energy Policy* (blog). July 12, 2024. Available at: https://kleinmanenergy.upenn.edu/commentary/blog/communicating-policy-raising-ira-awareness-in-philadelphia/.

Carlisle, Stephanie. 2020. "I've Been Polluting the Plant for Years. I'm Not an Oil exec—I'm an Architect." Fast Company, January 3, 2020. Available at: https://www.fastcompany.com/90435650/these-are-the-last-years-of-design-as-we-know-it.

Carlisle, Stephanie, and Nicholas Pevzner. 2013. "Introduction: Rethinking Infrastructure." *Scenario Journal* 4.

Carlisle, Stephanie, and Nicholas Pevzner. 2021. "The Thin Thread of Carbon." In *Non Extractive Architecture, Volume 1: On Designing Without Depletion*, edited by Space Caviar. Sternberg Press.

Carlisle, Stephanie. 2017. "Why Architects Must Confront Embodied Carbon." In R. Stahl and M. Rudolph (Eds.), *Embodied Energy and Design: Making Architecture Between Metrics and Narratives*. Lars Müller.

Carlock, Greg, Emily Mangan, and Sean McElwee. 2018. "A Green New Deal: A Progressive Vision for Environmental Sustainability and Economic Stability." Data for Progress. Available at: https://www.dataforprogress.org/green-new-deal-report.

Carr, Sara Jensen. 2021. *The Topography of Wellness: How Health and Disease Shaped the American Landscape*. University of Virginia Press.

Cayer, Aaron, Peggy Deamer, Sben Korsh, Eric Peterson, and Manuel Shvartzberg. 2016. *Asymmetric Labors: The Economy of Architecture in Theory and Practice*. Architecture Lobby.

Ceschin, Fabrizio, and İdil Gaziulusoy. 2016. "Design For Sustainability: An Evolutionary Review." *Proceedings of the Design Research Society Conference (DRS 2016)*, 3731–3748.

Brighton, UK. Available at: https://dl.designresearchsociety.org/cgi/viewcontent.cgi?article=1265&context=drs-conference-papers.

Ceschin, Fabrizio, and İdil Gaziulusoy. 2019. *Design for Sustainability: A Multi-Level Framework from Products to Socio-Technical Systems*. Routledge.

Cha, J. Mijin. 2024. *A Just Transition for All: Workers and Communities for a Carbon-Free Future*. MIT Press.

Cha, J. Mijin, Vivian Price, Dimitris Stevis, and Todd E. Vachon. 2021. "Workers and Communities in Transition: Report of the Just Transition Listening Project."

Chapman, Jonathan. 2005. *Emotionally Durable Design: Objects, Experiences and Empathy*. Earthscan.

Charbonnier, Pierre. 2022. "The Birth of War Ecology." *GREEN* 2(1): 70–76. https://doi.org/10.3917/green.002.0076.

Cheng, Irene, Charles L. Davis II, and Mabel O. Wilson. (Eds.). 2020. *Race and Modern Architecture: A Critical History from the Enlightenment to the Present*. University of Pittsburgh Press.

Choi, Ehoi 2020. *Sustainability's Image Problem*. Library Stack for the Oslo Architecture Triennale.

Christophers, Brett. 2024. *The Price Is Wrong: Why Capitalism Won't Save the Planet*. Verso Books.

Churkina, Galina, Alan Organschi, Christopher P. O. Reyer, Andrew Ruff, Kira Vinke, Zhu Liu, Barbara K. Reck, T. E. Graedel, and Hans Joachim Schellnhuber. 2020. "Buildings as a Global Carbon Sink." *Nature Sustainability* 3(4): 269–76. https://doi.org/10.1038/s41893-019-0462-4.

Coe, Neil M. 2021. "Afterword: Towards a Political Ecology of Labour?" *Area* 53 (3): 450–3.

Cohen, Daniel Aldana. 2019. "A Green New Deal for Housing." *Jacobin*, February 8, 2019. Available at: https://jacobin.com/2019/02/green-new-deal-housing-ocasio-cortez-climate.

Cohen, Daniel Aldana, and Mark Paul. 2020. "The Case for Social Housing." The Appeal, November 2.

Colgan, Jeff D., Jessica F. Green, and Thomas N. Hale 2021. "Asset Revaluation and the Existential Politics of Climate Change." *International Organization* 75(2): 586–610.

Cooke, Bill, and Uma Kothari (Eds.). 2001. *Participation: The New Tyranny?* Zed Books.

Cooley, Mike. 1980. *Architect or Bee? The Human/Technology Relationship*. Spokesman Books.

Commoner, Barry. 1990. *Making Peace with the Planet*. Pantheon Books.

Connolly, William E. 2017. *Facing the Planetary: Entangled Humanism and the Politics of Swarming*. Duke University Press.

COP28, IRENA, and GRA. 2023. "Tripling Renewable Power and Doubling Energy Efficiency by 2030: Crucial Steps towards 1.5°C." International Renewable Energy Agency. Available at: https://www.irena.org/Publications/2023/Oct/Tripling-renewable-power-and-doubling-energy-efficiency-by-2030.

Corner, James. 1999. *Recovering Landscape: Essays in Contemporary Landscape Theory*. Princeton Architectural Press.

Corvellec, Hervé, and Herman I. Stål. 2019. "Circular Business Models: Depoliticizing Waste." *Organization & Environment* 32(4).

Costanza-Chock, Sasha. 2020. *Design Justice: Community-Led Practices to Build the Worlds We Need*. MIT Press.

Creese, Walter L. 1990. *TVA's Public Planning: The Vision, the Reality*. University of Tennessee Press.
Cullenward, Danny, and David G. Victor. 2020. *Making Climate Policy Work*. John Wiley.
Cutler, Phoebe. 1985. *The Public Landscape of the New Deal*. 1st ed. Yale University Press.
Daggett, Cara. 2018. "Petro-Masculinity: Fossil Fuels and Authoritarian Desire." *Millennium* 47(1): 25–44.
Deamer, Peggy, and Phillip Bernstein. (Eds.). 2010. *Building (in) the Future: Recasting Labor in Architecture*. Princeton Architectural Press.
Deamer, Peggy, 2013. *Architecture and Capitalism: 1845 to the Present*. Routledge.
Deamer, Peggy (Ed.). 2015. *The Architect as Worker: Immaterial Labor, the Creative Class, and the Politics of Design*. Bloomsbury.
Deamer, Peggy. 2020. *Architecture and Labor*. Routledge.
Dharia, Namita Vijay. 2022. *The Industrial Ephemeral: Labor and Love in Indian Architecture and Construction*. Vol. 7. University of California Press.
Dilnot, Clive, Tony Fry, and Susan C. Stewart. 2015. *Design and the Question of History*. Bloomsbury Academic.
Di Paola, Roberto. 2021. "Milan's Private Vertical Forests vs. Horizontal Urban Greening." In I. Anguelovski and J. J. T. Connolly (Eds.), *The Green City and Social Injustice: 21 Tales from North America and Europe*. Routledge.
Dragsted, Pelle, and William Banks. 2025. "'Abundance' Requires Redistribution." *Compact Magazine*. https://www.compactmag.com/article/abundance-requires-redistribution/.
Dümpelmann, Sonja. 2022. *Frederick Law Olmsted: Landscape Architect and Environmental Planner*. Thames & Hudson.
Dunham-Jones, E. and Williamson, J. 2009. *Retrofitting Suburbia: Urban Design Solutions for Redesigning Suburbs*. John Wiley.
Dunham-Jones, E., and J. Williamson. 2021. *Case Studies in Retrofitting Suburbia: Urban Design Solutions for Greening Suburbs*. John Wiley.
Dutta, A. 2006. *The Bureaucracy of Beauty: Design in the Age of Its Global Reproducibility*. Routledge.
Ehn, P. 1990. *Work-Oriented Design of Computer Artifacts*. Lawrence Erlbaum Associates.
Eisenson, Matthew, Jacob Elkin, Andy Fitch, Matthew Ard, Kaya Sittinger, and Samuel Lavine. 2024. "Rebutting 33 False Claims About Solar, Wind, and Electric Vehicles." Sabin Center for Climate Change Law, April 2024 (revised February 2025). Available at: https://scholarship.law.columbia.edu/sabin_climate_change/217.
Ellen MacArthur Foundation 2013. *Towards the Circular Economy: Economic and Business Rationale for an Accelerated Transition*. Cowes: Ellen MacArthur Foundation.
Elzenbaumer, B., and Giuliani, C. 2014. "Designers' Inquiry: A Report of a Study on the Everyday Realities of Seven Graphic Designers." *Ephemera: Theory & Politics in Organization* 14(3): 657–68.
Ember. 2025. *Global Electricity Review 2025*. Available at: ember-climate.org/reports (accessed March 25, 2026).
Escobar, Arturo. 2018. *Designs for the Pluriverse*. Duke University Press.
Escobar, Arturo. 2020. *Pluriversal Politics: The Real and the Possible*. Duke University Press.
Estes, Nick. 2019a. *Our History Is the Future: Standing Rock versus the Dakota Access Pipeline, and the Long Tradition of Indigenous Resistance*. Verso Books.
Estes, Nick. 2019b. "Indigenous Resistance Is Post-Apocalyptic." Dissent Magazine, July 31.

Estes, Nick, Melanie Yazzie, Jennifer Nez Denetdale, and David Correia. 2021. *Red Nation Rising: From Bordertown Violence to Native Liberation*. PM Press.
Estevez, Isabel, and Riofrancos, Thea. 2025. "Global Green Industrial Policy: Navigating Power Dynamics for a Pro-Working-Class, Pro-Development Green Transformation." Climate and Community Institute, September 2025, https://climateandcommunity.org/research/global-green-industrial-policy.
Estreal, B., and M. Ramirez. 2024. "Design or Decline? A Decolonial Cease and Desist." *Diseña* (25).
European Commission. 2005. *Directive 2005/32/EC of the European Parliament and of the Council of 6 July 2005 Establishing a Framework for the Setting of Ecodesign Requirements for Energy-Using Products*. Official Journal L 191, July 22, 2005.
European Commission. 2009. *Directive 2009/125/EC of the European Parliament and of the Council of 21 October 2009 establishing a framework for the setting of ecodesign requirements for energy-related products*. Official Journal L 285, October 31, 2009: 10–35.
Ezell, Stephen. 2024. "How Innovative Is China in Nuclear Power?" https://itif.org/publications/2024/06/17/how-innovative-is-china-in-nuclear-power/.
Felman, Mackenzie, John Ikerd, Seth Warkins, Charlie Mitchell, Johny Bowman, Calla Rose Ostrander, and Greg Carlock. 2020. "Regenerative Farming and the Green New Deal." *Data For Progress*. Available at: https://www.dataforprogress.org/memos/regenerative-agriculture-and-the-green-new-deal.
Ferro, Sérgio. 2024. *Architecture from Below: An Anthology*. Edited by Silke Kapp and Mariana Moura. Translated by Ellen Heyward and Ana Naomi de Sousa. MACK.
Fertik, Ted. 2025. "The Anti-Climate Common Sense: How a New Climate Common Sense Was Built and How Trump's Assault Threatens Its Unraveling." *Phenomenal World*. Available at: https://www.phenomenalworld.org/analysis/the-anti-climate-common-sense/ (accessed February 3, 2026).
Fisher, Mark. 2022. *Capitalist Realism: Is There No Alternative?* John Hunt.
Fleming, Billy. 2019. "Design and the Green New Deal." *Places Journal*.
Fleming, Billy. 2023. "Red Design and the Green New Deal." *Social Text 60* 41(3): 1–25.
Flyvbjerg, Bent, and Dan Gardner. 2023. *How Big Things Get Done: The Surprising Factors That Determine the Fate of Every Project, from Home Renovations to Space Exploration and Everything in Between*. Penguin Random House.
Förster, Kim, Aleksandr Bierig, Nerea Calvillo Gonzalez, Daniel Barber, Kiel Moe, Jiat-Hwee Chang, Hannah le Roux, Isabelle Doucet, and Paulo Tavares. 2022. *Environmental Histories of Architecture*. Canadian Centre for Architecture (CCA).
Forty, A., 1987. *Objects of Desire: Design and Society from Wedgwood to IBM*. Pantheon Books.
Fox, Sarah. 2024. "'Hey ChatGPT, Finish This Building…': A Worker-Led AI Agenda for the Construction Industry." *Architectural Design* 94(3): 38–45.
Fox, Sarah E., Vera Khovanskaya, Clara Crivellaro, Niloufar Salehi, Lynn Dombrowski, Chinmay Kulkarni, Lilly Irani, and Jodi Forlizzi. 2020. "Worker-Centered Design: Expanding HCI Methods for Supporting Labor."
Franco, Michelle Arevalos, 2022. "Invisible Labor: Precarity, Ethnic Division, and Transformative Representation in Landscape Architecture Work." *Landscape Journal* 41(1): 95–111.
Franco, Michelle Arevalos. 2024. "Care's Repair, Landscape's Labor." *Landscape Research* 49(4): 584–94.

Fraser, Nancy. 2009. *Scales of Justice: Reimagining Political Space in a Globalizing World*. Vol. 31. Columbia University Press.

Fraser, Nancy. 2022. *Cannibal Capitalism: How Our System Is Devouring Democracy, Care, and the Planet—and What We Can Do About It*. Verso Books.

Freitag, Charlotte, Mike Berners-Lee, Kelly Widdicks, Bran Knowles, Gordon Blair, and Adrian Friday. 2021. "The Climate Impact of ICT: A Review of Estimates, Trends and Regulations." *arXiv Preprint arXiv:2102.02622*.

Fressoz, Jean-Baptiste 2024. *More and More and More: An All-Consuming History of Energy*. Allen Lane.

Friant, Martin Calisto, Walter J. V. Vermeulen, and Roberta Salomone. 2020. "A Typology of Circular Economy Discourses: Navigating the Diverse Visions of a Contested Paradigm." *Resources, Conservation and Recycling* 161: 104917.

Friant, Martin Calisto, and Walter V. J. Vermeulen 2024. "Visioning Four Different Circular Futures: What Could 2050 Look Like?" In Renato Passaro, Patrizia Ghisellini, Mario Pansera, Stefania Barca, and Martin Calisto Friant (Eds.), *Circular Economy for Social Transformation: Multiple Paths to Achieve Circularity*. Milano: Ledizioni, 436 p. : ill. - ISBN: 9791256001446 - Permalink: https://digital.casalini.it/9791256001446 - Casalini id: 5839691.

Friedman, Thomas L. 2007. "The Power of Green." *The New York Times*, April 15.

Friedman, Thomas L. 2008. *Hot, Flat, and Crowded: Why We Need a Green Revolution—And How It Can Renew America*. Farrar, Straus, and Giroux.

Fry, Tony. 2009. *Design Futuring: Sustainability, Ethics and New Practice*. Bloomsbury.

Fry, Tony. 2010. *Design as Politics*. Berg.

Fry, Tony. 2013. *Becoming Human by Design*. A&C Black.

Fry, Tony, 2014. *City Futures in the Age of a Changing Climate*. Routledge.

Fry, Tony, and Adam Nocek. 2020. *Design in Crisis: New Worlds, Philosophies and Practices*. Routledge.

Fullilove, M. T. 2004. *Root Shock: How Tearing Up City Neighborhoods Hurts America, and What We Can Do About It*. One World/Ballantine Books.

Furnish, Jim. 2015. *Toward a Natural Forest: The Forest Service in Transition (A Memoir)*. Oregon State University Press.

Fursman, Lindy. 2025. "The Climate Paradox: Why We Need to Reset Action on Climate Change." The Tony Blair Institute for Global Change. https://institute.global/insights/climate-and-energy/the-climate-paradox-why-we-need-to-reset-action-on-climate-change.

Gadanho, Pedro. 2022. *Climax Change! How Architecture Must Transform in the Age of Ecological Emergency*. Actar.

Geels, F. W. 2002. "Technological Transitions as Evolutionary Reconfiguration Processes: A Multi-Level Perspective and a Case-Study." *Research Policy* 31(8–9): 1257–74.

Geels, F. W. 2011. "The Multi-Level Perspective on Sustainability Transitions: Responses to Seven Criticisms." *Environmental Innovation and Societal Transitions* 1(1): 24–40.

Geels, F. W. 2020. *Micro-foundations of the Multi-Level Perspective on Socio-Technical Transitions: Developing a Multi-Dimensional Model of Agency*… Technological Forecasting and Social Change, 152.

Gehl, Jan. 1987. *Life Between Buildings: Using Public Space*. Translated by J. Koch. Van Nostrand Reinhold.

Gehl, J. 2010. *Cities for People*. Island Press.

GibsonGraham, J. K. 2006. *A Postcapitalist Politics*. University of Minnesota Press.

Gilman, Nils. 2025. "The Coming Ecological Cold War: Decarbonization Isn't just About Technology and Markets—It's a Geopolitical Revolution." *Foreign Policy*, September 1. Available at: https://foreignpolicy.com/2025/09/01/the-coming-ecological-cold-war/ (accessed February 4, 2026).

Girard, Valentin, Maud Rio, and Romain Couillet. 2024. "Computing, Complexity and Degrowth: Systemic Considerations for Digital De-Escalation." In Proceedings of the 2024 Workshop on Computing within Limits (LIMITS '24), June 18–19, 2024, Online, 1–11.

Goh, Kian. 2020. "Planning the Green New Deal: Climate Justice and the Politics of Sites and Scales." *Journal of the American Planning Association* 86(2): 188–95.

Goh, Kian. 2021. *Form and Flow: The Spatial Politics of Urban Resilience and Climate Justice*. MIT Press.

Goldstein, Jesse. 2018. *Planetary Improvement: Cleantech Entrepreneurship and the Contradictions of Green Capitalism*. MIT Press.

Gong, Weila, and Joanna I. Lewis. 2024. "The Politics of China's Just Transition and the Shift Away from Coal." *Energy Research & Social Science* 115: 103643.

Gordon Nembhard, J. 2014. *Collective Courage: A History of African American Cooperative Economic Thought and Practice*. University Park: Penn State University Press.

Graeber, David, and David Wengrow. 2021. *The Dawn of Everything: A New History of Humanity*. Allen Lane.

Griffith, S. and NREL. 2017. *100% Clean Electrification in the United States by 2035: A Scenario Analysis for the Decarbonization of Electricity, Transportation, Industry, and Buildings*. Golden, CO: National Renewable Energy Laboratory (NREL). Available at: https://www.nrel.gov/docs/fy17osti/68783.pdf (accessed January 26, 2026).

Griffith, Saul, and Sam Calisch. 2020. "Mobilizing for a Zero Carbon America: Jobs, Jobs, Jobs, and More Jobs." Rewiring America [report]. July 2020.

Gropius, Walter. 1919. The Bauhaus Manifesto. Available at: https://gropius.house/location/bauhaus-manifesto/.

Grubert, Emily, and Sara Hastings-Simon. 2022. "Designing the Mid-transition: A Review of Medium-term Challenges for Coordinated Decarbonization in the United States." *Wiley Interdisciplinary Reviews: Climate Change* 13(3): e768.

Grubler, Arnulf, Charlie Wilson, and Gregory Nemet. 2016. "Apples, Oranges, and Consistent Comparisons of the Temporal Dynamics of Energy Transitions." *Energy Research & Social Science* 22: 18–25.

Gutiérrez, Arnulf, 2020. "When Design Goes South: From Decoloniality, Through Declassification to Dessobons." In *Design in Crisis*, 56–73. Routledge.

Gutman, R. 1988. *Architectural Practice: A Critical View*. Princeton Architectural Press.

Haffner, Jeanne, ed. *Landscapes of Housing: Design and Planning in the History of Environmental Thought*. Routledge, 2021.

Hanauer, N. and D. M. Farrell. 2019. "A Green New Deal for Whom?," *Democracy: A Journal of Ideas*, 55. Available at: https://democracyjournal.org/magazine/55/green-new-deal/ (accessed January 26, 2026).

Haraway, D. 1991. *Simians, Cyborgs, and Women: The Reinvention of Nature*. Routledge.

Haraway, D. 2016. *Staying with the Trouble: Making Kin in the Chthulucene*. Duke University Press.

Hare, N. 1970. "Black Ecology." *The Black Scholar* 1(6).

Harper, P., and G. Boyle (Eds.). 1976. *Radical Technology*. London: Wildwood House.

Harry, Steven J., Tomas Maltby, and Kacper Szulecki. 2024. "Contesting Just Transitions: Climate Delay and the Contradictions of Labour Environmentalism." *Political Geography* 112: 103114.

Harvey, David. 1996. *Justice, Nature and the Geography of Difference*. Blackwell.

Harvey, David. 2000. *Spaces of Hope*. University of California Press.

Harvey, David. 2004. *Paris, Capital of Modernity*. Routledge.

Harvey, Hal, and Justin Gillis. 2022. *The Big Fix: Seven Practical Steps to Save Our Planet*. First Simon&Schuster. Simon & Schuster.

Harvey, Hal, Robbie Orvis, Jeffrey Rissman, Michael O'Boyle, Chris Busch, and Sonia Aggarwal. 2018. *Designing Climate Solutions: A Policy Guide for Low-Carbon Energy*. Springer.

Hatherley, Owen. 2014. "Imagining the Socialist City." Jacobin Online. https://jacobin.com/2014/10/imagining-the-socialist-city.

Hatherley, Owen. 2025 "A City Built for Sunshine" Jacobin Online https://jacobin.com/2025/03/a-city-built-for-sunshine.

Hausfather, Z. 2025. "An Assessment of Current Policy Scenarios over the 21st Century and the Reduced Plausibility of High-Emissions Pathways." *Dialogues on Climate Change* 2(1 April): 26–32.

Hawken, Paul. 2017. *Drawdown: The Most Comprehensive Plan Ever Proposed to Reverse Global Warming*. Penguin.

Hawken, Paul, Amory B. Lovins, and L. Hunter Lovins. 1999. *Natural Capitalism: The Next Industrial Revolution*. Little, Brown.

Hayden, D. 1982. *The Grand Domestic Revolution: A History of Feminist Designs for American Homes, Neighborhoods, and Cities*. MIT Press.

Healy, N., and J. Barry 2017. "Political economy of critical interstices: Energy transitions and the just transition." *Environmental Innovation and Societal Transitions* 25: 106–16.

Heffron, Raphael J. 2021. *Achieving a Just Transition to a Low-Carbon Economy*. Springer.

Heglar, Mary Annaïse. 2019. "Climate Change Isn't the First Existential Threat." *ZORA*. Available online at: https://Zora. Medium. Com/Sorry-Yall-but-Climate-Change-Aint-the-First-Existential-Threat-B3c999267aa0 (accessed January 12, 2022).

Helveston, John, and Jonas Nahm. 2019. "China's Key Role in Scaling Low-Carbon Energy Technologies." *Science* 366(6467): 794–6.

Hernandez, Rebecca R., Alona Armstrong, Jennifer Burney, Greer Ryan, Kara Moore-O'Leary, Ibrahima Diédhiou, Steven M. Grodsky, Leslie Saul-Gershenz, Rob Davis, and Jordan Macknick. 2019. "Techno–Ecological Synergies of Solar Energy for Global Sustainability." *Nature Sustainability* 2(7): 560–8.

Hickel, Jason. 2020. *Less Is More: How Degrowth Will Save the World*. Random House.

Hill, Dan, and Mazzucato, Mariana. 2024. *Modern Housing: An environmental common good. Council on Urban Initiatives* (CUI DP 2024–02). UCL Institute for Innovation and Public Purpose Working Paper Series: IIPP WP 2024–03. Available at: https://www.ucl.ac.uk/bartlett/public-purpose/publications/2024/mar/modern-housing-environmental-common-good.

Hilton, Isabel 2024. "How China Became the World's Leader on Renewable Energy." *Yale Environment* 360.

Hochhäusl, S. 2021. "A Collective Endeavour: Martin Wagner and the Making of Collective Housing in Interwar Berlin." In J. Haffner (Ed.), *Landscapes of Housing: Design and Planning in the History of Environmental Thought*. Routledge.

Hochschild, Arlie Russell. 2018. *Strangers in Their Own Land: Anger and Mourning on the American Right*. New Press.

Hosey, L. 2012. *The Shape of Green: Aesthetics, Ecology, and Design*. Washington, DC: Island Press.

Houston, C. H., and J. P. Davis. 1934 "Lily-White Reconstruction (The TVA)." *The Crisis* 41(10): 290–1.

Huber, Matthew T. 2022. *Climate Change as Class War: Building Socialism on a Warming Planet*. Verso Books.

Huber, Matthew, T., and Fred Stafford. 2023. "Socialist Politics and the Electricity Grid." *Catalyst* 6(4) Winter: 59–92.

Hulme, M. 2011. "Reducing the Future to Climate: A Story of Climate Determinism and Reductionism." Osiris, [online] Available at: https://www.journals.uchicago.edu/doi/abs/10.1086/661274.

Hutton, J. 2019. *Reciprocal landscapes: Stories of material movements*. Routledge.

Ibañez, Daniel, Jane Hutton, and Kiel Moe. 2019. *Wood Urbanism: From the Molecular to the Territorial*. Actar.

IEA, 2026. *Energy Storage Outlook 2026*. Paris: International Energy Agency. Available at: iea.org/reports (accessed March 25, 2026).

Illich, I. 1973. *Tools for Conviviality*. Harper & Row.

Indigenous Peoples Principles and Protocols for Just Transition, 2024. Indigenous Peoples' Perspectives, Knowledge, and Lived Experiences Summit, Geneva, Switzerland, 8–10 October 2024. Available at: https://www.sirgecoalition.org/statements/indigenous-peoples-reach-unanimous-agreement-on-defining-the-just-transition-and-provide-principles-and-protocols-to-eliminate-harm-from-renewable-energy-and-green-development (accessed October 3, 2025).

Ioannou, O. 2023. "Integrating Circularity in Education: The CBE Hub Approach." In M. Berger and K. Irvin (Eds.), *Repair: Sustainable Design Futures*. Routledge.

IPCC. 2018: *Global Warming of 1.5°C. An IPCC Special Report on the impacts of global warming of 1.5°C above pre-industrial levels and related global greenhouse gas emission pathways, in the context of strengthening the global response to the threat of climate change, sustainable development, and efforts to eradicate poverty* [Masson-Delmotte, V., P. Zhai, H.-O. Pörtner, D. Roberts, J. Skea, P. R. Shukla, A. Pirani, W. Moufouma-Okia, C. Péan, R. Pidcock, S. Connors, J. B. R. Matthews, Y. Chen, X. Zhou, M. I. Gomis, E. Lonnoy, T. Maycock, M. Tignor, and T. Waterfield (Eds.)]. Cambridge University Press.

IPCC. 2022: *Climate Change 2022: Impacts, Adaptation, and Vulnerability*. Contribution of Working Group II to the Sixth Assessment Report of the Intergovernmental Panel on Climate Change [H.-O. Pörtner, D.C. Roberts, M. Tignor, E. S. Poloczanska, K. Mintenbeck, A. Alegría, M. Craig, S. Langsdorf, S. Löschke, V. Möller, A. Okem, and B. Rama (Eds.)]. Cambridge University Press.

IPCC. 2023. *Climate Change 2023: Synthesis Report. Contribution of Working Groups I, II and III to the Sixth Assessment Report of the Intergovernmental Panel on Climate Change* [Core Writing Team, H. Lee, and J. Romero (Eds.)]. IPCC, Geneva, Switzerland, 35–115.

Irani, Lilly C., and M Six Silberman. 2016. "Stories We Tell About Labor: Turkopticon and the Trouble with Design." In *Proceedings of the 2016 CHI Conference on Human Factors in Computing Systems*, 4573–86. ACM.

Irwin, Terry. 2015. "Transition Design: A Proposal for a New Area of Design Practice, Study, and Research." *Design and Culture* 7(2): 229–46.

Irwin, Terry, Gideon Kossoff, and Cameron Tonkinwise. 2015. "Transition Design Provocation." *Design Philosophy Papers* 13(1): 3–11.

Irwin, Terry, Cameron Tonkinwise, and Gideon Kossoff. 2022. "Transition Design: An Educational Framework for Advancing the Study and Design of Sustainable Transitions." *Cuadernos Del Centro de Estudios En Diseño y Comunicación. Ensayos* (105): 31–72.

Iturbe, Elisa. 2019. "Architecture and the Death of Carbon Modernity." *Log* (47): 10–23.

Jacobs, Jane. 1961. *The Death and Life of Great American Cities*. Random House.

Jacobson, Mark Z. 2023. *No Miracles Needed: How Today's Technology Can Save Our Climate and Clean Our Air*. Cambridge University Press.

Jäger, A. 2026. *Hyperpolitics: Extreme Politicization Without Political Consequences*. Verso.

Jamieson, Dale. 2014. *Reason in a Dark Time: Why the Struggle against Climate Change Failed--and What It Means for Our Future*. Oxford University Press.

Jarzombek, Mark. 2006. "Sustainability: Fuzzy Systems and Wicked Problems." *Log* (8): 7–12.

Jenkins, Jesse D., Max Luke, and Samuel Thernstrom. 2018. "Getting to Zero Carbon Emissions in the Electric Power Sector." *Joule* 2(12): 2498–2510.

Jenkins, Jesse D., Erin N Mayfield, Eric D. Larson, Stephen W. Pacala, and Chris Greig. 2021. "Mission Net-Zero America: The Nation-Building Path to a Prosperous, Net-Zero Emissions Economy." *Joule* 5(11): 2755–61.

Jones, T. 1975. "Apartheid Ecology in America: On Building the Segregated Society." *Black World* 24(7).

Jones, Peter Blundell, Doina Petrescu, and Jeremy Till. 2005. *Architecture and Participation*. Taylor & Francis.

Joseph, Jomy. 2024. "In Another Life Another Being: On Design and the Wages of Decoloniality." *Diseña* 25: 284–305.

Julier, Guy. 2017. *Economies of Design*. Sage.

Kallipoliti, Lydia. 2024. *Histories of Ecological Design: An Unfinished Cyclopedia*. Actar.

Kaminer, Tahl. 2007. "Autonomy and Commerce: The Integration of Architectural Autonomy." *Arq: Architectural Research Quarterly* 11(1): 63–70.

Kaminer, Tahl. 2016. "Capital A." In *Asymmetric Labors: The Economy of Architecture in Theory and Practice*, 16–20. The Architecture Lobby.

Katznelson, Ira. 2013. *Fear Itself: The New Deal and the Origins of Our Time*. Liveright Publishing Corp.

Katz, Peter. 1994. *The New Urbanism. Toward an Architecture of Community*. McGraw-Hill.

Kimmerer, Robin. 2013. *Braiding Sweetgrass: Indigenous Wisdom, Scientific Knowledge and the Teachings of Plants*. Milkweed editions.

Kirk, Andrew 2007. *Counterculture Green: The Whole Earth Catalog and American Environmentalism*. University Press of Kansas

Klinger, Julie Michelle. 2018. *Rare Earth Frontiers: From Terrestrial Subsoils to Lunar Landscapes*. Cornell University Press.

Knuth, S. 2018. "'Breakthroughs' for a Green Economy? Financialization and Clean Energy Transition." *Energy Research & Social Science* 41: 220–9.

Knuth, S., I. Behrsin, A. Levenda, and J. McCarthy. 2022. "New Political Ecologies of Renewable Energy." *Environment and Planning E: Nature and Space* 5(4): 1477–1502.

Kossoff, Gideon. 2015. "Holism and the Reconstitution of Everyday Life: A Framework for Transition to a Sustainable Society." *Design Philosophy Papers* 13(1): 25–38.

Krasny, Elke. 2019. "Architecture and Care." In Angelika Fitz, Elke Krasny, and Architekturzentrum Wien (Eds.), *Critical Care: Architecture and Urbanism for a Broken Planet*, 33–41. MIT Press.

Kreikamp, E. 2022 "Shared Ideals: World Garden Cities and the Amsterdam School," World Garden Cities. Available at: https://www.worldgardencities.com/images/uploads/articles/Shared%20Ideals.pdf (accessed January 25, 2026).

Krivý, Maroš, and L. Ma. 2018. "The Limits of the Livable City: From Homo Sapiens to Homo Cappuccino." *Avery Review* 30.

Kuittinen, M., A. Organschi, and S. Carlisle. 2022. *Carbon: A Field Manual for Building Designers*. New Haven, CT: Yale School of Architecture/Actar Publishers.

ILO. 2025. 2025 Report of the Committee of Experts on the Application of Conventions and Recommendations.

LaFarge, Pier. 2024. "Pier LaFarge in Conversation with David Roberts—Should We Put Utilities in Charge of Distributed Energy?" Volts December 18, 2024. Available at: https://www.volts.wtf/p/should-we-put-utilities-in-charge.

Lamb, Z., and Khirfan, L. 2022. "Against Climate Haussmannization: Transformation Through and in Urban Design." *Journal of Planning Literature* 37(3): 416–25.

Larson, Eric. 2020. *Net-Zero America: Potential Pathways, Infrastructure, and Impacts*. Princeton University Press.

Latour, Bruno. 2008. "A Cautious Prometheus? A Few Steps Toward a Philosophy of Design (with special attention to Peter Sloterdijk)," keynote lecture, Design History Society Annual Conference, Falmouth, September 3–6. Available at: http://www.bruno-latour.fr/sites/default/files/112-DESIGN-CORNWALL-GB.pdf (accessed January 22, 2026).

Law, Kara Lavender, Natalie Starr, Theodore R. Siegler, Jenna R. Jambeck, Nicholas J. Mallos, and George H. Leonard. 2020. "The United States' Contribution of Plastic Waste to Land and Ocean." *Science Advances* 6(44).

Leadbeater, C., and Miller, P. (2004) *The Pro-Am Revolution: How Enthusiasts Are Changing Our Economy and Society*. Demos.

Leiserowitz, A., Carman, J., Rosenthal, S., Cutler, N., Kotcher, J., Botoruz, M., Goldberg, M., and Marlon, J. 2023. *Climate Change in the American Mind: Politics & Policy, Fall 2023*. New Haven, CT: Yale Program on Climate Change Communication, Yale University. Available at: https://climatecommunication.yale.edu/publications/climate-change-in-the-american-mind-politics-policy-fall-2023/ (accessed January 26, 2026).

Lennon, Myles. 2017. "Decolonizing Energy: Black Lives Matter and Technoscientific Expertise amid Solar Transitions." *Energy Research & Social Science* 30: 18–27.

Lennon, Myles. 2021. "Energy Transitions in a Time of Intersecting Precarities: From Reductive Environmentalism to Antiracist Praxis." *Energy Research & Social Science* 73: 101930.

Lennon, Myles. 2025. *Subjects of the Sun Solar Energy in the Shadows of Racial Capitalism*. Duke University Press.

Leopold, L. 2007. *The Man Who Hated Work and Loved Labor: The Life and Times of Tony Mazzocchi*. Chelsea Green.

Liebreich, Michael. 2023. "Net Zero Will Be Harder Than You Think—And Easier." Part I: Harder https://about.bnef.com/insights/clean-energy/liebreich-net-zero-will-be-harder-than-you-think-and-easier-part-i-harder/.

Liebreich, Michael. 2024. "Net Zero Will Be Harder Than You Think – And Easier. Part II: Easier." February 22, 2024. https://about.bnef.com/insights/clean-energy/liebreich-net-zero-will-be-harder-than-you-think-and-easier-part-ii-easier/.

Lilley, D., and G. T. Wilson. 2017. "Design Interventions for Sustainable Behavior." In T. Bhamra and R. L. Cooper (Eds.), *Routledge Handbook of Sustainable Product Design*, 123–44. Routledge. https://doi.org/10.4324/9781315693309.

Little, Peter C. 2022. *Burning Matters: Life, Labor, and E-Waste Pyropolitics in Ghana.* Oxford University Press.
Little, Peter C. 2023. *Critical Zones of Technopower and Global Political Ecology: Platforms, Pathologies, and Plunder.* Lexington Books.
Liu, Alan. 2009. *The Laws of Cool: Knowledge Work and the Culture of Information.* University of Chicago Press.
Long, J., J. L. Rice, and A. M. Levenda. 2020. "From Sustainable Urbanism to Climate Urbanism." *Urban Studies* 57(1): 39–55.
Lovins, Amory. 1973. *World Energy Strategies: Facts, Issues, and Options.* Friends of the Earth.
Lovins, Amory. 1976. "Energy Strategy: The Road Not Taken?" *Foreign Affairs* 55(1): 65–96.
Luke, Nikki. 2023. "Just Transition for All? Labor Organizing in the Energy Sector beyond the Loss of 'Jobs Property.'" *Annals of the American Association of Geographers* 113(1): 94–109.
MacKaye, Benton. 1921. "An Appalachian Trail: A Project in Regional Planning." *Journal of the American Institute of Architects* 9(10): 325–33.
Maher, Neil M. 2008. *Nature's New Deal.* Oxford University Press.
Maldonado, T. 1972. *Design, Nature, and Revolution: Toward a Critical Ecology.* Translated by Mario Domandi. New York: Harper & Row.
Malik, A. 2023. "The Political Economy of Distributed Solar in Pakistan: Inequality, Grid Defection, and Energy Justice." *Energy Research & Social Science* 103: 103190.
Malm, Andreas. 2016. *Fossil Capital: The Rise of Steam Power and the Roots of Global Warming.* Verso Books.
Malterre-Barthes, Charlotte. 2025. *A Moratorium on New Construction: Critical Spatial Practice 14.* Sternberg Press.
Mann, Charles C. 2005. *1491: New Revelations of the Americas before Columbus.* Knopf.
Mann, Michael. 2022. *The New Climate War: The Fight to Take Back Our Planet.* PublicAffairs.
Manzini, E., and F. Jégou. 2003. "Sustainable Everyday." *Design Philosophy Papers* 1(4): 188–91.
Manzini, Ezio. 2015. *Design, When Everybody Designs: An Introduction to Design for Social Innovation.* MIT Press.
Manzini, Ezio. 2019. *Politics of the Everyday*, Bloomsbury Academic.
Martin, Reinhold. 2018. "On the Uses and Disadvantages of Architecture for History" e-flux. Available at: https://www.e-flux.com/architecture/history-theory/225181/on-the-uses-and-disadvantages-of-architecture-for-history.
Martin, Reinhold. 2020. "Abolish Oil." *Places Journal.* June 2020. Sternberg Press.
Martinez-Diaz, L. N. Gordon, and M. McBride. 2025. *Climate Clarity: On the Future of Climate Action in the United States.* September 2025 Carnegie Endowment for International Peace.
Marvel, Kate. 2025. *Human Nature: Nine Ways to Feel About Our Changing Planet.* New York: Ecco.
Massey, D. 2005. *For Space.* Sage.
Mastini, Riccardo, Giorgos Kallis, and Jason Hickel 2021. "A Green New Deal Without Growth?" *Ecological Economics* 179.
Mattson, Kevin. 2020. *A History of the Appalachian Trail: The People, the Places, and the Politics of Hiking America's First National Scenic Trail.* University of California Press.
Mazria, E. 1979. *The Passive Solar Energy Book.* Rodale Press.
Mazzocchi, T. 1993. "A Superfund for Workers." *Earth Island Journal* 9(1): 40–1.
Mazzucato, Mariana. 2015. *The Green Entrepreneurial State.* Routledge.

Mazzucato, Mariana. 2021. *Mission Economy: A Moonshot Guide to Changing Capitalism.* Penguin UK.

Mazzucato, M., and D. Rodrik. 2023. Industrial Policy with Conditionalities: A Taxonomy and Sample Cases. UCL *Institute for Innovation and Public Purpose, Working Paper Series* (IIPP WP 2023–07). Available at: https://www.ucl.ac.uk/bartlett/public-purpose/wp2023-07.

Mazzucato, M., and R. Collington. 2023. *The Big Con: How the Consulting Industry Weakens Our Businesses, Infantilizes Our Governments, and Warps Our Economies.* Allen Lane (Penguin).

McAlevey, Jane F. 2016. *No Shortcuts: Organizing for Power in the New Gilded Age.* Oxford University Press.

McBride, J. 2018. "The Green New Deal and the Legacy of Public Power." *The Breakthrough Institute.* Available at: https://thebreakthrough.org/issues/energy/the-green-new-deal-and-the-legacy-of-public-power (accessed January 26, 2026).

McDonough, W., and M. Braungart. 2002. *Cradle to Cradle: Remaking the Way We Make Things.* North Point Press.

McKibben, Bill. 2025 *Here Comes the Sun: A Last Chance for the Climate and a Fresh Chance for Civilization.* W. W. Norton.

McIlroy, Damian, Seán Brennan, and John Barry. 2022. "Just Transition: A Conflict Transformation Approach." In *Handbook of Critical Environmental Politics*, 416–30. Edward Elgar.

McKie, Ruth E., Carlos Milani, Paul K. Gellert, Omar M. Faruque, K. Hochstetler, Guy Edwards, D. McElwee Pamela, Jonathan Walz, Prakash Kashwan, and Timmons Roberts. 2023. "Climate Obstruction in the Global South: Future Research Trajectories." *PLOS Climate* 2(7): e0000241.

Mehaffy, M., and N. A. Salingaros 2013. "Toward Resilient Architectures: Why Green Often Isn't." *Metropolis Magazine*, April 4. Available at: https://metropolismag.com/viewpoints/toward-resilient-architectures-2-why-green-often-isnt/.

Mertins-Kirkwood, H. 2016. *Towards a just transition for Canadian workers*, Canadian Centre for Policy Alternatives [online]. Available at: https://www.policyalternatives.ca/newsroom/news-release/just-transition-canadian-workers (accessed January 23, 2026).

Meyer, Robinson. 2021. "Unfortunately, I Care About Power Lines Now." *The Atlantic* (blog). July 28, 2021. Available at: https://www.theatlantic.com/science/archive/2021/07/america-is-bad-at-building-power-lines-lets-fix-that-transmission-climate/619591/.

Meyer, Robinson. 2025. "The Electricity Affordability Crisis Is Coming." Heatmap. July 23, 2025. Available at: https://heatmap.news/politics/electricity-price-crisis#.

Mies, M. (1986) *Patriarchy and Accumulation on a World Scale: Women in the International Division of Labour.* Zed Books.

Mirletz, Heather, Henry Hieslmair, Silvana Ovaitt, Taylor L. Curtis, and Teresa M. Barnes. 2023. "Unfounded Concerns about Photovoltaic Module Toxicity and Waste Are Slowing Decarbonization." *Nature Physics* 19(10): 1376–78.

Moe, Kiel. 2017. *Empire, State & Building.* Actar D, Inc.

Moe, Kiel. 2018. "Iatrogenic architecture: Unreliable narratives of sustainability." *Harvard Design Magazine*, 44. Available at: https://www.harvarddesignmagazine.org/articles/iatrogenic-architecture-unreliable-narratives-of-sustainability/

Moore, Jason. 2015. *Capitalism in the Web of Life: Ecology and the Accumulation of Capital.* Verso Books.

Morena, Edouard, Dunja Krause, and Dimitris Stevis (Eds.). 2020. *Just Transitions: Social Justice in the Shift Towards a Low-Carbon World.* Pluto Press.

Mulgan, Geoff. 2014. *Design in Public and Social Innovation: What Works and What Could Work Better*. London: Nesta. Available at: https://www.nesta.org.uk/report/design-in-public-and-social-innovation/ (accessed January 26, 2026).

Mulvaney, Dustin. 2019. *Solar Power: Innovation, Sustainability, and Environmental Justice*. University of California Press.

Mulvaney, Dustin. 2020. *Sustainable Energy Transitions: Socio-Ecological Dimensions of Decarbonization*. Palgrave Macmillan. Available at: https://doi.org/10.1007/978-3-030-48912-0

Mumford, Lewis. 1938. *The Culture of Cities*. New York: Harcourt, Brace and Company.

Muzaffar, Ijlal. 2024. *Modernism's Magic Hat: Architecture and the Illusion of Development without Capital*. University of Texas Press.

Natural Resources Conservation Service "A Brief History of NRCS." n.d. Available at: https://www.nrcs.usda.gov/about/history/brief-history-nrcs (accessed October 20, 2024).

National Renewable Energy Laboratory (NREL). n.d. "WINDExchange: Wind Energy Community Benefits Guide." Available at: https://windexchange.energy.gov/community-benefits-guide (accessed October 21, 2024).

Nelson, G. D. 2019. "An Appalachian Trail: A Project in Regional Planning." *Places Journal*.

Newell, Peter, and Dustin Mulvaney. 2013. "The Political Economy of the 'just transition." *Geographical Journal* 179(2): 132–40.

Newell, Peter, and Andrew Simms. 2021. "How Did We Do That? Histories and Political Economies of Rapid and Just Transitions." *New Political Economy* 26(6): 907–22.

NoiseCat, J. B. 2025. *We Survived the Night*. Alfred A. Knopf.

Obach, Brian K. 2004. *Labor and the Environmental Movement: The Quest for Common Ground*. MIT Press. Available at: https://doi.org/10.7551/mitpress/4080.001.0001.

Ocasio-Cortez, A., and B. Sanders. Ramirez Reintroduce The Green New Deal for Public Housing Act." 2021. Representative Ocasio-Cortez. March 21, 2021. http://ocasio-cortez.house.gov/media/press-releases/ocasio-cortez-sanders-ramirez-reintroduce-green-new-deal-public-housing-act.

Oatley, Thomas, and Mark Blyth. 2021. "The Death of the Carbon Coalition." *Foreign Policy*, February 12, 2021.

Oldfield, Philip. 2019. *The Sustainable Tall Building: A Design Primer*. London: Routledge.

Oldfield, Philip. 2022. "Seven Principles for Designing Low-Carbon Skyscrapers." *Dezeen*, November 21. Available at: https://www.dezeen.com/2022/11/22/principles-low-carbon-skyscrapers-philip-oldfield/ (accessed January 25, 2026).

Oreskes, Naomi, and Erik M. Conway. 2011. *Merchants of Doubt: How a Handful of Scientists Obscured the Truth on Issues from Tobacco Smoke to Global Warming*. Bloomsbury Publishing USA.

Papanek, Victor. 1972. *Design for the Real World*. Pantheon Books.

Paprocki, Kasia, Daniel Aldana Cohen, Rebecca Elliott, and Liz Cohen Kozlov. 2019. "The Useful Discomfort of Critical Climate Social Science." *Social Science Research Council Items*. May 7, 2019. Available at: https://items.ssrc.org/from-our-fellows/the-useful-discomfort-of-critical-climate-social-science/ (accessed October 3, 2025).

Passaro, Renato, Patrizia Ghisellini, Mario Pansera, Stefania Barca, and Martin Calisto Friant (Eds.). 2024. *Circular Economy for Social Transformation: Multiple Paths to Achieve Circularity*. LED Edizioni Universitarie di Macerata, Deliverable D.1.1 of the EU H2020 JUST2CE project (Grant n° 101003491).

Patel, Raj, and Jim Goodman. 2019. *A Green New Deal for Agriculture*. Jacobin.

Patterson, J. J. 2023. "Backlash to Climate Policy." *Global Environmental Politics* 23(1): 68–90.
Penin, Lara. (Ed.). 2021. *The Disobedience of Design. A Gui Bonsiepe Reader*. Bloomsbury.
Penin, Lara, and Rashid Owoyele. 2026. "Worker-centred Service Design: Countering the Invisibility of Workers." In L. Penin, A. Prendiville, and D. Sangiorgi (Eds.), *The Bloomsbury Handbook of Service Design: Plural Perspectives and a Critical Contemporary Agenda*, Bloomsbury.
Pevzner, Nicholas. 2019. "The Great New Deal, Landscape, and Public Imagination." *Landscape Architecture* 109(7): 68–81.
Pevzner, Nicholas, Yekang Ko, and Kirk Dimond. 2021. "Power Player: Designing for Just and Multifunctional Energy Landscapes." *Landscape Architecture Magazine*. June 8, 2021. https://landscapearchitecturemagazine.org/2021/06/08/power-player/.
Piketty, Thomas. 2020. "Brahmin left vs merchant right: Rising inequality and the changing structure of political conflict". World Inequality Lab Working Paper. Available at: https://wid.world/document/piketty2020brahminleft/ (accessed February 19, 2026).
Pirani, S. 2024. "Socialism and electricity: renewables and decentralisation versus nuclear." *Capitalism Nature Socialism* 35(3): 1–20.
P2P Foundation. 2015. *Peer to Peer and the Commons: A Path Towards Transition*. P2P Foundation.
Price, R. 1973. *Maroon Societies: Rebel Slave Communities in the Americas*. Anchor Press/Doubleday.
Pursell, Carroll. 1993. "The Rise and Fall of the Appropriate Technology Movement in the United States, 1965–1985." *Technology and Culture* 34(3): 629–37.
Putnam, R. D. 2000. *Bowling Alone: The Collapse and Revival of American Community*. Simon & Schuster.
Räthzel, Nora, Dimitris Stevis, and David Uzzell (Eds.). 2021. *The Palgrave Handbook of Environmental Labour Studies*. Springer International. Available at: https://doi.org/10.1007/978-3-030-71909-8.
Riach, J. 2014. "Zaha Hadid Defends Qatar World Cup Role Following Migrant Worker Deaths." *The Guardian*, February 25. Available at: https://www.theguardian.com/world/2014/feb/25/zaha-hadid-qatar-world-cup-migrant-worker-deaths (accessed January 23, 2026).
Rich, Damon. 2021. "What Would Happen If All the Details Were Put Back in the Story? Interview with Damon Rich." In H. Harriss, R. Hyde, and R. Marcaccio (Eds.), *Architects After Architecture: Innovative Practices for a World in Crisis*, 268–77. Routledge.
Riofrancos, Thea. 2020. *Resource Radicals: From Petro-Nationalism to Post-Extractivism in Ecuador*. Duke University Press.
Riofrancos, Thea. 2021. "The Rush to 'Go Electric' Comes with a Hidden Cost: Destructive Lithium Mining." *The Guardian*, June 14, 2021, sec. Opinion. Available at: https://www.theguardian.com/commentisfree/2021/jun/14/electric-cost-lithium-mining-decarbonasation-salt-flats-chile.
Riofrancos, Thea, Alissa Kendall, Matthew Haugen, Kira McDonald, Batul Hassan, Margaret Slattery, and Xan Lillehei. 2023. "Achieving Zero Emissions with More Mobility and Less Mining." Climate and Community Project. Available at: http://www.climateandcommunity.org/more-mobility-less-mining.
Riofrancos, Thea. 2025a. *Extraction: The Frontiers of Green Capitalism*. W.W. Norton.
Riofrancos, Thea. 2025b. "The Perils of Climate Alarmism." *Journal of Democracy* 36(1): 169–74.

Ritchie, Hannah. 2024. "Low-Carbon Technologies Need Far Less Mining Than Fossil Fuels." *Our World in Data*. Retrieved from https://ourworldindata.org/low-carbon-technologies-need-far-less-mining-fossil-fuels.

Ritchie, Hannah. 2025. *Clearing the Air – A Hopeful Guide to Solving Climate Change*. Chatto and Windus.

Rodríguez-Labajos, B., I. Yánez, P. Bond, L. Greyl, S. Munguti, G. U. Ojo, and W. Overbeek, 2019. "Not So Natural an Alliance? Degrowth and Environmental Justice Movements in the Global South." *Ecological Economics* 157(C): 175–84.

Rosenbaum, L. 2014. "Zaha Brouhaha: Hadid and the Conscience of Architects." *CultureGrrl* [blog], August 27. Available at: https://www.artsjournal.com/culturegrrl/2014/08/zaha-hadid-and-the-conscience-of-architects.html (accessed January 23, 2026).

Rosenow, Jan 2024. "Have We Been Duped by the Primary Energy Fallacy?" https://medium.com/@jan.rosenow/have-we-been-duped-by-the-primary-energy-fallacy-167f53c58961.

Rudofsky, Bernard. 1964. *Architecture Without Architects: A Short Introduction to Non-Pedigreed Architecture*. UNM Press.

Saad, B., 2020. Betraying the Profession: Design Labour Against Technocracy. Failed Architecture, [online] Available at: https://failedarchitecture.com/betraying-the-profession-design-labour-against-technocracy/ (accessed January 22, 2026).

Sadler, Simon. 2008. "An Architecture of the Whole." *Journal of Architectural Education* 61: 4.

Sadler, Simon. 2015. "Mandalas or Raised Fists? Hippie Holism, Panther Totality, and Another Modernism." In Andrew Blauvelt (Ed.), *Hippie Modernism: The Struggle for Utopia*, 114–25. Walker Art Center.

Satgar, V. (Ed.). 2018. *The climate crisis: South African and global democratic eco-socialist alternatives*. Wits University Press.

Schmelzer, Matthias, Andrea Vetter, and Aaron Vansintjan. 2022. *The Future Is Degrowth: A Guide to a World beyond Capitalism*. Verso Books.

Scholtz, Trebor. 2016. Platform cooperativism. *Challenging the Corporate Sharing Economy*. Rosa Luxemburg Foundation, 435.

Schumacher, E. F. 1974. *Small Is Beautiful: Economics as If People Mattered*. Blond & Briggs.

Schumacher, Patrik. "The End of Architecture." *Khōrein: Philosophy and Social Theory Journal* (University of Belgrade, Institute for Philosophy and Social Theory), 2024/2025. Available at: https://patrikschumacher.com/the-end-of-architecture/.

Schwartz, John 2018. "Will We Survive Climate Change." *New York Times* November 19.

Schwarz, Moritz, Philippe Benoit, and Alex Clark. 2023. "Decarbonising State Owned Power Companies: A Framework for Applying Policy Actions." *University of Oxford Smith School Working Paper*: 23–01.

Scoones, Ian, Melissa Leach, and Peter Newell. 2015. *The Politics of Green Transformations*. Taylor & Francis.

Scott, Felicity. 2007. *Architecture or Techno-Utopia: Politics After Modernism*. MIT Press.

Scott, Felicity. 2016. *Outlaw Territories: Environments of Insecurity/Architectures of Counterinsurgency*. Zone Books.

Timothy D. Searchinger, Liqing Peng, Jessica Zionts, and Richard Waite. 2023. "The Carbon Costs of Global Wood Harvests." *Nature* 620(7972): 110–15.

Sennett, Richard. 2008. *The Craftsman*. Yale University Press.

Shin, Jae, and Rich, Damon 2019. "HECTOR" Lecture Given to the Harvard Graduate School of Design, https://www.youtube.com/watch?v=63HsDx00F1o

Sharma, Vishal, Neha Kumar, and Bonnie Nardi. 2023. "Post-Growth Human–Computer Interaction." *ACM Transactions on Computer-Human Interaction* 31(1): 1–37.

Sijmons, Dirk. 2014. *Landscape and Energy: Designing for Transition.* nai010 uitgevers.

Simay, Philippe. 2024. *Bâtir avec ce qui reste: quelles ressources pour sortir de l'extractivisme ?* Éditions Terre Urbaine

Simms, Andrew, Ann Pettifor, Caroline Lucas, Charles Secrett, Colin Hines, Jeremy Leggett, Larry Elliott, Richard Murphy, and Tony Juniper. 2008. "A Green New Deal: Joined-up Policies to Solve the Triple Crunch of the Credit Crisis, Climate Change and High Oil Prices." New Economics Foundation. Available at: https://neweconomics.org/uploads/files/8f737ea195fe56db2f_xbm6ihwb1.pdf.

Simon, Robert M., and David J. Hayes. 2017. "America's Clean Energy Success, by the Numbers" Center for American Progress [report]. June 29, 2017.

Simon, Herbert A. 2019. *The Sciences of the Artificial, Reissue of the Third Edition with a New Introduction by John Laird.* MIT Press.

Sivaram, Varun. 2025. "We Need a Fresh Approach to Climate Policy. It's Time for Climate Realism." Available at: https://www.cfr.org/article/we-need-fresh-approach-climate-policy-its-time-climate-realism.

Skocpol, Theda. 2003. *Diminished Democracy: From Membership to Management in American Civic Life.* University of Oklahoma Press.

Smil, Vaclav. 2019. *Energy in World History.* Routledge.

Smil, Vaclav. 2022. *How the World Really Works: The Science Behind How We Got Here and Where We're Going.* First United States edition. Viking.

Smith, Adrian. 2005. "The Alternative Technology Movement: An Analysis of Its Framing and Negotiation of Technology Development." *Human Ecology Review* 12 (2): 106.

Smith, Judy. 1978. *Something Old, Something New, Something Borrowed, Something Due: Women and Appropriate Technology.* Butte, MT: National Center for Appropriate Technology.

Smith, Judy. 1983. "Sex Role Separation in Appropriate Technology." In *Women and Technology: Deciding What's Appropriate: Proceedings of a Conference on Women and Technology*, San Diego, CA: San Diego State University.

Smith, Noah 2025 "Nonprofits Are Sapping the Progressive Project: We Need to Bring Back State Capacity." Available at: https://www.noahpinion.blog/p/nonprofits-are-sapping-the-progressive?utm_source=substack&utm_medium=email.

Soper, Kate. 2020. *Post-Growth Living: For an Alternative Hedonism.* Verso Books.

Soules, Matthew. 2021. *Icebergs, Zombies, and the Ultra-Thin: Architecture and Capitalism in the 21st Century.* Chronicle Books.

Sovacool, Benjamin K. 2016. "How Long Will It Take? Conceptualizing the Temporal Dynamics of Energy Transitions." *Energy Research & Social Science* 13: 202–15.

Spann Edward, K. 1996. *Designing Modern America: The Regional Planning Association of America and Its Members.* Columbus: Ohio State University Press.

Spektor, Franchesca, Sarah E. Fox, Ezra Awumey, Christine A. Riordan, Hye Jin Rho, Chinmay Kulkarni, Marlen Martinez-Lopez, Betsy Stringam, Ben Begleiter, and Jodi Forlizzi. 2023. "Designing for Wellbeing: Worker-Generated Ideas on Adapting Algorithmic Management in the Hospitality Industry." In *Proceedings of the 2023 Designing Interactive Systems Conference (DIS '23)*, 623–37. ACM.

Spirn, Anne Whiston. 1984. *Granite Garden.* Basic Books.

Steffen, Alex. 2017. "On Climate, Speed Is Everything." *The Nearly Now* [blog], December 7. Available at: https://thenearlynow.com/on-climate-speed-is-everything-5bc47203bc7a (accessed January 23, 2026).
Stein, Sam. 2019. *Capital City: Gentrification and the Real Estate State*. Verso Books.
Stein, Jesse Adams, and Chantel Carr (Eds.). 2025. *Designing Through Planetary Breakdown: Locating Material Knowledge and Practical Skill*. Taylor & Francis.
Stevis, Dimitris. 2013. "Global Labour Unions and Just Transition to a Green Economy." *International Environmental Agreements: Politics, Law and Economics* 15(1): 29–43.
Stevis, Dimitris, and Romain Felli. 2015. "Global Labour Unions and Just Transition to a Green Economy." *Global Labour Journal* 6(1): 25–42.
Stevis, Dimitris, Edouard Morena, and Dunja Krause. (Eds.). 2019. *Just Transitions: Social Justice in the Shift Towards a Low-Carbon World*. Pluto Press.
Stevis, Dimitris. 2023. *Just Transitions: Promise and Contestation*. Cambridge University Press.
Stokes, Leah, C. 2020. *Short Circuiting Policy: Interest Groups and the Battle over Clean Energy and Climate Policy in the American States*. Oxford University Press.
Stylianou, Nikos, Jana Tauschinski, and Edward White White. 2025. "How Xi sparked China's electricity revolution." *Financial Times*, May 12, viewed January 27, 2026. Available at: https://www.ft.com/content/[article-id-if-available] (reprinted/quoted at https://climateenergyfinance.org/wp-content/uploads/2025/05/How-Xi-sparked-Chinas-electricity-revolution-FT-12-May-2025.pdf).[1]
Sultana, Farhana. 2022. "The Unbearable Heaviness of Climate Coloniality." *Political Geography* 99.
Sweeney, Sean, and John Treat. 2018. "Trade Unions on the Front Lines of Just Transition to a Green Economy." *International Journal of Labour Research* 38(2): 327–53.
Sweeney, Sean Treat, John, and HongPing Shen, Irene. 2020. "Transition in Trouble? The Rise and Fall of 'Community Energy' in Europe" (Working Paper #13).
Swilling, Mark, and Eve Annecke. 2012. *Just Transitions: Explorations of Sustainability in an Unfair World*. UCT Press.
Szumilas-Kowalzyk, Hanna, and Nicholas Pevzner. 2019. "Getting Beyond Visual Impact: Designing Renewable Energy as a Positive Landscape Addition." In *Proceedings of the 2019 Visual Resource Stewardship Conference*. Argonne National Laboratory.
Táíwò, Olúfẹ́mi O. 2022. *Reconsidering Reparations*. Oxford University Press.
Teles, Steven. 2025. *Varieties of Abundance*. Niskanen Center. https://www.niskanencenter.org/abundance-varieties/
Temple, James. 2018 "China's Giant Transmission Grid Could be the Key to Cutting Climate Emissions." *MIT Technology Review*, November 8. Available at: https://www.technologyreview.com/2018/11/08/138280/chinas-giant-transmission-grid-could-be-the-key-to-cutting-climate-emissions/ (accessed January 26, 2026).
Thomas, Katie Lloyd, Tilo Amhoff, and Nick Beech. 2015. *Industries of Architecture*. Routledge.
The Red Nation. 2021. *The Red Deal: Indigenous Action to Save Our Earth*. Common Notions.
Till, Jeremy. 2009. *Architecture Depends*. Vol. 55. MIT Press.
Todd, Nancy, J. 1977. *The Book of the New Alchemists*. E.P. Dutton.
Todd, Nancy J., and John Todd. 1994. *From Eco-Cities to Living Machines: Principles of Ecological Design*. North Atlantic Books.

Todd, N. J. 2012. *A Safe and Sustainable World: The Promise of Ecological Design.* Island Press.

Toensmeier, Eric. 2016. *The Carbon Farming Solution: A Global Toolkit of Perennial Crops and Regenerative Agriculture Practices for Climate Change Mitigation and Food Security.* Chelsea Green.

Tonkinwise, Cameron. 2003. "Interminable Design: Techne and Time in the Design of Sustainable Service Systems." *5th European Academy of Design, Barcelona, Spain.*

Tonkinwise, Cameron. 2015. "Radical Sustainable Innovation." In Peter Stebbing, and Ursula Tischner (Eds.), *Changing Paradigms: Designing for a Sustainable Future.* Cumulus Think Tank.

Tonkinwise, Cameron. 2019. "Design's (dis) Orders: Mediating Systems-Level Transition Design." *Cuadernos del Centro de Estudios en Diseño y Comunicación. Ensayos* (73): 84–95.

Tonkinwise, Cameron. 2022. "Before Design, More-than-Design: Elucidating 'Ontological Design.'" *Design and Culture* 14(3): 341–59.

Tonkinwise, Cameron. 2024. "The Structure of Structural Change Making a Habit of Being Alienated as a Designer." In *Routledge Handbook of Sustainable Design*, 459–71. Routledge.

Tooze, Adam. 2024. "Chartbook Carbon Notes 10: In China, Clean Energy Is Now THE Driver of Overall Economic Growth." https://adamtooze.substack.com/p/chartbook-carbon-notes-10-in-china.

Tooze, Adam. 2025a. "Climbs the China Learning Curve." Sinica Podcast July 16, 2025. Available at: https://www.sinicapodcast.com/p/adam-tooze-climbs-the-china-learning.

Tooze, Adam. 2025b. "Chartbook 347 The Trouble with Transitioning: Putting Energy History Back on Its Feet." Carbon notes #19. Available at: https://adamtooze.substack.com/p/chartbook-347-the-trouble-with-transitioning.

Tooze, Adam. 2025c. "Electrostates, Petrostates and the New Cold War" [Lecture]. LRB Autumn Lecture, The New School, Tishman Auditorium, New York, NY, October 27.

Trappmann, Vera, Daniel Eversberg, and Felix Schulz. 2025. "German Trade Unions and Decarbonisation: A Transition to Green Growth, a-growth or Degrowth?" *Industrial Relations Journal* 56(4): 322–33.

Turner, Fred. 2010. *From Counterculture to Cyberculture: Stewart Brand, the Whole Earth Network, and the Rise of Digital Utopianism.* University of Chicago Press.

Ukeles, Mierle Laderman. 1969. *Maintenance Art Manifesto 1969! Proposal for an exhibition "CARE"* https://queensmuseum.org/wp-content/uploads/2016/04/Ukeles-Manifesto-for-Maintenance-Art-1969_5-pages.pdf.

Unger, Roberto. 2005. *What Should the Left Propose?* Verso Books.

U.S. Congress. House. *H.Res.109—Recognizing the Duty of the Federal Government to Create a Green New Deal.* H.R.109. 116th Congress, 1st session. Introduced in House February 7, 2019. Available at: https://www.congress.gov/bill/116th-congress/house-resolution/109/text.

U.S. Congress. Senate. *Green New Deal for Public Housing Act.* S.1218. 117th Congress, 1st session. Introduced in Senate April 19, 2021. Available at: https://www.congress.gov/bill/117th-congress/senate-bill/1218/text.

U.S. Congress. House. *Green New Deal for Public Schools Act of 2021.* H.R.4442. 117th Congress, 1st session. Introduced in House July 16, 2021. Available at: https://www.congress.gov/bill/117th-congress/house-bill/4442/text.

U.S. Congress. House. *Homes for All Act of 2019*. H.R.5244. 116th Congress, 1st session. Introduced in House November 21, 2019. Available at: https://www.congress.gov/bill/116th-congress/house-bill/5244/text.

U.S. Congress. House. *Green New Deals for Cities Act of 2021*. H.R.2644. 117th Congress, 1st session. Introduced in House April 19, 2021. Available at: https://www.congress.gov/bill/117th-congress/house-bill/2644/text.

U.S. Congress. House. *H.Res.109 - Recognizing the Duty of the Federal Government to Create a Green New Deal*. H.R.109. 116th Congress, 1st session. Introduced in House February 7, 2019. Available at: https://www.congress.gov/bill/116th-congress/house-resolution/109/text.

Usina-CTAH/Arantes, Pedro. 2025. "Architecture, Politics and Self-Management: *Mutirao* as Transformative Practice." In Malterre-Barthes, Charlotte *On Architecture and Work: The Political Economy of Space Vol. 3*. Hatje Cantz.

Vachon, Todd. 2021. "Just Transitions, Work Futures and the Role of Unions." In *Just Transition Listening Project Report*. Labor Network for Sustainability, 1–52. Available at: https://www.labor4sustainability.org/files/JTLP_report2021.pdf (accessed January 25, 2026).

Vaheesan, Sandeep. 2024. *Democracy in Power: A History of Electrification in the United States*. University of Chicago Press.

Valenzuela, Fernanda, and Steffen Böhm. 2017. "Against Wasted Politics: A Critique of the Circular Economy." *Ephemera: Theory & Politics in Organization* 17(1).

Van Alstyne, Andrew D. 2015. "The United Auto Workers and the Emergence of Labor Environmentalism." *WorkingUSA* 18(4): 613–27.

Van der Ryn, Sim. 2013. *Culture, Architecture and Nature: An Ecological Design Retrospective*. Routledge.

Vassillo, Chiara. 2024. "Circular Economy Transition in European Union Countries." In *Circular Economy for Social Transformation: Multiple Paths to Achieve Circularity*, edited by Renato Passaro, Patrizia Ghisellini, Mario Pansera, Stefania Barca, and Martin Calisto Friant. LED Edizioni Universitarie di Macerata (Deliverable D.1.1, EU H2020 JUST2CE project).

Velicu, Irina, and Stefania Barca. 2020. "The Just Transition and Its Work of Inequality." *Sustainability: Science, Practice and Policy* 16(1): 263–73.

Vermeulen, Walter J. V., Martin Calisto Friant, Kieran Campbell-Johnston, Pauline Deutz, Aodhan Newsholme, Małgorzata Pusz, Natacha Klein, Tomás B Ramos, Anna Diaz, and Kaustubh Thapa. 2024. "Policy Considerations for a Circular Economy." In *Circular Economy Realities*, 173–202. Routledge.

Veselova, Emīlija, and İdil Gaziulusoy. 2022. "Mediating the Needs of Human and Natural Nonhuman Stakeholders: Towards a Design Methodological Framework." *Design and Culture* 14(2): 149–83.

Vezzoli, Carlo, and Ezio Manzini. 2008. *Design for Environmental Sustainability*. Springer.

Wagner, Jason, Daniel A. Cohen, and Ruthy Gourevitch. 2025. *Transforming the Housing Sector with Green Industrial Policy*. Climate and Community Institute. Available at: https://climateandcommunity.org/research/transforming-the-housing-sector-with-green-industrial-policy/ (accessed January 26, 2026).

Wainwright, Hilary, and Dave Elliot. 1981. *The Lucas Plan: A New Trade Union Strategy for Industry*. Spokesman Books.

Waldheim, Charles. (Ed.) 2006. *The Landscape Urbanism Reader*. Princeton Architectural Press.

Walker, Ryan, John D. Sterman, Aaron Jones, and Theodore Jojola. 2013. *Reclaiming Indigenous Planning*. McGill-Queen's University Press.
Walter, Daan, Sam Butler-Sloss, and Katie Bond. 2025. "Rewiring the Energy Debate: The Electrotech Perspective." *The Electrotech Revolution*, May 28. Available at: https://electrotech-revolution.substack.com (accessed January 26, 2026).
Wang, Dan. 2025. *Breakneck: China's Quest to Engineer the Future*. W.W. Norton.
Wang, Seaver. 2021. "Perfectly Safe Energy Doesn't Exist and That's Fine." Breakthrough Institute Blog, May 11. https://thebreakthrough.org/issues/energy/perfectly-safe-energy-doesnt-exist.
Wang, Seaver, Z. Hausfather, S. Davis, J. Lloyd, E.B. Olson, L. Liebermann, ... and J. McBride. 2022. "Future Demand for Electricity Generation Materials Under Different Climate Mitigation Scenarios." *Joule* 6(8): 1856–73.
Wang, Seaver, Peter Cook, Adam Stein, Juzel Lloyd, and Charlyne Smith. 2024. "Updated Mining Footprints and Raw Material Needs for Clean Energy: Challenges and Opportunities for Managing Energy Transition Mining Impacts." *The Breakthrough Institute*. https://thebreakthrough.org/issues/energy/updated-mining-footprints-and-raw-material-needs-for-clean-energy.
Wang, Seaver, Ted Nordhaus, and Vijay Ramachandran. 2025. Greenwashing with Chinese Characteristics: China's "Green Electrostate" Is Neither. The Breakthrough Institute, December 11. Available at: https://thebreakthrough.org/issues/energy/greenwashing-with-chinese-characteristics (accessed February 4, 2026).
Ward, Colin. 1991. *Reflected in Water: A Crisis of Social Responsibility*. Cassell.
Wark, McKenzie. 2013. "Against Social Reductionism" Public Seminar. December 5. Available at: https://publicseminar.org/2013/12/against-social-determinism/
Wark, McKenzie. 2015. *Molecular Red: Theory for the Anthropocene*. Verso Books.
Watson, David. 1998. "Social Ecology and the Problem of Technology." In Andrew Light (Ed.), *Social Ecology After Bookchin*. Guilford Press.
Watson, Julia. 2019. *Lo-TEK: Design by Radical Indigenism*. Cologne: Taschen.
Weber, Isabelle. 2021. *How China Escaped Shock Therapy: The Market Reform Debate*. Routledge.
Weizman, Eyal. 2017. *Forensic Architecture: Violence at the Threshold of Detectability*. Princeton University Press.
Weller, Richard. 2006. "An Art of Instrumentality: Thinking Though Landscape Urbanism." In Charles Waldheim (Ed.), *The Landscape Urbanism Reader*, 71. Princeton Architectural Press.
Weller, Richard, and Billy Fleming. 2016. "Has Landscape Architecture Failed?" *Landscape Architecture Foundation News Blog*.
White, Damian. 2008. *Bookchin – A Critical Appraisal*. Pluto Press.
White, Damian, Alan Rudy, and Brian Gareau. 2016. *Environments, Natures and Social Theory: Towards a Critical Hybridity*. Bloomsbury.
White, Damian F., and J. Timmons Roberts. 2019. "Post Carbon Transition Futuring: For a Reconstructive Turn in the Environmental Social Sciences?" In K. Legun, J. C. Keller, M. Carolan, and M. M. Bell (Eds.), *The Cambridge Handbook of Environmental Sociology*. Cambridge University Press.
White, Richard. 1995. "Are You an Environmentalist or Do You Work for a Living?" In W. Cronon (Ed.), *Uncommon Ground: Toward Reinventing Nature*. W.W. Norton.
Whyte, Kyle. 2017. "Indigenous Climate Change Studies: Indigenizing Futures, Decolonizing the Anthropocene." *English Language Notes* 55(1): 153–62.

Wilbert, Chris, and Damian F. White. 2011. *Autonomy, Solidarity, Possibility: The Colin Ward Reader*. AK Press.
Willis, Anne-Marie. 2018. *The Design Philosophy Reader*. Bloomsbury.
Winant, Gabriel. 2019. “Professional-Managerial Chasm.” n+1 (October 10, 2019).
Wizinsky, Matthew. 2022. *Design after Capitalism: Transforming Design Today for an Equitable Tomorrow*. MIT Press.
Wolff, Jane, 2007. “Redefining Landscape.” In Tim Culvahouse (Ed.), *The Tennessee Valley Authority: Design and Persuasion*. Princeton Architectural Press.
World Commission on Environment and Development. 1987. *Our Common Future*. Oxford University Press.
Woodhouse, Edward J., and James W. Patton. 2004 “Design by Society: Science and Technology Studies and the Social Shaping of Design.” *Design Issues* 20(3): 1–12.
Wright, Erik Olin. 1985. *Classes*. Verso Books.
Wright, Erik Olin. 2010. *Envisioning Real Utopias*. Verso Books.
Xu, L. 2022. “Towards Green Innovation by China’s Industrial Policy: Evidence from Made in China 2025.” *Frontiers in Environmental Science* 10: 924250.
Xue, Xiaokang, and Mathias Larsen. 2025. *China’s Green Leap Outward: The Rapid Scale- up of Overseas Chinese Clean-tech Manufacturing Investments*. Net Zero Industrial Policy Lab. September.
Yergin, Daniel. 2025. “The Energy Transition Needs a Reality Check.” *Foreign Affairs*, February 25. Available at: https://press.spglobal.com/2025-02-25-The-Energy-Transition-Needs-a-Reality-Check (accessed January 26, 2026).
York, Richard, and Shannon Elizabeth Bell. 2019. “Energy Transitions or Additions?: Why a Transition from Fossil Fuels Requires More Than the Growth of Renewable Energy.” *Energy Research & Social Science* 51(2019): 40–3.
Zografos, Christos, and Paul Robbins. 2020. “Green Sacrifice Zones, or Why a Green New Deal Cannot Ignore the Cost Shifts of Just Transitions.” *One Earth* 3(5): 543–63.
Zonta, Michela. 2024. “Increasing Affordable Housing Stock Through Modular Building.” *Center for American Progress. Report*. February 6, 2024.

Index

www.ingramcontent.com/pod-product-compliance
Ingram Content Group UK Ltd.
Pitfield, Milton Keynes, MK11 3LW, UK
UKHW020245290926
15470UKWH00002B/744

9 781350 332553